Semi-active Suspension Control

Emanuele Guglielmino • Tudor Sireteanu
Charles W. Stammers • Gheorghe Ghita
Marius Giuclea

Semi-active Suspension Control

Improved Vehicle Ride and Road Friendliness

Emanuele Guglielmino, PhD
Italian Institute of Technology (IIT)
Via Morego, 30
16163 Genoa
Italy

Charles W. Stammers, PhD
Department of Mechanical Engineering
University of Bath
Bath BA2 7AY
UK

Tudor Sireteanu, PhD
Gheorghe Ghita, PhD

Institute of Solid Mechanics
Romanian Academy
C-tin Mille Street
010141 Bucharest
Romania

Marius Giuclea, PhD
Department of Mathematics
Academy of Economic Studies 6
Piata Romana
010374 Bucharest
Romania

ISBN 978-1-84996-761-7 e-ISBN 978-1-84800-231-9

DOI 10.1007/978-1-84800-231-9

British Library Cataloguing in Publication Data
A catalogue record for this book is available from the British Library

Cover design: eStudio Calamar S.L., Girona, Spain

Printed on acid-free paper

9 8 7 6 5 4 3 2 1

springer.com

Naturae phaenomena ex principiis mechanicis eodem argumentandi genere derivare liceret.

Newton

Preface

The fundamental goals of a car suspension are the isolation of the vehicle from the road and the improvement of road holding by means of a spring-type element and a damper.

The inherent limitations of classical suspensions have motivated the investigation of controlled suspension systems, both semi-active and active. In a semi-active suspension the damper is generally replaced by a controlled dissipative element and no energy is introduced into the system. In contrast, an active suspension requires the use of a fully active actuator, and a significant energy input is generally required. Due to their higher reliability, lower cost and comparable performance semi-active suspensions have gained wide acceptance throughout the automotive engineering community.

This book provides an overview of vehicle ride control employing smart semi-active damping systems. In this context the term smart refers to the ability to modify the control logic in response to measured vehicle ride and handling indicators.

The latest developments in vehicle ride control stem from the integration of diverse engineering disciplines, including classical mechanics, hydraulics, biomechanics and control engineering as well as software engineering and analogue and digital electronics.

This book is not intended to be a general-purpose text on vehicle dynamics and vehicle control systems (traction control, braking control, engine and emissions control *etc.*). The focus of this work is on controlled semi-active suspension systems for ride control and road friendliness using a multidisciplinary mechatronic approach.

If effective control of a suspension is to be achieved, which makes the most of the potentialities of a semi-active damping device, it is paramount to understand the interactions between mechanical, hydraulic and electromagnetic sub-systems.

The book analyses the different facets of the technical problems involved when designing a novel smart damping system and the technical challenges involved in its control. The emphasis of this work is not only on modelling and control algorithm design, but also on the practical aspects of its implementation. It describes the practical constraints encountered and trade-offs pursued in real-life engineering practice when designing and testing a novel smart damping system. Hence sound mathematical modelling is balanced by large sections on experimental implementation as well as case studies, where a variety of automotive applications are described, covering different applications of ride control, namely semi-active suspensions for a saloon car, seat suspensions for vehicles not equipped with a primary suspension and control of heavy-vehicle dynamic tyre loads to reduce road damage and improve handling.

Within the book issues such as road holding, passenger comfort and human body response to vibration are thoroughly analysed. Appropriate control-oriented dampers models are described, along with their experimental validation. Vehicle ride and human body models are illustrated and robust algorithms are designed.

The book is centered around two types of semi-active dampers: friction dampers and magnetorheological dampers. The former can be viewed as an out-of-the-box non-conventional damper while the latter can be thought as a conventional controllable damper (it is used in several cars). Based on these two types of dampers in the course of the book it is shown how to design a semi-active damping system (using a friction damper) and how to implement an effective semi-active control system on a well-established damper (the magnetorheological damper).

The book can be fruitful reading for mechanical engineering students (at both undergraduate and postgraduate level) interested in vehicle dynamics, electrical and control engineering students majoring in electromechanical and electrohydraulic control systems. It should be valuable reading for R&D and design engineers working in the automotive industry and automotive consultants. It can be of interest also to engineers, physicists and applied mathematicians working in the broad area of noise and vibration control, as many concepts can potentially be applied to other fields of vibration control.

The book is structured as follows:

Chapter 1 is a general introduction to active, semi-active and passive suspensions and introduces the fundamental concepts of vehicle ride and handling dynamics.

Chapter 2 focusses on dampers modelling (including hysteresis modelling) and reviews the main vehicle ride and road surface models.

Chapter 3 analyses the human body response to vibration via appropriate human body models based on recent studies in the field of biomechanics.

Chapter 4 is dedicated to control algorithms. After a brief qualitative overview of the fundamentals of modern control theory, the main semi-active suspensions algorithms are introduced. The focus is on an algorithm known as balance logic, which is analysed from a mathematical viewpoint. Emphasis is also placed on robust algorithm design and on techniques to increase the reliability of the systems (*e.g.*, anti-chattering algorithms).

Chapter 5 details the design of a semi-active suspension system based on a friction damper.

Chapter 6 illustrates the design of a magnetorheological-based semi-active suspension.

Chapter 7 offers a comprehensive overview of the applications with a number of case studies including a friction damper-based suspension unit for a saloon car, a magnetorheological damper-based seat suspension for vehicles not equipped with primary suspensions which uniquely rely on this suspension mounted underneath the driver seat to provide ride comfort and semi-active suspension for heavy vehicles where the emphasis is not only on ride comfort but also on road damage reduction.

Disclaimer

All the experimental work and numerical simulations presented in this book are the result of academic research carried out at the University of Bath (UK) and at the Institute of Solid Mechanics of the Romanian Academy (Romania). All devices described in the book are purely experimental prototypes.

Therefore in no circumstances shall any liability be accepted for any loss or damage howsoever caused to the fullest possible extent of the law that may result from the reader's acting upon or using the content contained in the publication.

Acknowledgements

The authors wish to express their gratitude to Springer (London) for the invitation to write this book. Many thanks to our Editor Oliver Jackson for his guidance throughout this work.

The book is the fruit of many years of research work. Most of the results presented were obtained at the Department of Mechanical Engineering, University of Bath (UK), which we would like to thank.

We are deeply grateful to Professor Kevin Edge, Pro-Vice-Chancellor for Research of the University of Bath, for his contribution to the hydraulic control of friction dampers.

We would like to thank the University for providing the technical facilities and would also like to express our gratitude to the technical staff of the Department of Mechanical Engineering without whom the manufacture of devices and test rigs and technical trouble-shooting would not have been possible.

We wish to express our appreciation to the Royal Society of London for sponsoring a decade of collaboration with the Romanian Academy in Bucharest, where valuable contributions to this text were made.

Thanks are also due to Dr Georgios Tsampardoukas for kindly providing valuable material on trucks.

We finally wish to thank all the people whose input, support and sympathetic understanding was instrumental to the successful completion of such an undertaking.

Contents

1

Introduction

1.1 Introduction

Today's vehicles rely on a number of electronic control systems. Some of them are self-contained, stand-alone controllers fulfilling a particular function while others are co-ordinated by a higher-level supervisory logic. Examples of such vehicle control systems include braking control, traction control, acceleration control, lateral stability control, suspension control and so forth. Such systems aim to enhance ride and handling, safety, driving comfort and driving pleasure. This book focuses on semi-active suspension control. The thrust of this work is to provide a comprehensive overview of theoretical and design aspects (including several case studies) of vehicle semi-active systems based on smart damping devices.

Isolation from the forces transmitted by external excitation is the fundamental task of any suspension system. The problem of mechanical vibration control is generally tackled by placing between the source of vibration and the structure to be protected, suspension systems composed of spring-type elements in parallel with dissipative elements. Suspensions are employed in mobile applications, such as terrain vehicles, or in non-mobile applications, such as vibrating machinery or civil structures. In the case of a vehicle, a classical car suspension aims to achieve isolation from the road by means of spring-type elements and viscous dampers (shock absorbers) and contemporarily to improve road holding and handling.

The elastic element of a suspension is constituted by a spring (coil springs but also air springs and leaf springs), whereas the damping element is typically of the viscous type. In such a device the damping action is obtained by throttling a viscous fluid through orifices; depending on the physical properties of the fluid (mainly its viscosity), the geometry of the orifices and of the damper, a variety of force versus velocity characteristics can be obtained. This technology is very reliable and has been used since the beginning of the last century (Bastow, 1993). However it is possible to achieve a damping effect by other means, as subsequently discussed.

Spring rate and damping are chosen according to comfort, road holding and handling specifications. A suspension unit ought to be able to reduce chassis acceleration as well as dynamic tyre force within the constraint of a set working space. Depending upon the type of vehicle, either the former or the latter criterion is emphasised. In applications different from automotive ones (*e.g.*, rotating machinery, vibration mitigation in civil structures) the comfort criterion is not usually an issue, but other specifications exist, *e.g.*, on the maximum value of some quantities (displacements, velocities *etc.*).

Passive suspensions have inherent limitations as a consequence of the trade-off in the choice of the spring rate and damping characteristics, in order to achieve acceptable behaviour over the whole range of working frequencies. As is known from linear systems theory a one-degree-of-freedom (1DOF) spring–mass–damper system (modelled by a second-order linear differential equation) having high damping performs well in the vicinity of the resonant frequency and poorly far from it, whilst a low-damped system behaves conversely (Rao, 1995).

The necessity of compromising between these conflicting requirements has motivated the investigation of controlled suspension systems, where the elastic and the damping characteristics are controlled closed-loop. By using an external power supply and feedback-controlled actuators, controlled suspension systems can be designed which outperform any passive system.

The external energy needed to generate the required control forces of a smart suspension is an important issue that must be considered in controller design. The controllers must be designed so as to achieve an acceptable trade-off between control effectiveness and energy consumption. From this point of view, the control strategies can be grouped in two main categories: active and semi-active.

Usually, the active control strategies need a substantial amount of energy to produce the required control forces. A fully active system can potentially provide higher performance than its passive counterpart. However in many engineering applications this goal can be achieved only at the expense of a complex and costly system, with large energy consumption and non-trivial reliability issues. In particular when designing an active control system two important aspects must be taken into consideration: the potential failure of the power source, and the injection of a large amount of mechanical energy into the structure that has the potential to destabilise (in the bounded input/bounded output sense) the controlled system. Hence a careful hazard and failure-modes analysis must be carried out and a fail-safe design adopted.

Semi-active control devices offer reliability comparable to that of passive devices, yet maintaining the versatility and adaptability of fully active systems, without requiring large power sources. In a semi-active suspension the amount of damping can be tuned in real time. Hence most semi-active devices produce only a modulation of the damping forces in the controlled system according to the control strategy employed. In contrast to active control devices, semi-active control devices cannot inject mechanical energy into the controlled system and, therefore, they do not have the potential to destabilise it. Examples of such devices are variable orifice dampers, controllable friction devices and dampers with controllable fluids (*e.g.*, electrorheological and magnetorheological fluids).

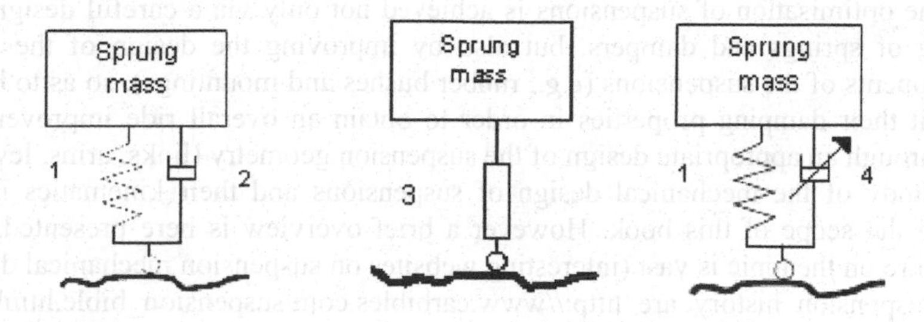

Fig. 1.1. Illustration of passive and controlled vehicle suspensions (1 - passive spring, 2 - passive damper, 3 - actuator, 4 - controllable damper)

The above discussed control solutions are illustrated schematically in Figure 1.1 for the simplest (1DOF) model of a primary vehicle suspension.

As previously stated, a suspension algorithm is designed to reduce chassis acceleration as well as dynamic tyre force. Chassis acceleration is related to ride and comfort, and tyre force to road holding and handling.

Dynamic tyre force reduction results in better handling of the vehicle, as the cornering force, tractive and braking efforts developed by the tyre are related to normal load, which can be controlled by semi-active methods. Road holding and handling performance can be quantified by the consideration of the forces and moments applied to the chassis and to the tyres.

Comfort is more difficult to quantify and, although standards exist, its assessment is a controversial issue, because it is an inherently subjective matter. In Section 1.5 a survey of comfort assessment criteria is presented to shed light on this matter.

1.2 Historical Notes on Suspensions

Suspensions, as many other vehicle systems, followed relatively closely the evolution of the transportation technology. For centuries carts were not equipped with any sort of suspension at all. Only later, in the eigth century, was a primitive suspension based on an iron chain system developed. Metal springs were first developed in the 17th century and shortly afterwards leaf springs. Various designs were developed until the last century, which saw the development of the concept of suspension based on a spring and a damper.

Early vehicle ride studies date back to the 1920s and 1930s (Lanchester, 1936). Investigation on handling and steering dynamics followed later in the 1950s as reported by Milliken WF and Milliken DL (1995) as well as the application of random vibration theory to vehicle studies. The advent of digital computers with greater processing power and the development of multi-body vehicle models of ever increasing complexity has contributed to produce more and more sophisticated designs.

The optimisation of suspensions is achieved not only via a careful design and tuning of springs and dampers, but also by improving the design of the other components of the suspensions (*e.g.*, rubber bushes and mountings), so as to better exploit their damping properties in order to obtain an overall ride improvement, and through an appropriate design of the suspension geometry (links, arms, levers). The study of the mechanical design of suspensions and their kinematics is not within the scope of this book. However a brief overview is here presented. The literature on the topic is vast (interesting websites on suspension mechanical design and suspension history are http://www.carbibles.com/suspension_bible.html and http://www.citroenet.org.uk/miscellaneous/suspension/suspension1.html).

Essentially suspensions can be categorised into two large families: dependent and independent suspensions, the difference being whether the two suspension units (on either the front or the rear of the vehicle) are linked or not. Whilst it is very common to have rear dependent suspension, most front suspensions are of the independent type. Sometimes suspensions are linked by an anti-roll bar, which is essentially a torsion spring that helps reduce roll while negotiating a bend.

As far as the kinematics of the suspension is concerned car manufacturers developed a variety of designs, including the so-called double wishbone system and the multi-link suspensions (used on the Audi A4 for instance).

It was said that suspension unit is composed of a damper and a spring. Springs are typically of coil type but leaf springs are still common, particularly on trucks.

A classical design is the so-called MacPherson strut, named after Earle S. MacPherson who designed it in the 1940s. It is a very compact design where the damper is mounted within the coil spring. Hydropneumatic suspension is another type of suspension developed by Citroën, which has worked on controlled suspensions for many years (Curtis, 1991). Another type of suspension is the Hydragas suspension employed for instance on the MGF Roadster (Moulton and Best, 1979a and 1979b; Rideout and Anderson, 2003).

Controlled suspensions (both active and semi-active) have appealed to automotive engineers for many decades. Semi-active dampers have been developed by damper manufacturers such as ZF Sachs (ABC -Active Body Control- and CDC http://www.zf.com/content/en/import/zf_konzern/startseite/f_e/nutzen_fuer_unsere_kunden/variable_daempfungssysteme/Variable_Daempfungssysteme.html). At present many vehicles offer some kind of controlled suspensions. Active suspensions were first developed for Formula 1 cars: Lotus's was the first car to be equipped with an active system in 1983 (Baker, 1984; Milliken, 1987). Besides racing cars, active systems have been studied and developed for a long time also for road vehicles (typically saloon cars). Hillebrecht *et al.* (1992) 15 years ago discussed the trade-off between customer benefit and technological challenge from the angle of a car manufacturer. Mercedes have worked for many years on active suspensions. The Mercedes CL Coupe (Cross, 1999) is equipped with a fully integrated suspension and traction control. The Citroën BX model was fitted with a self-leveller system and the Xantia Activa is equipped with active anti-roll bars. Toyota worked on controlled suspensions, for example in the Toyota Celica (Yokoya *et al.*, 1990) as well as Volvo (Tiliback and Brood, 1989). Most recently plenty of high-segment cars are equipped with semi-active suspensions (Mercedes, Lamborghini and Ferrari vehicles, to name but a few).

Magnetorheological-based semi-active suspensions are used on a number of high-segment market cars which employ the Delphi MagneRide™ (http://delphi.com/manufacturers/auto/other/ride/magneride/) system based on magnetorheological dampers. The system is fitted on a few vehicles including some Cadillac models (Imaj, Seville, SRX, XLR, STS, DTS), the Chevrolet Corvette and most recently the Audi TT, the Audi R8 and the Ferrari 599 GTB.

Another interesting type of suspension worth mentioning is the Bose® linear electromagnetic suspension designed by Dr Amar Bose, which is based on a linear electric motor and power amplifier instead of a spring and a damper (http://www.automobilemag.com/features/news/0410_bose_suspension/).

1.3 Active and Semi-active Suspensions in the Scientific Literature

A vast amount of work on controlled suspension systems is present in the technical and scientific literature. The first paper dealing with active suspensions dates back to the 1950s (Federspiel-Labrosse, 1954). One of the first reviews of the state of the art of controlled suspensions was carried out by Hedrick and Wormely (1975). Another one was produced in 1983 by Goodall and Kortüm who surveyed the active suspension technology. A few years later Sharp and Crolla (1987) and Crolla and Aboul Nour (1988) produced comparative reviews of advantages and drawbacks of various types of suspensions. Another historical review and also an attempt to present some design criteria was given by Crolla (1995). A first choice in the design of a fully active suspension is the type of actuation. The actuator can be hydraulic, pneumatic or electromagnetic, or a hybrid solution. Williams *et al.* (1996) analysed the merits of an oleo-pneumatic actuator, Martins *et al.* (1999) proposed a hybrid electromagnetic-controlled suspension. An active suspension employing a hydraulic actuator, pressure-controlled rather than flow-controlled, has been proposed by Satoh *et al.* (1990).

Active suspensions are a challenging field for control engineers. All the main control techniques developed in the past 30 years have been applied to the problem of controlling vehicle suspensions. An overview of this research now follows.

At the outset it must be stressed that one of the main problems in the design of control algorithms is the identification of the vehicle and suspension parameters. Errors in their knowledge can spoil the performance of the most sophisticated controllers, designed with very refined mathematical techniques. Majjad (1997) and Tan and Bradshaw (1997) addressed the problem of the identification of car suspension parameters. The necessity of trading off among the conflicting requirements of the suspensions in terms of comfort and road holding led to the use of optimisation techniques. In 1976 Thompson studied a quarter car model and employed optimal linear state feedback theory for designing a controlled suspension; Chalasani (1987) optimised active ride performance using a full car model. An H_∞ algorithm for active suspensions was proposed by Sammier *et al.* (2000).

The driving conditions in a car change greatly depending upon the road and the speed. This suggests the necessity of some forms of adaptive control. Hac (1987) implemented this kind of scheme. Other types of adaptive controls were proposed over the years. Ramsbottom *et al.* (1999) and Chantranuwathana and Peng (1999) studied an adaptive robust control scheme for active suspensions.

Robust control, the control philosophy for dealing with systems with uncertain parameters, has been investigated by many researchers as well. Mohan and Phadke (1996) studied a variable structure controller for a quarter car. Park and Kim (1998) extended this study to a full 7DOF ride model. Sliding mode control was investigated by Yagtz *et al.* (1997) and by Kim and Ro (1998). A mixed sliding mode–fuzzy controller was proposed by Al-Houlu *et al.* (1999).

Active suspension systems have been studied also for off-road vehicles (Crolla *et al.*, 1987). Stayner (1988) proposed an active suspension for agricultural vehicles. Active devices have been investigated also for rail applications as reported by Goodall *et al.* (1981).

Semi-active suspensions were firstly introduced in the 1970s (Crosby and Karnopp, 1973; Karnopp *et al.*, 1974) as an alternative to the costly, highly complicated and power-demanding active systems. Similar work was performed by Rakheja and Sankar (1985) and Alanoly and Sankar (1987) in terms of active and semi-active isolators. A comparative study with passive systems was carried out by Margolis (1982) and by Ahmadian and Marjoram (1989). The most attractive feature of that work was that the control strategies were based only upon the measurement of the relative displacement and velocity. A review can be found in Crolla (1995).

A control scheme known as skyhook damping, based on the measurement of the absolute vertical velocity of the body of the car (the aim is to achieve the same damping force as that produced by a damper connected to an ideal inertial reference in the sky), was proposed in the 1970s by Karnopp and is still employed in a number of variations (Alleyne *et al.*, 1993). Yi and Song (1999) proposed an adaptive version of the skyhook control. Some authors (Chang and Wu, 1997), in order to improve comfort, designed a suspension based on a biological, neuromuscular-like control system. Recently Liu *et al.* (2005) studied four different semi-active control strategies based on the skyhook and balance control strategies.

The reduction of the dynamic tyre force is a challenging field. Cole *et al.* (1994) did extensive work on it, both theoretical and experimental. Groundhook control logic was also investigated by Valasek *et al.* (1998) to reduce dynamic tyre forces.

As far as the applications of semi-active suspensions are concerned, they have been envisaged not only for saloon cars but also for other types of vehicles. Besinger *et al.* (1991) studied an application of semi-active dampers on trucks. A skyhook algorithm for train applications was investigated by Ogawa *et al.* (1999). Miller and Nobes (1988) studied a semi-active suspension for military tanks. A study was performed by Margolis and Noble (1991) in order to control heave and roll motions of large off-road vehicles.

As with active systems, a variety of control schemes have been proposed for semi-active suspensions: adaptive schemes (Bellizzi and Bouc, 1989), optimal control (Tseng and Hedrick, 1994), LQG (Linear Quadratic Gaussian) schemes (Barak and Hrovat, 1988) as well as robust algorithms (Titli *et al.*, 1993). Crolla and Abdel Hady (1988) proposed a multivariable controller for a full vehicle model. Preview control was proposed by Hac (1992) and Hac and Youn (1992), receding horizon control was investigated by Ursu *et al.*, (1984) and H_∞ optimal control by Moran and Nagai (1992).

An interesting solution has been proposed by Groenewald and Gouws (1996) who suggested improving ride and handling by using closed-loop control to adjust the tyre pressure. By controlling tyre pressure it is possible to control wheel-hop resonance and therefore improve the so-called secondary ride, *i.e.*, the behaviour close to the wheel-hop resonance (Shaw, 1999) as well as improving the lifespan of tyres and reducing fuel consumption.

During the last few years there has been a tremendous amount of activity on the applications of artificial intelligence techniques to suspension systems. Neural networks and fuzzy logic (Vemuri, 1993; Agarwal, 1997) have attracted the attention of many researchers in the field (Moran and Nagai, 1994; Watanabe and Sharp, 1996 and 1999; Ghazi Zadeh *et al.* 1997; Yoshimura *et al.*, 1997).

This brief survey has shown that a great deal of research has been and is still being carried out on designing cheap and reliable controlled suspension systems for vehicles.

1.4 Comfort in a Vehicle

Whilst road holding and handling can be objectively quantified by the analysis of the dynamic equations of a vehicle, this is not the case with comfort, as it is an inherently subjective matter. Ride quality, driving pleasure and driving fun are concerned with passenger comfort and driver feeling in a moving vehicle. Vibration transmitted to passengers originates from a host of causes, including, amongst others, road uneveness, aerodynamic forces and engine- and powertrain-induced vibration. Road irregularities are indeed the major source of vibration. In a comfortable vehicle, vibration must stay within some boundaries. In order to establish these boundaries, it is firstly necessary to assess and quantify how to measure comfort.

There is no generally accepted method to assess human sensitivity to vibration; human response is quite subjective and dependent on several factors. Firstly it must be highlighted that the road forces transmitted to tyres are asymmetrical. In the occurrence of a bump, vertical upward acceleration can reach several g while if a pothole is encountered, the vertical downward acceleration cannot be larger than 1 g. This is also a reason why hydraulic dampers are designed with non-symmetrical characteristics for the bound and rebound strokes.

The human body presents asymmetric reactions to vibration as well: the body reacts differently if a vertical acceleration of a given magnitude is applied upward or downward. People can better withstand an increase rather than a decrease in the gravitational force (as can be experienced in a fast elevator, for instance). Likewise motions with low roll centre (*e.g.*, rolling and pitching of a ship) are more troublesome and likely to induce motion sickness than those with a high roll centre.

From these considerations it follows that comfort tests carried out with sinusoidal inputs are not sufficient, but only useful for comparison and benchmarking purposes. However even if a multi-harmonic input is applied, it is difficult to weigh the various frequencies to which the body is more, or less, sensitive.

Early studies associated the feeling of comfort with the frequency of vibration of the walking pace. A review of these early works is provided by Demic (1984). Thus early car suspensions were designed according to this criterion. Further studies (Dieckmann, 1958) proved that different frequency bands are uncomfortable for different organs. Frequencies lower than 1 Hz are related to symptoms like motion sickness; frequencies in the range 5–6 Hz are troublesome for the stomach, while frequencies around 20 Hz are pernicious for head and neck.

Several criteria have been proposed over the past 30 years to assess comfort based on the nature of the vibration. Some of them are general purpose whilst others are employed in specialist fields such as off-road and military vehicles (Pollock and Craighead, 1986).

One first criterion, relevant to the automotive field, is Janeway's comfort criterion (SAE Society of Automotive Engineers, 1965). It relates the comfort to vertical vibration amplitude and permits to find the largest allowed chassis displacement for each frequency. In essence it states that within the range 1–6 Hz the maximum allowable amplitude is the one resulting in a peak jerk value of not more than 12.6 m/s^3. In other words, if X is the maximum allowed displacement amplitude and ω the angular frequency, Janeway's criterion states that:

$$X\omega^3 = 12.6\,. \tag{1.1}$$

At higher frequencies the criterion does not refer to jerk but to acceleration and velocity. It affirms that in the range 6-20 Hz, acceleration peak value should not exceed 0.33 m/s^2, whilst between 20 and 60 Hz the maximum velocity should stay below 2.7 mm/s. The major limitation of the SAE criterion is that it applies only to vertical sinusoidal disturbances and it does not give any indication of the situation in the case of multi-harmonic inputs.

In the same period Steffens (1966) proposed an empirical formula to determine the amplitude of vibration causing discomfort as a function of the frequency:

$$X[\text{cm}^2] = 7.62\cdot 10^{-3}(1+\frac{125}{f^2})\,, \tag{1.2}$$

where again vertical displacement (rather than acceleration) is used to assess comfort.

The most general criterion however is the standard ISO 2631 (1978); it is a general standard applicable not only to vehicles but to all vibrating environments. It defines the exposure limits for body vibration in the range 1–80 Hz, defining limits for reduced comfort, for decreased proficiency and for preservation of health. A subsequent addendum to the norm also takes into consideration the frequencies in the particular range 0.1–1 Hz. It relates discomfort to root mean square (RMS) acceleration as a function of frequency for different exposure times.

If vibrations are applied in all three directions (foot-to-head, side-to-side, back-to-chest), the corresponding boundaries apply for each component. Subsequently to the ISO standard, the British standard BS 6841 was published in 1987 in address what were perceived in Britain as the weakest points of the ISO standard and in 1998 Griffin thoroughly reviewed both of them, analysing merits and their weakest points.

Another criterion proposed is the vibration dose value (Griffin, 1984) which provides an indication based on the integral of the fourth power of the frequency-weighted acceleration. The vibration dose (VD) value is calculated as:

$$\mathrm{VD} = \int_0^t a^4 \mathrm{d}t. \tag{1.3}$$

If the acceleration has components along two or three axes, the total dose value is the algebraic sum of the values for each axis. A value above 15 m^4/s^7 is considered to cause discomfort. This criterion is independent of the type of waveform; besides the fourth power of weighted acceleration emphasises the peak value (which is the other important parameter of a waveform —together with the RMS value— to assess comfort).

Some studies have considered the energy absorbed by the body in a vibrating environment. An initial study by Zeller (1949) relates the comfort to the maximum specific kinetic energy absorbed by the body over a period for a sinusoidal input, and establishes some boundaries for the maximum allowed absorbable energy. The energy is calculated as $E = 0.5v^2_{max}/T = a^2_{max}/8\pi^2 f$. The disturbance considered here is again sinusoidal, T being its period.

A further integral criterion is known as the absorbed power method (Lee and Pradko, 1969); it calculates the power absorbed by the body when it experiences vibration. It has been used by the army to test human tolerance in military vehicles. The criterion is expressed by:

$$\mathrm{Power} = \int_0^f (A^2 w)\mathrm{d}f, \tag{1.4}$$

the parameter A being the acceleration spectrum along an axis and w a frequency weighting function. The tolerance limit is taken to be 6 W.

Another criterion, known as the DISC rating value (Leatherwood *et al.*, 1980) assesses the discomfort with the formula:

$$\mathrm{DISC} = -0.44 + 1.65(\mathrm{DVERT}^2 + \mathrm{DLAT}^2), \tag{1.5}$$

where

$$\mathrm{DVERT} = -1.75 + 0.857\mathrm{CFV} - 0.102\mathrm{CFV}^2 + 0.00346\mathrm{CFV}^3 + 33.4\mathrm{GV}, \tag{1.6}$$

CFV being the centre frequency of vertical axis applied vibration band and GV the peak acceleration within this band (DLAT is defined by an analogous expression).

This survey has shown that the problem of assessing comfort is in some way quantifiable, but the main issue is the choice of the most suitable comfort criteria for a particular application.

1.4.1 Comfort Assessment

From the previous survey it is clear that comfort assessment is an ill-posed problem from a quantitative standpoint. In this work a choice has been made on how to assess it. For the purpose of this work the assessment will be based on RMS, peak and jerk values for sinusoidal inputs, on the response to a pseudo-random input and to a bump as well as on the spectral characteristics of the non-linear acceleration response.

In terms of sinusoidal input, the simplest method to compare passive and semi-active suspension response is through the peak value of chassis accelerations. In a linear case it is straightforward: for an output displacement expressed by $x(t)=X\sin(2\pi ft)$, the peak values of the higher-order derivatives (velocity, acceleration and jerk) are:

$$\dot{x}_{\mathrm{MAX}} = 2\pi fX, \tag{1.7}$$

$$\ddot{x}_{\mathrm{MAX}} = 4\pi^2 f^2 X, \tag{1.8}$$

$$\dddot{x}_{\mathrm{MAX}} = 8\pi^3 f^3 X. \tag{1.9}$$

An assessment based on peak jerk values is a possibility investigated by Hrovat and Hubbard (1987). Another possible approach to estimate comfort is in the frequency domain, carrying out a Fourier analysis of the chassis acceleration amplitude spectrum. Actually this analysis is more appropriate to assess the degree of non-linearity in terms of harmonic distortion. However, as a rule of thumb, a large harmonic content may possibly indicate higher peak acceleration and jerk amplitudes (even if strictly speaking the analysis of the phase spectrum would be necessary), although there is no precise relationship between spectrum and comfort.

As far as the bump input is concerned, the relevant quantities to minimise, comfortwise, are the peak value of the acceleration and the number of oscillations after the bump.

1.5 Introduction to Controlled Dampers

The kernel of a semi-active system is the controllable damper. Therefore it is of paramount importance to gain an understanding of the different types of dampers, their working principles, and how to model them. Consider a simple spring–mass–damper system, depicted in Figure 1.2.

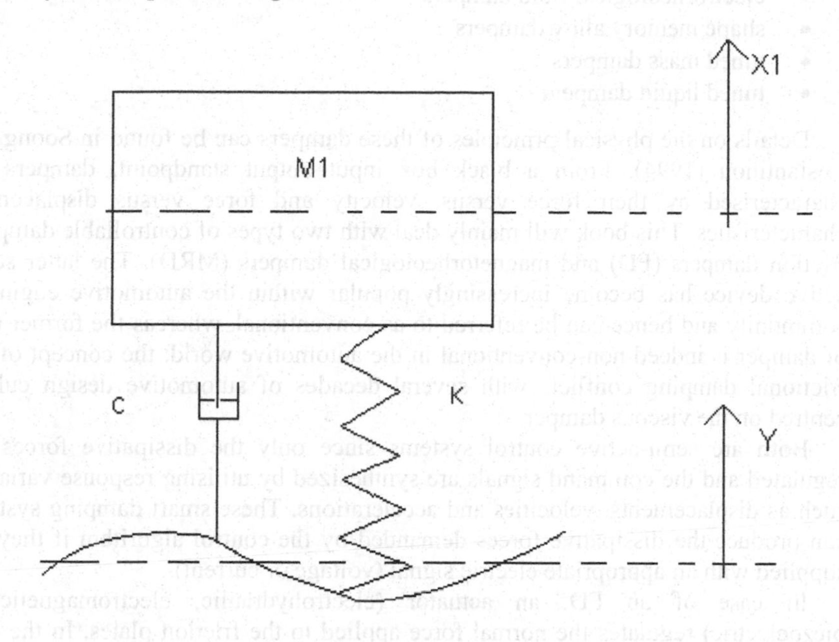

Fig. 1.2. 1DOF spring–mass–damper system

A spring is an elastic element which stores potential energy and whose position-dependent characteristic can be expressed by the functional relation $F=k(x)$, x being the displacement across it. Examples of springs employed in automotive engineering are coil springs, air springs, leaf springs and torsion bars. Hooke's law $F=kx$, k being the spring stiffness (or spring rate) expressed in N/m, represents the particular case of a linear spring.

A damper is a device which dissipates energy through an internal mechanism (*e.g.*, by throttling a viscous flow through an orifice). Typically a damper has a velocity-dependent characteristic expressed by $F=c(\dot{x})$, $\dot{x}$ being the velocity across it. The particular case $F=c\dot{x}$ represents an ideal linear viscous damper and c is the damping coefficient, expressed in Ns/m.

A generic nonlinear semi-active element has a characteristic expressed by the functional relation $F=f(x,\dot{x})$ or also $F=f(x,\dot{x},\ddot{x})$.

A wide range of dampers exist based on a variety of dissipating mechanisms (deformation of viscoelastic solids, throttling of fluids, frictional sliding, yielding of metals, and so forth). The following is a list of some common types of dampers employed in engineering applications.

- viscous dampers
- viscoelastic dampers
- friction dampers

- magnetorheological fluid dampers
- electrorheological fluid dampers
- shape memory alloy dampers
- tuned mass dampers
- tuned liquid dampers

Details on the physical principles of these dampers can be found in Soong and Costantinou (1994). From a black box input output standpoint, dampers are characterised by their force versus. velocity and force versus displacement characteristics. This book will mainly deal with two types of controllable dampers: friction dampers (FD) and magnetorheological dampers (MRD). The latter semi-active device has become increasingly popular within the automotive engineers community and hence can be referred to as conventional, whereas the former type of damper is indeed non-conventional in the automotive world: the concept of dry frictional damping conflicts with several decades of automotive design culture centred on the viscous damper.

Both are semi-active control systems since only the dissipative forces are regulated and the command signals are synthesized by utilising response variables such as displacements, velocities and accelerations. These smart damping systems can produce the dissipative forces demanded by the control algorithm if they are supplied with an appropriate electric signal (voltage or current).

In case of an FD, an actuator (electrohydraulic, electromagnetic or piezoelectric) regulates the normal force applied to the friction plates. In the case of an MRD, an electromagnetic circuit is used to modulate the intensity of the magnetic field applied to the magnetorheological fluid.

By developing sufficiently accurate analytical models to portray the dynamic behaviour of these smart damping devices, the vehicle and the controller can be modelled as a dynamic system having as inputs the external excitation (produced by road unevenness) and the control algorithm command signals.

1.6 Introduction to Friction Dampers

Viscous dampers are the most widely used type of dampers in automotive engineering. However in principle it is possible to achieve a damping effect by other means. A possible alternative is frictional damping. Friction in automotive engineering is usually associated with braking systems or friction-based power transmission systems (*e.g.*, belt transmissions). However it can also be used in vibration control.

A friction damper is a device which conceptually is composed of a plate fixed to a moving mass and a pad pressing against it. A sketch showing the physical principle of a friction device is depicted in Figure 1.3.

An external normal force F_n is applied to a mass by the pad and consequently, because of the relative motion between the pad and the plate and of the presence of friction (which can be represented by a friction coefficient μ or a more elaborated model), a damping force F_d is produced.

Frictional damping is one of the oldest techniques of achieving a damping effect. It is worthwhile remarking that historically early cars were equipped with friction dampers before the advent of the technology of viscous dampers (Bastow, 1993). Leaf spring suspensions too exploit the damping properties of friction arising from the sliding motion among leaves when they are bent.

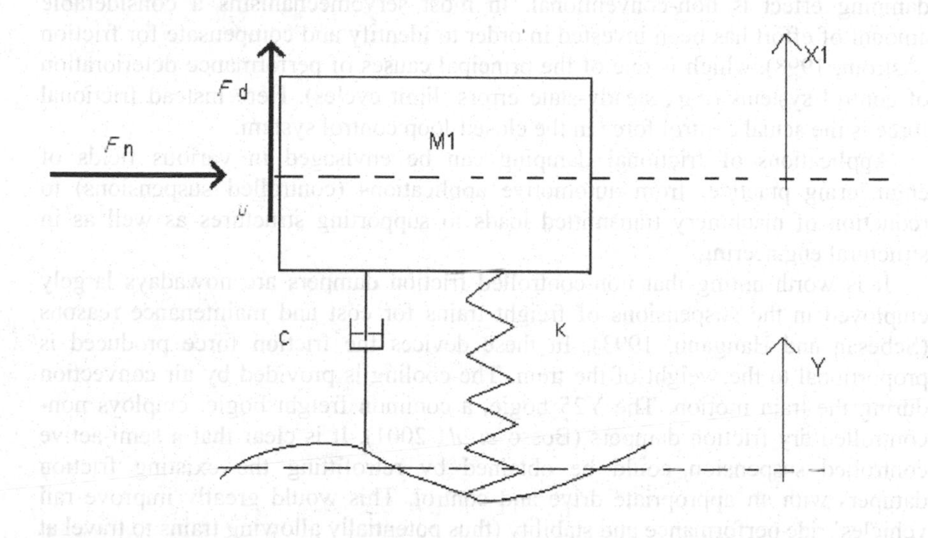

Fig. 1.3. Principle of a friction damper (copyright Elsevier (2003), reproduced from Guglielmino E, Edge KA, Controlled friction damper for vehicle applications, Control Engineering Practice, Vol 12, N 4, pp 431–443 and used by permission)

The pure dry (or lubricated) friction characteristic is of no practical use because of its harshness, but if friction force is modulated by employing modern control techniques, performance can improve tremendously. Anti-locking brakes (ABS) are a good example of closed-loop control using friction force.

Controlling friction entails revisiting an early-day damping technology, improving it in the light of the latest advances in mechatronics and control, and proving through simulation and experimental studies that a controlled semi-active friction damper has potentially superior performance to a conventional viscous damper and could be successfully used in a variety of mobile (and also non-mobile) applications where it can potentially replace a conventional viscous damper for the purpose of reducing vibration.

The challenge is therefore the control of the friction force so as to obtain appropriate damping characteristics for an automotive application. If a friction damper is properly controlled, it can emulate viscous- and spring-type characteristics or any combination of the two, and it is possible to create any other type of generalised damping (*e.g.*, proportional to acceleration). It is also interesting to remark that, if opportunely driven, this device can produce a high damping at low velocity, unlike a passive viscous damper (where at low velocity the damping is low) and can clamp a mass if contingencies require it.

This device is similar in its physical principle to a controlled brake, even if its aim is different. In an ABS the target is to prevent slip so as to minimise the braking distance, whereas in a controlled friction damper the interest is in minimising displacements, velocities and/or accelerations so as to improve vehicle ride and handling. The choice of using dry friction as a means of achieving a damping effect is non-conventional. In most servomechanisms a considerable amount of effort has been invested in order to identify and compensate for friction (Åstrom, 1998), which is one of the principal causes of performance deterioration of control systems (*e.g.*, steady-state errors, limit cycles). Here instead frictional force is the actual control force in the closed-loop control system.

Applications of frictional damping can be envisaged in various fields of engineering practice, from automotive applications (controlled suspensions) to reduction of machinery transmitted loads to supporting structures as well as in structural engineering.

It is worth noting that non-controlled friction dampers are nowadays largely employed in the suspensions of freight trains for cost and maintenance reasons (Sebesan and Hanganu, 1993). In these devices the friction force produced is proportional to the weight of the train. The cooling is provided by air convection during the train motion. The Y25 bogie, a common freight bogie, employs non-controlled dry friction dampers (Bosso *et al.*, 2001). It is clear that a semi-active controlled suspension could be obtained by retrofitting the existing friction dampers with an appropriate drive and control. This would greatly improve rail vehicles' ride performance and stability (thus potentially allowing trains to travel at faster speeds) and also reduce mechanical stress and damage to the railway tracks, with great economical benefit both for train and infrastructure operators. In a study by O'Neill and Wale (1994) it was pointed out how, in order to reach higher speeds and have better ride in rail transport, semi-active systems can be a lower-cost solution than making any modification to the tracks and other infrastructures.

Friction dampers are widely employed in non-mobile applications as well, most importantly in civil engineering *e.g.*, vibration mitigation in buildings and civil structures (Nishitani *et al.*, 1999). Friction dampers have been also proposed for turbine blade vibration control (Sanliturk *et al.*, 1995). Applications to vehicle driveshaft vibration reduction have also been considered (Wang *et al.*, 1996).

In the course of the book it will be shown that friction force can be electro-hydraulically controlled. The damper is force-controlled; in an electrohydraulically actuated scheme this translates to pressure control, and hence the flow required in order to set up the working pressure in the hydraulic circuit is negligible, since the actuator load (the pad) does not move. As a consequence it consumes less energy compared to a controlled viscous damper. Chapter 5 will deal with FDs in detail.

1.7 Introduction to MR Dampers

A magnetorheological damper (MRD) is not very different from a conventional viscous damper. The key difference is the magnetorheological (MR) oil and the presence of a solenoid embedded inside the damper which produces a magnetic field.

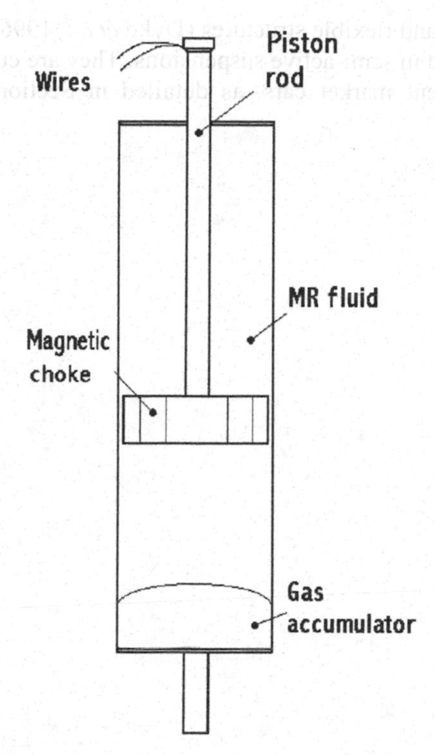

Fig. 1.4. Principle of a magnetorheological damper (copyright Inderscience (2005), reproduced from Guglielmino E, Stammers CW, Stancioiu D, Sireteanu T, Conventional and non-conventional smart damping systems, Int J Vehicle Auton Syst, Vol. 3, N 2/3/4, used by permission)

An MR oil is a particular type of fluid which contains micron-sized ferromagnetic particles in suspension. A detailed analysis of MR fluids properties is given in Agrawal *et al.* (2001). As a consequence of the polarising magnetic field, particles tend to form chains, which modifies the value of the oil yield stress. In such a state rheological properties of oil change and the fluid passes from the liquid state to the semi-solid state. Hence by controlling the solenoid current, continuously variable damping can be produced without employing moving parts such as valves or variable orifices. The energy requirements are extremely low. For control, it is only necessary to supply the solenoid with a conventional battery.

Likewise it is important to remark that MR fluid rheological properties are virtually temperature and contamination independent. Therefore MRDs are rugged and reliable devices, capable of providing excellent performance over a wide variety of operating conditions.

It is clear that an appropriate control logic is crucial to take full advantage of the potential offered by an MRD. It will be shown later that similarities exist between FD and MRD control as their static characteristics are somehow similar.

The main domains of application are automotive and structural. In the latter, they are employed for earthquake protection and for damping wind-induced oscillations of bridges and flexible structures (Dyke *et al.*, 1996). In the automotive field they are employed in semi-active suspensions. They are currently present on a number of high-segment market cars, as detailed in Section 1.2. Chapter 6 is dedicated to MRDs.

2

Dampers and Vehicle Modelling

2.1 Introduction

The heart of a semi-active suspension is the controllable damper. Its accurate modelling is crucial for suspension analysis and design. In many practical applications the damper characteristic exhibits a strong non-linearity, which must be taken into account in simulation studies in order to obtain realistic results when investigating system performance.

A damper is identified by its force versus velocity characteristics (damping characteristics or hydraulic characteristics), which can be expressed by the functional relation

$$F_d = f(\dot{x}), \tag{2.1}$$

F_d being the damping force generated and $\dot{x}$ the velocity across it. Other damping elements have a more general non-linear characteristics expressed by the functional relation

$$F_d = f(x, \dot{x}). \tag{2.2}$$

Such a characteristic is typical of viscoelastic materials, but it could well represent a controlled semi-active damper, whilst generalised semi-active damping devices can be made to have a characteristic expressed by

$$F_d = f(x, \dot{x}, \ddot{x}) \tag{2.3}$$

with acceleration-dependent damping too.

A certain amount of hysteresis is always present in a damper characteristic, depending upon its internal dissipation mechanism. As introduced in Chapter 1 a damper can be viewed from an energy standpoint as a device that dissipates energy through an internal mechanism (*e.g.*, by throttling a viscous flow through an orifice). In order to fully identify a damper, besides its damping characteristics it is customary to define also its force versus displacement characteristics (damper work characteristics), the area of which gives a measure of the energy dissipated over a complete cycle.

This chapter presents the mathematical techniques necessary to model real dampers with hysteresis in their characteristics, and subsequently reviews the main car and truck ride models, developed for suspension studies. The last part of the chapter deals with road modelling.

Figures 2.1 and 2.2 plot the characteristics of two types of dampers: an ideal linear viscous damper and an ideal Coulomb friction damper. These are idealised characteristics as no hysteresis is present in the force versus velocity characteristics (real dampers always contain a certain amount of hysteresis in their force versus velocity map).

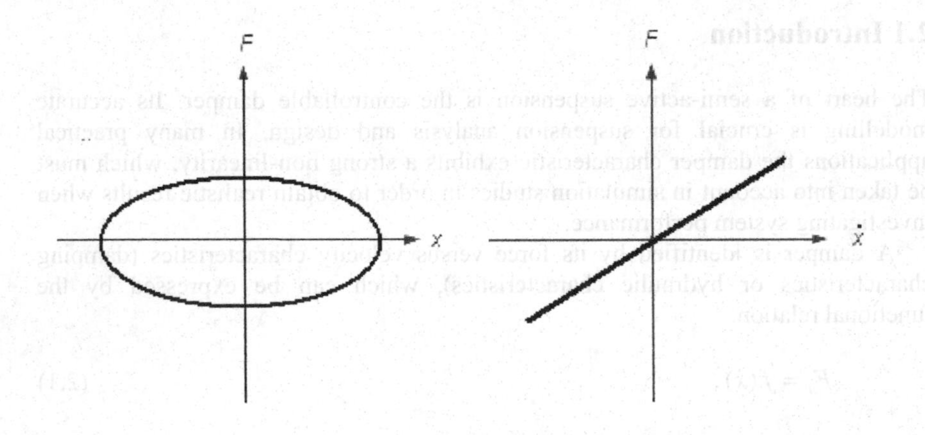

Fig. 2.1. Linear viscous damper characteristics

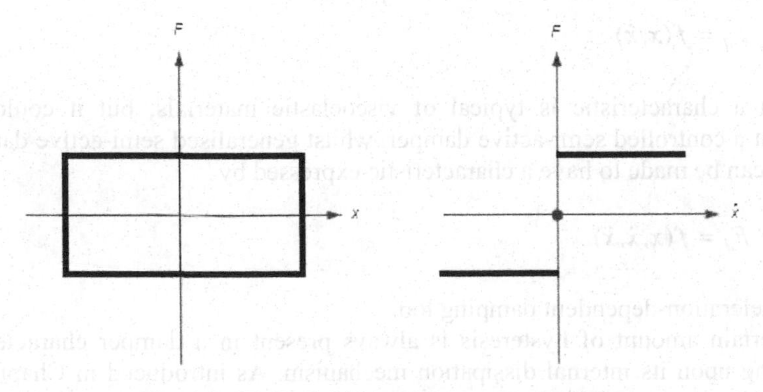

Fig. 2.2. Coulomb friction damper characteristics

2.2 Phenomenology of Hysteresis

Hysteresis occurs in a variety of physical systems; the most noteworthy examples are ferromagnetic materials, constituting the core of motors, generators, transformers and a wide range of other electrical devices. In automotive applications hysteresis is present not only in damper characteristics but also, for instance, in tyre characteristics and in all viscoelastic and viscoplastic materials in general.

From a control systems standpoint, hysteresis must not always be regarded as a non-linearity hampering controller performance or making its design more difficult. Under particular circumstances hysteresis can be beneficially exploited in control systems, namely in on–off control algorithms (Gerdes and Hedrick, 1999) for reducing chatter which may occur when in a control loop the difference between setpoint and feedback (*i.e.*, the error) is close to zero. In such systems the required amount of hysteresis is typically generated within the control software in microprocessor-based programmable architectures or in hardware by employing electronic devices or exploiting the inherent hysteresis present in actuator characteristics (*e.g.*, valves).

Virtually no material or device employed in mechanical and structural systems is perfectly elastic, and restoring forces generated as a result of deformations are not perfectly conservative. Likewise internal dissipation within viscous fluids in dampers result in a hysteretic characteristic.

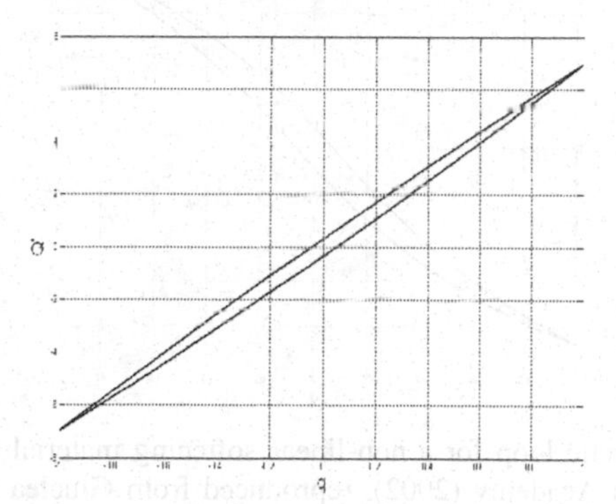

Fig. 2.3. Typical hysteretic loop for a linear material (copyright Publishing House of the Romanian Academy (2002), reproduced from Giuclea M, Sireteanu T, Mita AM, Ghita G, Genetic algorithm for parameter identification of Bouc–Wen model, Rev Roum Sci Techn Mec Appl, Vol 51, N 2, pp 179–188, used by permission)

A wide variety of micromechanisms contribute to energy dissipation in cyclically loaded materials and in viscous fluids. This behaviour macroscopically results in a hysteretic loop when the material or device is subject to a sinusoidal displacement Q'. However, it is not possible — except in very special cases — to quantitatively predict the macroscopic hysteretic behaviour starting from physical models of the microscopic behaviour. Therefore, from an engineering standpoint, an experimental assessment of the hysteresis loop and the corresponding energy losses is required to characterise a material. Figures 2.3, 2.4 and 2.5 depict typical hysteretic loops for a linear material, a non-linear hardening material and a non-linear softening material, respectively.

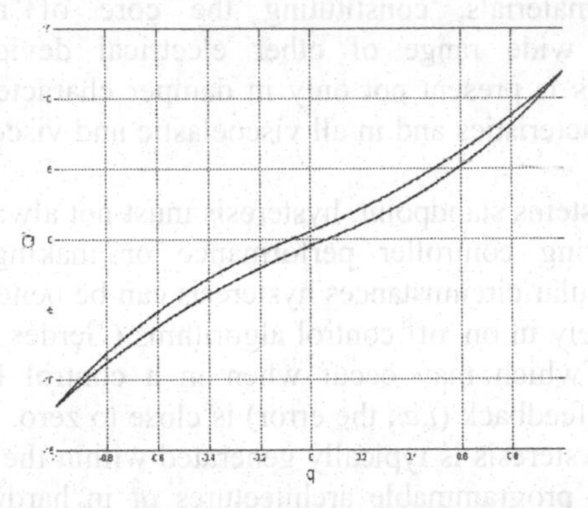

Fig. 2.4. Typical hysteretic loop for a non-linear hardening material (copyright Publishing House of the Romanian Academy (2002), reproduced from Giuclea M, Sireteanu T, Mita AM, Ghita G, Genetic algorithm for parameter identification of Bouc–Wen model, Rev Roum Sci Techn Mec Appl, Vol 51, N 2, pp 179–188, used by permission)

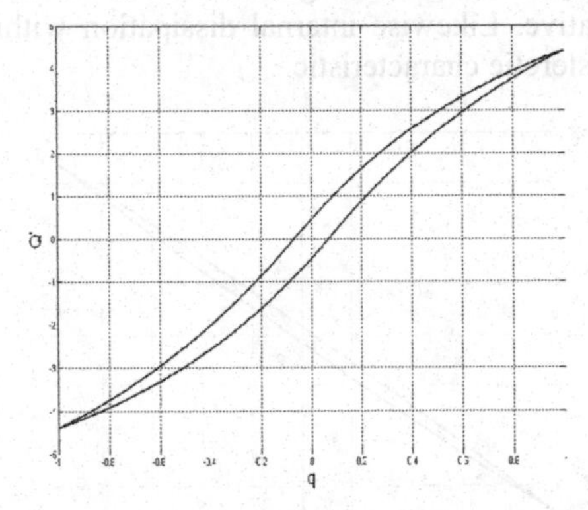

Fig. 2.5. Typical hysteretic loop for a non-linear softening material (copyright Publishing House of the Romanian Academy (2002), reproduced from Giuclea M, Sireteanu T, Mita AM, Ghita G, Genetic algorithm for parameter identification of Bouc–Wen model, Rev Roum Sci Techn Mec Appl, Vol 51, N 2, pp 179–188, used by permission)

The area enclosed by the loop is a measure of the dissipated energy. The area, and hence the energy dissipated per cycle, can be calculated by the following contour integral:

$$E = \oint Q' \, \mathrm{d}q \, . \tag{2.4}$$

Depending upon the material, the hysteresis loop can be either very thin (and generally elliptical in shape), resulting in a very small amount of energy dissipation (Figure 2.6) or larger, hence producing a more significant energy consumption.

Thin loops are likely to occur in elastic materials, such as steel, when they are cyclically loaded within their elastic range. Conversely, materials loaded in their inelastic range exhibit wider hysteresis loops, as portrayed in Figure 2.7. Composite materials too dissipate significant amounts of energy.

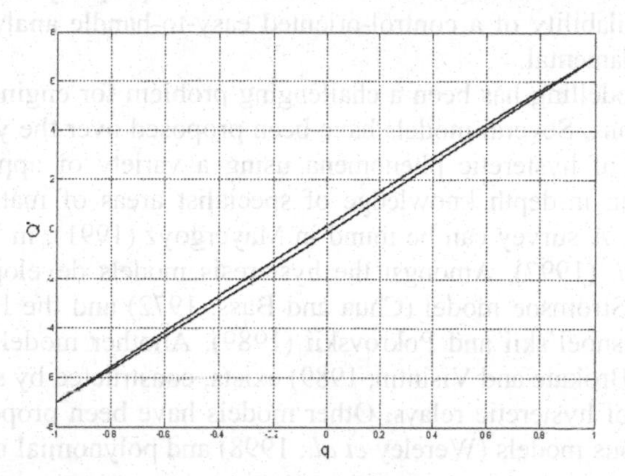

Fig. 2.6. Deformation in the elastic range (copyright Publishing House of the Romanian Academy (2002), reproduced from Giuclea M, Sireteanu T, Mita AM, Ghita G, Genetic algorithm for parameter identification of Bouc–Wen model, Rev Roum Sci Techn Mec Appl, Vol 51, N 2, pp 179–188, used by permission)

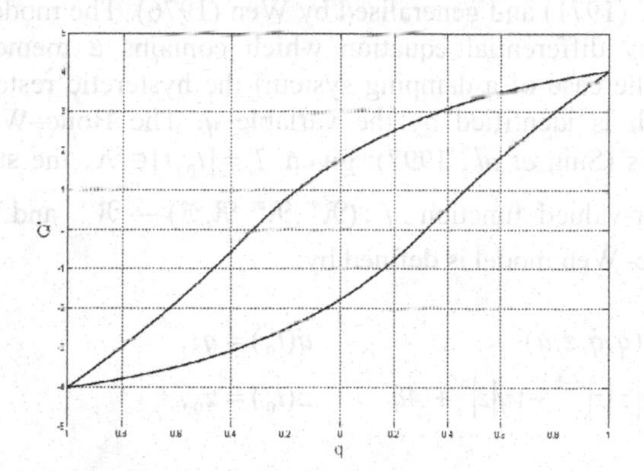

Fig. 2.7. Deformation in the plastic range (copyright Publishing House of the Romanian Academy (2002), reproduced from Giuclea M, Sireteanu T, Mita AM, Ghita G, Genetic algorithm for parameter identification of Bouc–Wen model, Rev Roum Sci Techn Mec Appl, Vol 51, N 2, pp 179–188, used by permission)

A loss function $L(E)$ can therefore be defined for a particular material or damping device under a sinusoidal load. Measuring the area enclosed by the hysteretic loop, and dividing it by the cycle period T, the average energy loss per cycle can be obtained

$$L(E) = \frac{1}{T}\int_0^T Q' \dot{q} \mathrm{d}t \,. \tag{2.5}$$

2.3 Damper Hysteresis Modelling

The energy dissipation and the impact on control systems performance of hysteretic materials can be significant and in order to properly assess the system response the availability of a control-oriented easy-to-handle analytical model of hysteresis is fundamental.

Hysteresis modelling has been a challenging problem for engineers, physicists and mathematicians. Several models have been proposed over the years to capture different classes of hysteretic phenomena using a variety of approaches. Some models require an in-depth knowledge of specialist areas of mathematics to be fully appreciated. A survey can be found in Mayergoyz (1991), in Visintin (1994) and in Sain *et al.* (1997). Amongst the hysteresis models developed it is worth citing the Chua–Stromsoe model (Chua and Bass, 1972) and the Hysteron model proposed by Krasnoel′skii and Pokrovskii (1989). Another model, known as the Preisach model (Brokate and Visintin, 1989) exists, constructed by superposing the outputs of a set of hysteretic relays. Other models have been proposed, including hysteretic biviscous models (Wereley *et al.*, 1998) and polynomial models (Choi *et al.*, 2001).

A model having an appealing simplicity is the Bouc–Wen model, which has gained large consensus within the engineering community. A wide variety of hysteretic shapes can be represented by using this simple differential model proposed by Bouc (1971) and generalised by Wen (1976). The model is based on a nonlinear ordinary differential equation which contains a memory variable z, representing (in the case of a damping system) the hysteretic restoring force, the position of which is identified by the variable q. The Bouc–Wen equation is defined as follows (Sain *et al.*, 1997): given $T=[t_0,t]\in\Re$, the states $q(t)$, $z(t)$: $T\to\Re$, a vector-valued function $f:(\Re^m,\Re^m,\Re,\Re)\to\Re^m$ and the input $u(t)$: $T\to\Re$, the Bouc–Wen model is defined by:

$$\ddot{q}(t)=f(q,\dot{q},z,u) \qquad \ddot{q}(t_0)=q_0, \tag{2.6a}$$

$$\dot{z}=-\gamma|\dot{q}|\,z\,|z|^{n-1}-v\dot{q}|z|^{n}+A\dot{q} \qquad z(t_0)=z_0, \tag{2.6b}$$

where q is the imposed displacement of the device (or the material deformation). The quantities γ, v, A, n are loop parameters defining the shape and the amplitude of the hysteresis loop.

Figures 2.8, 2.9 and 2.10 depicts three typical hysteretic loops plotted for different sets of parameters and for an imposed harmonic displacement $q(t)=q_0\sin(2\pi ft)$, where $f=1$ Hz is the frequency and q_0 is the amplitude of the sinusoidal input.

By consideration of the three Figures 2.8, 2.9 and 2.10 it can be seen that the Bouc–Wen equation can describe both linear and non-linear (softening and hardening) hysteretic behaviour.

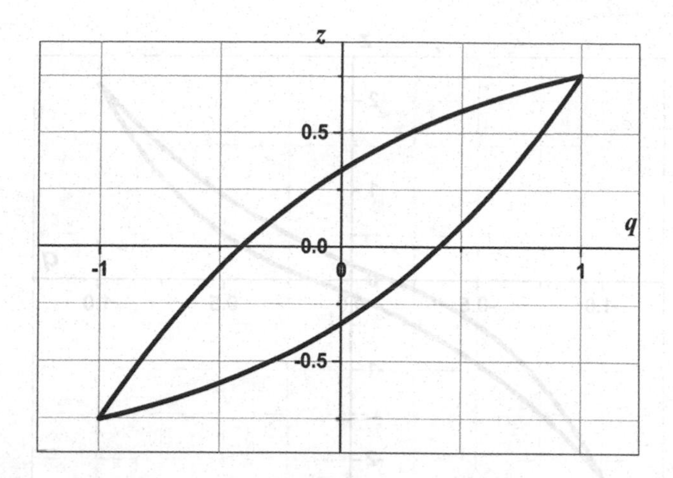

Fig. 2.8. Linear hysteretic behaviour, plotted for $\gamma = 0.9$, $\nu = 0$, $A = 1$, $n = 1$ (copyright Publishing House of the Romanian Academy (2002), reproduced from Giuclea M, Sireteanu T, Mita AM, Ghita G, Genetic algorithm for parameter identification of Bouc–Wen model, Rev Roum Sci Techn Mec Appl, Vol 51, N 2, pp 179–188, used by permission)

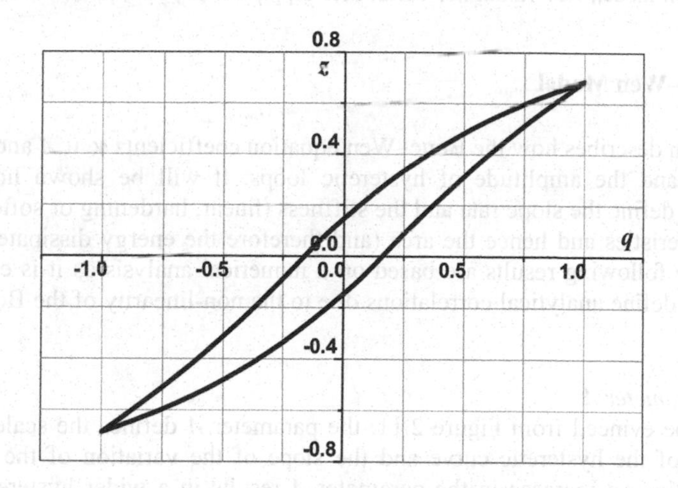

Fig. 2.9. Non-linear softening hysteretic behaviour, plotted for $\gamma = 0.75$, $\nu = 0.25$, $A = 1$, $n = 1$ (copyright Publishing House of the Romanian Academy (2002), reproduced from Giuclea M, Sireteanu T, Mita AM, Ghita G, Genetic algorithm for parameter identification of Bouc–Wen model, Rev Roum Sci Techn Mec Appl, Vol 51, N 2, pp 179–188, used by permission)

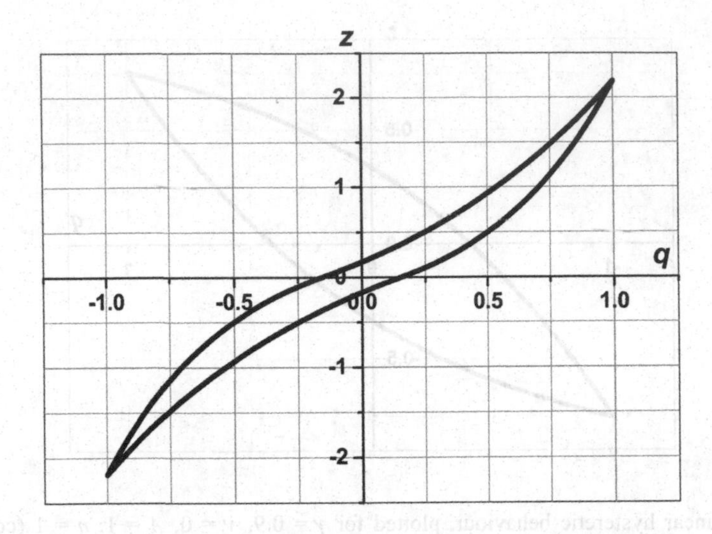

Fig. 2.10. Non-linear hardening hysteretic behaviour (plotted for $\gamma = 0.5$, $\nu = -1.5$, $A = 1$, $n = 1$) (copyright Publishing House of the Romanian Academy (2002), reproduced from Giuclea M, Sireteanu T, Mita AM, Ghita G, Genetic algorithm for parameter identification of Bouc–Wen model, Rev Roum Sci Techn Mec Appl, Vol 51, N 2, pp 179–188, used by permission)

2.3.1 Bouc–Wen Model

This section describes how the Bouc–Wen equation coefficients γ, ν, A and n affect the shape and the amplitude of hysteretic loops. It will be shown how these parameters define the slope rate and the stiffness (linear, hardening or softening) of the characteristics and hence the area (and therefore the energy dissipated over a cycle). The following results are based on a numerical analysis as it is extremely difficult to define analytical correlations due to the non-linearity of the Bouc–Wen equation.

2.3.1.1 Parameter A

As it can be evinced from Figure 2.11, the parameter A defines the scale and the amplitude of the hysteretic curve and the slope of the variation of the stiffness characteristic. An increase in the parameter A results in a wider hysteresis loop, and consequently in a larger energy dissipation.

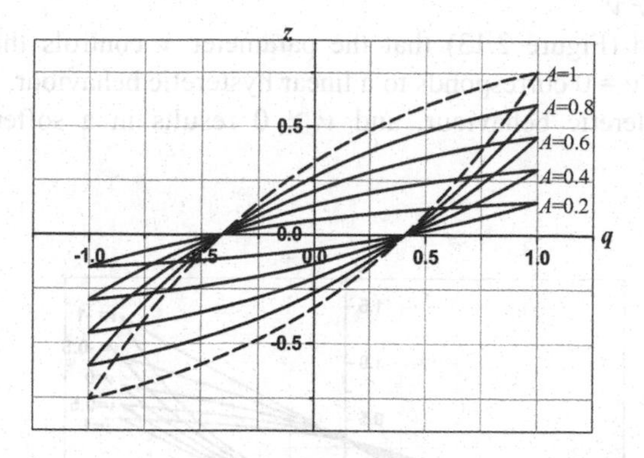

Fig. 2.11. Hysteretic loop dependence on the parameter A (plotted for $\gamma = 0.9$, $\nu = 0.1$, $n = 1$) (copyright Publishing House of the Romanian Academy (2002), reproduced from Giuclea M, Sireteanu T, Mita AM, Ghita G, Genetic algorithm for parameter identification of Bouc–Wen model, Rev Roum Sci Techn Mec Appl, Vol 51, N 2, pp 179–188, used by permission)

2.3.1.2 Parameter γ

The dependence of the hysteretic loops on the parameter γ is portrayed in Figure 2.12 (plotted for $\nu = 0.1$, $n = 1$, $A = 1$) and can be summarised by saying that the area of the loop increases if γ increases from 0 to a value γ_0 (between 2 and 3 in the numerical example of Figure 2.12, but in general dependent on the other coefficients), and for values of γ larger than γ_0 the energy dissipated per cycle slightly decreases.

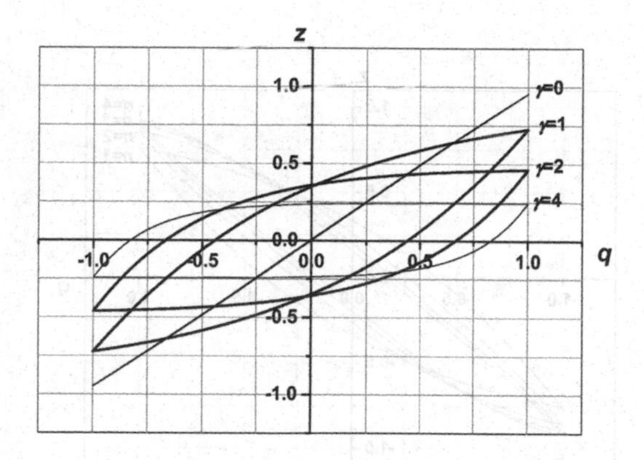

Fig. 2.12. Hysteretic loop dependence on the parameter γ (plotted for $\nu = 0.1$, $n = 1$, $A = 1$) (copyright Publishing House of the Romanian Academy (2002), reproduced from Giuclea M, Sireteanu T, Mita AM, Ghita G, Genetic algorithm for parameter identification of Bouc–Wen model, Rev Roum Sci Techn Mec Appl, Vol 51, N 2, pp 179–188 used by permission)

2.3.1.3 Parameter ν

It can be noticed (Figure 2.13) that the parameter ν controls the shape of the hysteretic curve: $\nu = 0$ corresponds to a linear hysteretic behaviour, $\nu < 0$ produces a hardening hysteretic behaviour, and $\nu > 0$ results in a softening hysteretic behaviour.

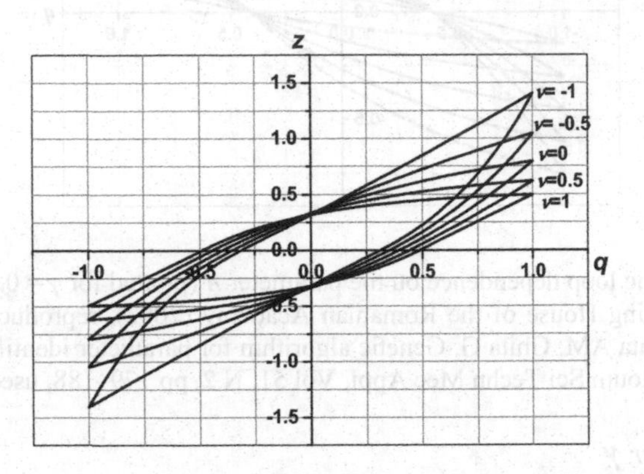

Fig. 2.13. Hysteretic loop dependence on the parameter ν (plotted for $\gamma = 0.9$, $n = 1$, $A = 1$) (copyright Publishing House of the Romanian Academy (2002), reproduced from Giuclea M, Sireteanu T, Mita AM, Ghita G, Genetic algorithm for parameter identification of Bouc–Wen model, Rev Roum Sci Techn Mec Appl, Vol 51, N 2, pp 179–188, used by permission)

2.3.1.4 Parameter n

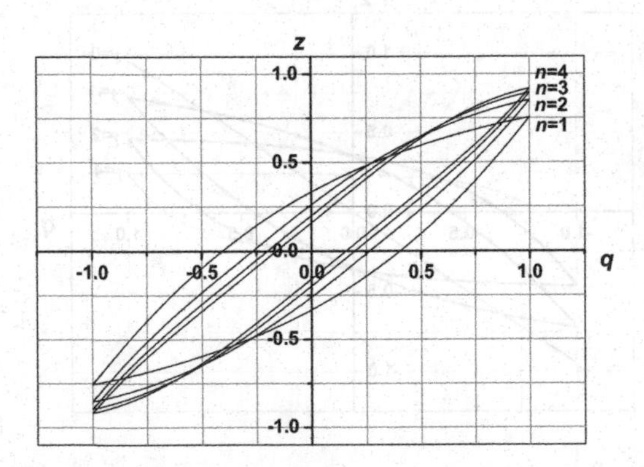

Fig. 2.14. Hysteretic loop dependence on the parameter n (plotted for $\gamma = 0.9$, $\nu = 0.1$, $A = 1$) (copyright Publishing House of the Romanian Academy (2002), reproduced from Giuclea M, Sireteanu T, Mita AM, Ghita G, Genetic algorithm for parameter identification of Bouc–Wen model, Rev Roum Sci Techn Mec Appl, Vol 51, N 2, pp 179–188, used by permission)

As Figure 2.14 (plotted for $\gamma = 0.9$, $\nu = 0.1$, $A = 1$) shows, the variation is significant for small values of n (between 1 and 2 in the numerical example of Figure 2.14), while for larger values ($n > 2$) its effect is negligible.

2.4 Bouc–Wen Parameters Identification

The Bouc–Wen model has the ability to portray a wide range of hysteretic behaviour and by an appropriate choice of the equation coefficients both the slope variation of the stiffness characteristic and the energy dissipated per cycle can be precisely established. The Bouc–Wen equation can be readily combined with plant differential equations to yield an overall dynamic model.

The shape of the Bouc–Wen hysteretic loop depends on the four parameters γ, ν, A and n whose physical meaning has been discussed above. However their identification is not straightforward as the dependence between z and the set of the four parameters is strongly nonlinear and not easy to investigate analytically; furthermore the parameter variation ranges are different. Parameter identification through least-square-based methods is a possible avenue, but may not be the best choice. Black-box optimisation methods based on artificial intelligence techniques such as genetic algorithms (GA) could also be beneficial. Such a method will be employed in Chapter 6 in the context of MRD parameters identification. The method is here briefly introduced. The reader interested in furthering the topic can refer to the textbook of Goldberg (1989).

A GA is a probabilistic search technique inspired from the evolution of species. Such an optimisation tool has an inherent parallelism and ability to avoid stagnation in local optima. It starts with a set of potential solutions called individuals and evolves towards better solutions with respect to an objective function. Genetic operators are defined (namely crossover, mutation and selection) and their application drives the solution towards the optimum. The main elements of a standard GA are genetic representation for potential solutions, an objective function, genetic operators, characteristic constants such as population size, probability of applying an operator, number of children and so forth.

The Bouc–Wen coefficients search problem can be stated as finding a set of parameters γ, ν, A and n such that the Bouc–Wen model given by Equation 2.6 determines a hysteretic curve which is a good approximation of an experimental one, knowing the imposed displacement $q(t)$ and a set of measured data $(q_i, z^*_i)_{i=1,\dots,n}$ corresponding to a complete cycle.

2.5 Vehicle Ride Models

A broad variety of vehicle mathematical models of increasing degree of complexity has been developed over the years by automotive engineers to provide reliable models for computer-aided automotive design and vehicle performance assessement.

From a purely mathematical standpoint vehicle models can be categorised as distributed models (*i.e.*, governed by partial differential equations) and lumped parameter models (*i.e.*, governed by ordinary differential equations). The former are mainly of interest to vehicle design rather than control algorithm design. Distributed models (typically solved numerically with finite-element-based methods) are widely employed in mechanical, thermal, aerodynamic analyses as well as crashworthiness analyses. For car dynamics (ride and handling) and control studies, lumped parameter models are usually employed. They typically aim to model either ride or handling dynamics or both. In this book only lumped parameter models will be considered.

A car can be thought as being composed of two main subsystems: the sprung mass (chassis) and the unsprung masses (wheels, axles and linkages), connected via a number of elastic and dissipative elements (suspensions, tyres *etc.*) and subjected to external inputs coming from the road profile, the steering system and other external disturbances (*e.g.*, wind gust).

The motion of a vehicle with the nonholonomic constraint of the road has six degrees of freedom (6DOF), classified as follows:

- longitudinal translation (forward and backward motion)
- lateral translation (side slip)
- vertical translation (bounce or heave)
- rotation around the longitudinal axis (roll)
- rotation around the transverse axis (pitch)
- rotation around the vertical axis (yaw)

Vehicle ride is essentially concerned with car vertical dynamics (bounce, pitch, roll) whereas handling is concerned with lateral dynamics (side slip, yaw, roll). Ride models are typically composed of interconnected spring–mass–damper systems and defined by a set of ordinary differential equations.

The most trivial representation of a vehicle suspension has 1DOF. In this simple model the chassis (body) is represented by a mass and the suspension unit by a spring and a damper. Tyre mass and stiffness are neglected as well as any cross-coupling dynamics.

By incorporating a wheel into the model, a more accurate representation having 2DOF (typically referred to as a quarter car model) can be developed. This model was (and still is) very popular in the automotive engineering community, especially before the widespread use of computer simulation, the reason being that the quarter car model, despite its simplicity, features the main variables of interest to suspension performance assessment: body acceleration, dynamic tyre force and suspension working space (Sharp and Hassan, 1986). A merit of the quarter car is that it permits to evaluate more straightforwardly the effects of modifications in control parameters because higher-order dynamics and cross-coupling terms with the other suspension units are not taken into account. A good suspension design should produce improvement of both vehicle road holding and passenger comfort (or possibly improvement of one without degradation of the other), although inherent trade-offs are unavoidable in the design of a passive suspension system.

The quarter car is a 2DOF system having two translational degrees of freedom. Another classical model can be obtained with only 2DOF: the half vehicle model having a translational degree of freedom and a rotational degree of freedom to describe, respectively, bounce and pitch motions or, analogously, bounce and roll motions (in the former case the model is referred to as a bicycle model). Its natural extension is a 4DOF model, which also includes tyre masses and elasticity. This model can be employed to study the vehicle pitch (or roll) behaviour. However the 4DOF model cannot take into account the cross-couplings between the right- and left-hand side of the car (or front and rear in the case of roll motion). These interactions can be taken into account only by using a 7DOF model (sometimes referred to as a full car model), which allows to represent bounce, roll and pitch motions.

The models mentioned above are classical ride models. Higher order ride models can be developed including further degrees of freedom, *e.g.*, accounting for seat and engine mounting elasticity. Driver and passengers can be modelled as well with springs, masses and damping elements. This is particularly important for accurate human comfort studies. Chapter 3 will deal with this topic in detail.

Analogously to ride vehicle models, also handling models having different degrees of complexity can be developed. The equivalent handling model of the quarter car is a linear single track model which describes lateral and yaw dynamic responses to handling manoeuvres (ignoring the effect of sprung and unsprung masses).

Models including both ride and handling dynamics are necessary when there is a need to accurately investigate the interaction between ride and handling (during a turning manoeuvre, for instance) and to study the limit of handling characteristics or elements such as anti-roll bars. Multibody techniques allow relatively easy development of complicated models with many degrees of freedom. In 1991, Zeid and Chang described a 64DOF model. Models with hundreds of degrees of freedom have been developed by automotive engineers. Such involved models, however, despite their sophistication, suffer from two main drawbacks: parameter uncertainty and long simulation running time. For these reasons they are not always the best choice for control design. Especially in the early stages of the design, a less complicated model is preferable, reserving the use of higher-order models for further refinements and optimisations.

Analogously to cars, several heavy and trailed vehicles (tractor/semi-trailer) models (Gillespie,1992; Wong, 1993) have been developed to examine their ride comfort, tractor–trailer interactions, dynamic tyre forces and road damage. A survey can be found in Jiang *et al.* (2001). These vehicle models usually include linear tyre models, linear or nonlinear suspension characteristics, tandem or single axles. Other truck ride models include suspended tractor cab and driver seat with linear or non-linear components.

The models described and the simulation results presented in this book are based on the use of MATLAB® and Simulink® software. In the following sections the classical vehicle and truck models will be briefly revised.

2.5.1 Quarter Car Model

The quarter car has for a long time been the *par excellence* model used in suspension design. It is a very simple model as it can only represent the bounce motion of chassis and wheel without taking into account pitch or roll vibration modes. However it is very useful for a preliminary design: it is described by the following system of second-order ordinary differential equations (Figure 2.15; tyre damping is not shown in the figure):

$$m_1\ddot{x}_1 = -2\xi\omega_1(\dot{x}_1 - \dot{x}_2) - k_s(x_1 - x_2), \tag{2.7a}$$
$$m_2\ddot{x}_2 = 2\xi\omega_1(\dot{x}_1 - \dot{x}_2) + k_s(x_1 - x_2) - k_t(x_2 - z_0) - c_t(\dot{x}_2 - \dot{z}_0). \tag{2.7b}$$

If relative displacement is defined as $x = x_1 - x_2$, Equations 2.7a and 2.7b can be rewritten in a more compact form:

$$m_1\ddot{x}_1 = -2\xi\omega_1\dot{x} - k_s\,x, \tag{2.8a}$$
$$m_2\ddot{x}_2 = 2\xi\omega_1\,\dot{x} + k_s\,x - k_t(x_2 - z_0) - c_t(\dot{x}_2 - \dot{z}_0). \tag{2.8b}$$

From the analysis of the quarter car model equations some fundamental properties of the passive suspensions can be analytically evinced and it is possible to quantify the compromise when reduction of both chassis acceleration, suspension working space (sometimes referred as rattle space) and tyre deflection is pursued: the

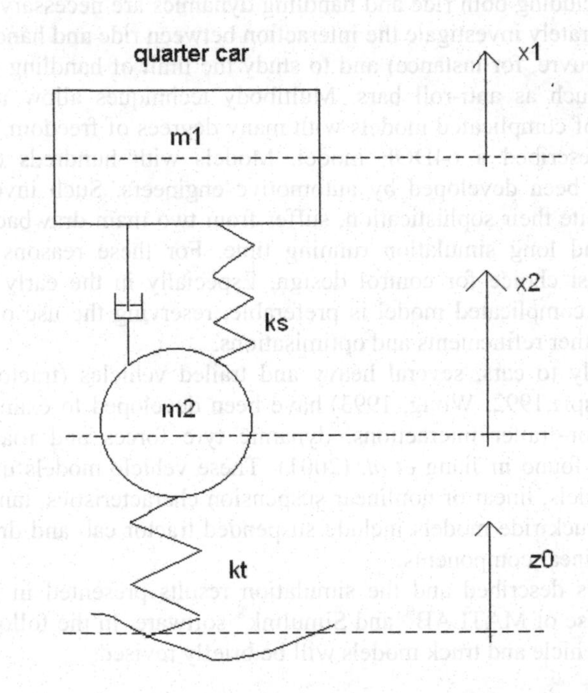

Fig. 2.15. Quarter car model (copyright Inderscience (2005) reproduced with minor modifications from Guglielmino, E, Stammers CW, Stancioiu D and Sireteanu T, Conventional and non-conventional smart damping systems, Int J Vehicle Auton Syst, Vol. 3, N 2/3/4, pp 216–229, used by permission)

quarter car is a dynamic system composed of two interconnected subsystems and as such is subject to constraint equations, independent of the type of interconnections. From the analysis of the quarter car model Hedrick and Butsuen (1988) showed that only three transfer functions can be independently defined and that invariant points (*i.e.*, values at specified frequencies depending only on k_t, m_1 and m_2 but not on k_s) exist at particular frequencies. In particular they showed that the acceleration transfer function has an invariant point at the wheel-hop frequency. Similarly, the suspension deflection transfer function has an invariant point at the rattle space frequency. The trade-off between passenger comfort and suspension deflection occurs because it is not possible to simultaneously keep both transfer functions small around the wheel-hop frequency in the low-frequency range.

2.5.2 Half Car Model

Pitch motion can be taken into account with the half vehicle model (also known as the bicycle model). The governing equations are the following (Figure 2.16), with α representing pitch:

$$\begin{cases} m\ddot{x} = -k_2(z-b\alpha-z_4)-k_1(z+a\alpha-z_3)-c_2(\dot{z}-b\dot{\alpha}-\dot{z}_4)-c_1(\dot{z}+a\dot{\alpha}-\dot{z}_3), \\ J\ddot{\alpha} = k_2(z-b\alpha-x_4)b-k_1(x+a\alpha-z_3)a-c_2(\dot{z}-b\dot{\alpha}-\dot{z}_4)b-c_1(\dot{z}+a\dot{\alpha}-\dot{z}_3)a, \\ m_1\ddot{z}_3 = k_1(z+a\alpha-z_3)-k_{01}(z_3-z_{01})+c_1(\dot{z}+a\dot{\alpha}-\dot{z}_3), \\ m_2\ddot{z}_4 = k_2(z-b\alpha-z_4)-k_{02}(z_4-z_{02})+c_2(\dot{z}-b\dot{\alpha}-\dot{z}_4), \end{cases} \tag{2.9}$$

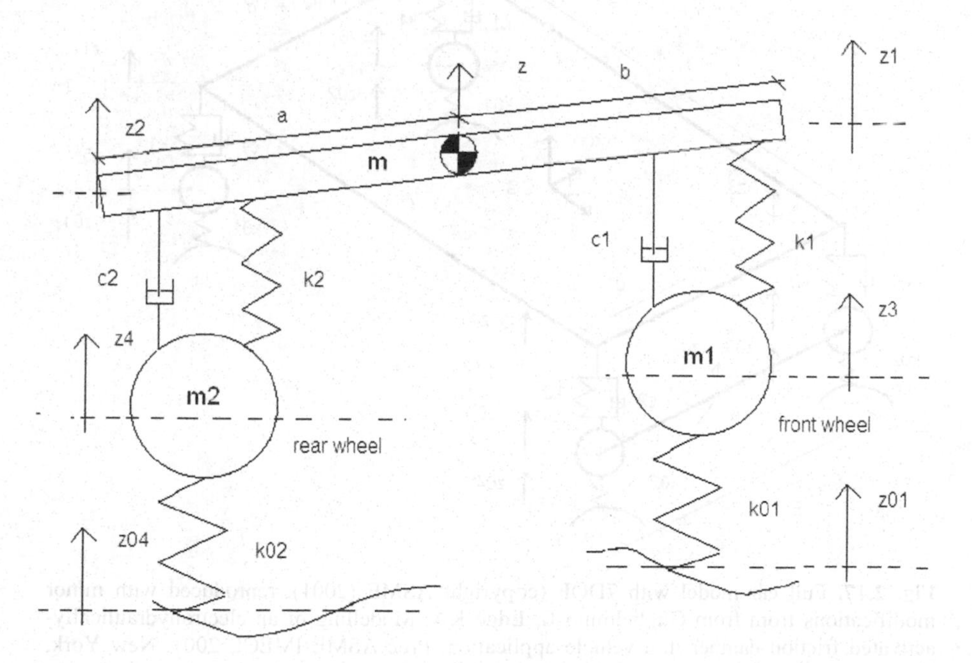

Fig. 2.16. Half car model with 4DOF

where m is the sprung mass, m_1, m_2 the front and rear unsprung masses, J the pitch inertia, and a and b the distances of the front and rear of the vehicle from its centre of gravity. Replacing α with the roll angle, pitch inertia J by roll inertia and b by half-track length this model can be also employed to describe roll motion.

2.5.3 Full Car Model

The 7DOF vehicle ride model (Sireteanu *et al.*, 1981) extends the half car model to the entire vehicle: 3DOF are used for the sprung mass (bounce, roll and pitch), while the unsprung masses have 4DOF (1DOF for each tyre), as depicted in Figure 2.17. The governing equations can be written compactly in matrix form (bold letters denote matrices and vectors):

$$\boldsymbol{M}\ddot{\boldsymbol{q}} + \boldsymbol{P}^T\boldsymbol{C}\boldsymbol{P}\dot{\boldsymbol{q}} + \boldsymbol{P}^T\boldsymbol{K}\boldsymbol{P}\boldsymbol{q} + \boldsymbol{F}_d = -\boldsymbol{P}^T\boldsymbol{K}_0\boldsymbol{z}_0 - \boldsymbol{P}^T\boldsymbol{C}_0\dot{\boldsymbol{z}}_0, \qquad (2.10)$$

with $\boldsymbol{q}\in\Re^7$, $\boldsymbol{z}_0\in\Re^4$, $\boldsymbol{F}_d\in\Re^7$, $\boldsymbol{M}\in\Re^{7x7}$, $\boldsymbol{K}\in\Re^{8x8}$, $\boldsymbol{C}\in\Re^{8x8}$, $\boldsymbol{K}_0\in\Re^{8x4}$, $\boldsymbol{C}_0\in\Re^{8x4}$ and $\boldsymbol{P}\in\Re^{8x7}$.

The vertical displacement vector $\boldsymbol{z}\in\Re^8$ is defined as:

$$\boldsymbol{z} = [z_1, z_2, z_3, z_4, z_5, z_6, z_7, z_8]^T \qquad (2.11)$$

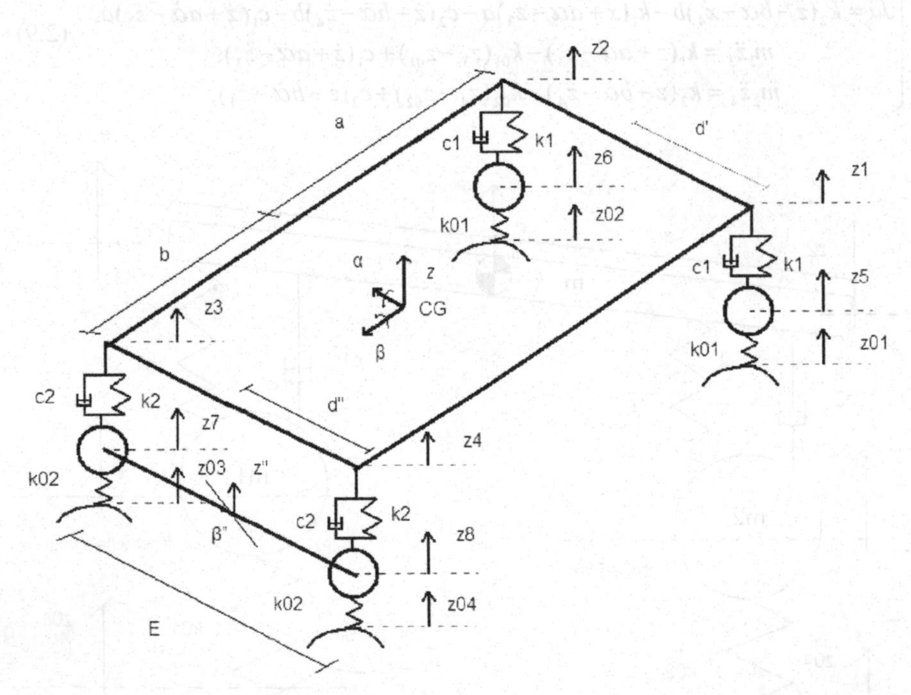

Fig. 2.17. Full car model with 7DOF (copyright ASME (2001), reproduced with minor modifications from from Guglielmino E, Edge KA, Modelling of an electrohydraulically-activated friction damper in a vehicle application, Proc ASME IMECE 2001, New York, used by permission)

(the vertical displacements are not all independent).

Let $\boldsymbol{q}$ be the vector of generalised co-ordinates:

$$\boldsymbol{q} = [q_1, q_2, q_3, q_4, q_5, q_6, q_7]^T \tag{2.12}$$

with the following choice of co-ordinates:

$$q_1 = z\,;\, q_2 = z_5\,;\, q_3 = z_6\,;\, q_4 = z''\,;\, q_5 = \alpha\,;\, q_6 = \beta\,;\, q_7 = \beta'', \tag{2.13}$$

z and $\boldsymbol{q}$ being related by the matrix $\boldsymbol{P}$, dependent upon the vehicle geometry:

$$z = \boldsymbol{Pq}. \tag{2.14}$$

Consider the vertical displacement vector and the matrix $\boldsymbol{P}$ being defined as:

$$\boldsymbol{P} = \begin{bmatrix} 1 & 0 & 0 & 0 & -a & d' & 0 \\ 1 & 0 & 0 & 0 & -a & -d' & 0 \\ 1 & 0 & 0 & 0 & b & d'' & 0 \\ 1 & 0 & 0 & 0 & b & -d'' & 0 \\ 0 & 1 & 0 & 0 & 0 & 0 & 0 \\ 0 & 0 & 1 & 0 & 0 & 0 & 0 \\ 0 & 0 & 0 & 1 & 0 & 0 & \frac{E}{2} \\ 0 & 0 & 0 & 1 & 0 & 0 & -\frac{E}{2} \end{bmatrix}, \tag{2.15}$$

where a and b are the distances of the front and rear of the vehicle from its centre of gravity, d' and d'' are, respectively, the front and rear half-track lengths and E the inter-wheel distance.

The road input vector z_0 is then defined as:

$$\boldsymbol{z_0} = [z_{01}, z_{02}, z_{03}, z_{04}]^T. \tag{2.16}$$

Equation 2.10 can be obtained using Lagrangian formalism. The Lagrange equations, expressed as a function of the kinetic energy are:

$$\frac{\mathrm{d}}{\mathrm{d}t}\frac{\partial T}{\partial \dot{q}_k} - \frac{\partial T}{\partial q_k} = g_k \qquad k=1,\dots,7 \tag{2.17}$$

where T is the total kinetic energy of the system, defined by the quadratic form

$$T = \frac{1}{2}\dot{q}^T M\dot{q}, \tag{2.18}$$

$\boldsymbol{M}$ being the mass matrix:

$$\boldsymbol{M} = \text{diag}(m, m_1, m_1, m_2, J_\alpha, J_\beta, J_{\beta''}), \tag{2.19}$$

where m is the sprung mass, m_1, m_2 the front and rear unsprung masses, J_α the pitch inertia and J_β and $J_{\beta''}$ the roll inertias of sprung mass and rear inter-axis bar.

The right-hand side of Equation 2.17 is defined as:

$$g_k = \sum_{i=1}^{8} f_i \frac{\partial x_i}{\partial q_k} \qquad k=1,\ldots, 7. \tag{2.20}$$

Defining the vector $\boldsymbol{f}$ of the forces applied to sprung and unsprung masses

$$\boldsymbol{f} = [f_1, f_2, f_3, f_4, f_5, f_6, f_7, f_8]^T. \tag{2.21}$$

the forces applied to sprung and unsprung masses are:

$$\boldsymbol{f} = -[\boldsymbol{K}\boldsymbol{z} + \boldsymbol{C}\dot{\boldsymbol{z}} + \boldsymbol{K_0}\boldsymbol{z_0} + \boldsymbol{C_0}\dot{\boldsymbol{z}}_{\boldsymbol{0}}], \tag{2.22}$$

where $\boldsymbol{K}$ is the stiffness matrix:

$$\boldsymbol{K} = \begin{bmatrix} k_1 & 0 & 0 & 0 & -k_1 & 0 & 0 & 0 \\ 0 & k_1 & 0 & 0 & 0 & -k_1 & 0 & 0 \\ 0 & 0 & k_2 & 0 & 0 & 0 & -k_2 & 0 \\ 0 & 0 & 0 & k_2 & 0 & 0 & 0 & -k_2 \\ -k_1 & 0 & 0 & 0 & k_1 + k_{01} & 0 & 0 & 0 \\ 0 & -k_1 & 0 & 0 & 0 & k_1 + k_{01} & 0 & 0 \\ 0 & 0 & -k_2 & 0 & 0 & 0 & k_2 + k_{02} & 0 \\ 0 & 0 & 0 & -k_2 & 0 & 0 & 0 & k_2 + k_{02} \end{bmatrix} \tag{2.23}$$

and $\boldsymbol{K_0}$ is the unsprung mass stiffness matrix

$$\boldsymbol{K}_0 = \begin{bmatrix} 0 & 0 & 0 & 0 \\ 0 & 0 & 0 & 0 \\ 0 & 0 & 0 & 0 \\ 0 & 0 & 0 & 0 \\ -k_{01} & 0 & 0 & 0 \\ 0 & -k_{01} & 0 & 0 \\ 0 & 0 & -k_{02} & 0 \\ 0 & 0 & 0 & -k_{02} \end{bmatrix}. \tag{2.24}$$

Analogous definitions can be given for matrices $\boldsymbol{C}$ and $\boldsymbol{C_0}$, the latter being the tyre damping matrix, usually negligible (tyre damping is not depicted in Figure 2.17).

Combining Equations 2.14 and 2.18 yields:

$$x_k = \sum_{i=1}^{7} p_{ik} q_k \quad k=1,\ldots, 8, \tag{2.25}$$

$$T = \frac{1}{2} \sum_{i=1}^{7} m_{kk} \dot{q}_k^2 . \tag{2.26}$$

Developing (2.17) yields:

$$g_k = \sum_{i=1}^{8} p_{ik} q_k = (P^T f) \quad k=1,\ldots, 7, \tag{2.27}$$

$$\frac{\mathrm{d}}{\mathrm{d}t} \frac{\partial T}{\partial \dot{q}_k} = (M\ddot{q})_k \qquad k=1,\ldots, 7, \tag{2.28}$$

hence

$$\boldsymbol{M\ddot{q}} = \boldsymbol{P}^T \boldsymbol{f} . \tag{2.29}$$

Taking into account Equation 2.22, the governing Equation 2.10 is obtained:

$$\boldsymbol{M\ddot{q}} + \boldsymbol{P}^T \boldsymbol{CP\dot{q}} + \boldsymbol{P}^T \boldsymbol{KPq} + \boldsymbol{F}_d = -\boldsymbol{P}^T \boldsymbol{K}_0 \boldsymbol{z}_0 - \boldsymbol{P}^T \boldsymbol{C}_0 \dot{\boldsymbol{z}}_0 . \tag{2.30}$$

In the model here described, front suspensions are taken to be independent and rear suspensions dependent (connected through a rigid axle). However the model can be easily modified to represent independent front suspensions or both front and rear dependent suspensions, by appropriate choices of the matrices ***M***, ***K*** and ***C***.

2.5.4 Half Truck Model

The half truck model is the equivalent of the half vehicle model for an articulated vehicle, and has seven degrees of freedom. Figure 2.18 depicts a schematic of the truck model (Tsampardoukas *et al.*, 2007)

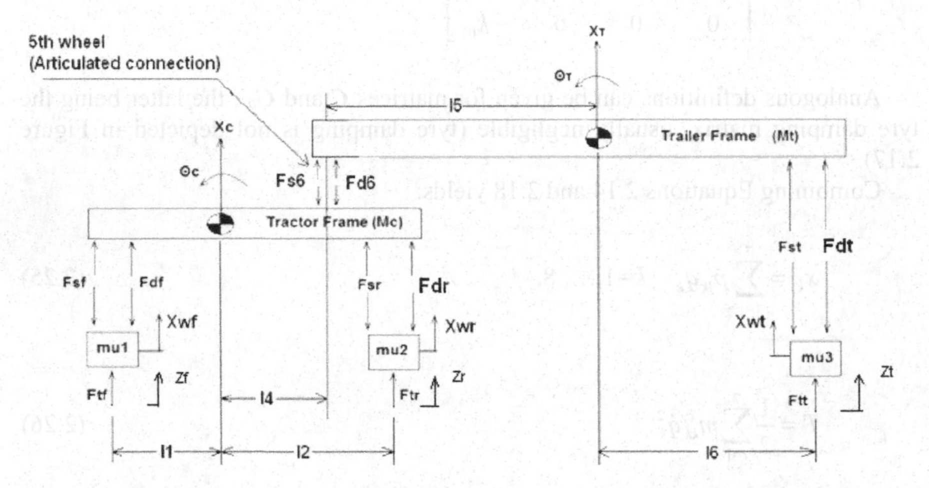

Fig. 2.18. Half truck model (copyright Elsevier, reproduced from Tsampardoukas G, Stammers CW and Guglielmino E, Hybrid balance control of a magnetorheological truck suspension, accepted for publication in Journal of Sound and Vibration, used by permission)

It is composed of two sprung masses, namely the tractor body (or frame) and the trailer body, together with three unsprung masses (the three wheels). The tractor and the trailer are linked through an articulated connection known as the fifth wheel. Two vibration modes are considered for each sprung unit (heave and pitch) and one for the unsprung masses (heave).

The governing equations can be readily obtained by consideration of forces and moments as follows:

$$\begin{cases} m_{\mathrm{c}}\ddot{x}_{\mathrm{c}} = F_{\mathrm{sf}} + F_{\mathrm{df}} + F_{\mathrm{sr}} + Fc_{\mathrm{rear}} - F_{\mathrm{s6}} - F_{\mathrm{d6}} \\ J_{\mathrm{c}}\ddot{\theta}_{\mathrm{c}} = F_{\mathrm{sr}} l_2 + Fc_{\mathrm{rear}} l_2 - F_{\mathrm{s6}} l_4 - F_{\mathrm{d6}} l_4 - F_{\mathrm{sf}} l_1 - F_{\mathrm{df}} l_1 \\ m_{\mathrm{t}}\ddot{x}_{\mathrm{t}} = F_{\mathrm{st}} + Fc_{\mathrm{trailer}} + F_{\mathrm{s6}} + F_{\mathrm{d6}} \\ J_{\mathrm{t}}\ddot{\theta}_{\mathrm{t}} = F_{\mathrm{st}} l_6 + Fc_{\mathrm{trailer}} l_6 - F_{\mathrm{s6}} l_5 - F_{\mathrm{d6}} l_5 \\ m_{\mathrm{u1}}\ddot{x}_{\mathrm{wf}} = F_{\mathrm{tf}} - F_{\mathrm{sf}} - F_{\mathrm{df}} \\ m_{\mathrm{u2}}\ddot{x}_{\mathrm{wr}} = F_{\mathrm{tr}} - F_{\mathrm{sr}} - F_{\mathrm{dr}} \\ m_{\mathrm{u3}}\ddot{x}_{\mathrm{wt}} = F_{\mathrm{tt}} - F_{\mathrm{st}} - F_{\mathrm{dt}} \end{cases} \tag{2.31}$$

where

$$F_{\mathrm{sf}} = k_f(x_{\mathrm{wf}} - x_{\mathrm{c}} + l_1\theta_{\mathrm{c}})$$
$$F_{\mathrm{df}} = c_f(\dot{x}_{\mathrm{wf}} - \dot{x}_{\mathrm{c}} + l_1\dot{\theta}_{\mathrm{c}})$$

$$F_{\mathrm{sr}} = k_r(x_{\mathrm{wr}} - x_{\mathrm{c}} - l_2\theta_{\mathrm{c}})$$
$$F_{\mathrm{dr}} = c_r(\dot{x}_{\mathrm{wr}} - \dot{x}_{\mathrm{c}} - l_2\dot{\theta}_{\mathrm{c}})$$

$$F_{\mathrm{st}} = k_t(x_{\mathrm{wt}} - x_{\mathrm{t}} - l_6\theta_{\mathrm{t}})$$
$$F_{\mathrm{dt}} = c_t(\dot{x}_{\mathrm{wt}} - \dot{x}_{\mathrm{t}} - l_6\dot{\theta}_{\mathrm{t}})$$

$$F_{\mathrm{s6}} = k_5(x_{\mathrm{c}} + l_2\theta_{\mathrm{c}} - x_{\mathrm{t}} + l_5\theta_{\mathrm{t}})$$
$$F_{\mathrm{d6}} = c_5(\dot{x}_{\mathrm{c}} + l_2\dot{\theta}_{\mathrm{c}} - \dot{x}_{\mathrm{t}} + l_5\dot{\theta}_{\mathrm{t}})$$

$$F_{\mathrm{tf}} = k_{\mathrm{tf}}(z_{\mathrm{f}} - x_{\mathrm{wf}})$$
$$F_{\mathrm{tr}} = k_{\mathrm{tr}}(z_{\mathrm{r}} - x_{\mathrm{wr}})$$
$$F_{\mathrm{tt}} = k_{\mathrm{tt}}(z_{\mathrm{t}} - x_{\mathrm{wt}})$$

$$rear\_relative\_velocity = (\dot{x}_{\mathrm{c}} + l_2\dot{\theta}_{\mathrm{c}} - \dot{x}_{\mathrm{wr}})$$

$$trailer\_relative\_velocity = (\dot{x}_{\mathrm{t}} + l_6\dot{\theta}_{\mathrm{t}} - \dot{x}_{\mathrm{wt}})$$

$$\begin{aligned} z_{\mathrm{f}} &= front\_wheel\_input \\ z_{\mathrm{r}} &= rear\_wheel\_input \\ z_{\mathrm{t}} &= trailer\_wheel\_input \end{aligned} \tag{2.32}$$

The equations can be implemented in Simulink®. Figure 2.19 shows a schematic of the Simulink® model. The meaning of the symbols in Equations 2.31 and 2.32 are listed in Table 2.1.

Table 2.1. Truck model symbols notation (copyright Elsevier, reproduced from Tsampardoukas G, Stammers CW and Guglielmino E, Hybrid balance control of a magnetorheological truck suspension, accepted for publication in Journal of Sound and Vibration, used by permission)

F_{sf}	Front tractor suspension spring force
F_{df}	Front tractor suspension damping force
F_{sr}	Rear tractor suspension spring force
F_{dr}	Rear tractor suspension damping force
F_{st}	Trailer suspension spring force
F_{dt}	Trailer suspension damping force
F_{s6}	Fifth wheel spring force
F_{d6}	Fifth wheel damping force
x_t	Trailer heave
x_c	Tractor heave
ϑ_c	Tractor pitch
ϑ_t	Trailer pitch
x_{wf}	Front tractor wheel heave
x_{wr}	Rear tractor wheel heave
x_{wt}	Trailer tractor wheel heave
z_f	Front tractor road input
z_r	Rear tractor road input
z_t	Trailer road input
F_{tf}	Excitation force (front tractor)
F_{tr}	Excitation force (rear tractor)
F_{tt}	Excitation force (trailer)

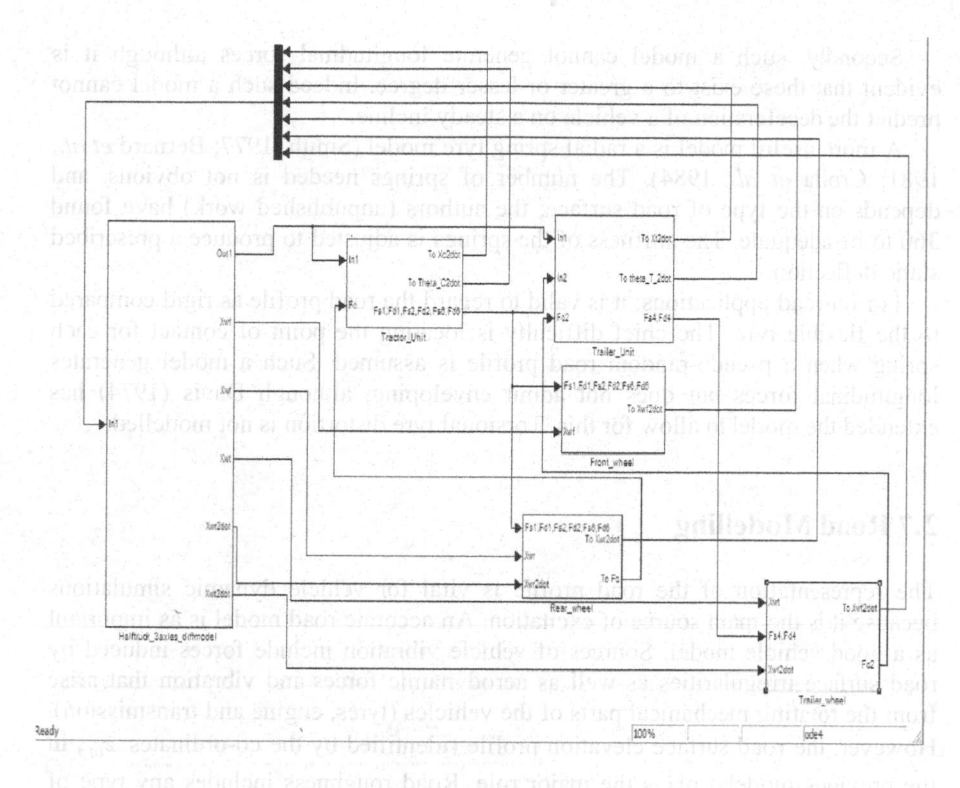

Fig. 2.19. Simulink® half truck model

2.6 Tyre Modelling

Tyres are made of rubber, *i.e.*, a viscoelastic material and in ride studies the vertical tyre stiffness can be approximated by a spring and some damping (either pure viscous or hysteretic damping, but often negligible). In the vehicle models described so far, the tyre has been represented as a spring (and a viscous damper term). This model although quite crude is acceptable for ride analysis. In handling, braking or traction studies more sophisticated models are required which account for road–tyre adhesion (both longitudinal and lateral) as well as rolling friction. A classical model is the so-called Pacejka magic formula (Pacejka and Bakker, 1991). At higher frequencies and for short road obstacles even more sophisticated models are required.

The majority of workers utilise a point contact model since it is easy to use. Such a model has at least two defects. Firstly, the point follows the slightest vertical excursion of the road and hence generates high-frequency inputs which in practice would not occur as the tyre contact patch bridges or envelops such points.

Secondly, such a model cannot generate longitudinal forces although it is evident that these exist to a greater or lesser degree. Indeed such a model cannot predict the deceleration of a vehicle on a steady incline.

A more useful model is a radial spring tyre model (Smith, 1977; Bernard *et al.*, 1981; Crolla *et al.*, 1984). The number of springs needed is not obvious, and depends on the type of road surface; the authors (unpublished work) have found 360 to be adequate. The stiffness of the springs is adjusted to produce a prescribed static deflection.

For on-road applications, it is valid to regard the road profile as rigid compared to the flexible tyre. The chief difficulty is locating the point of contact for each spring when a pseudo-random road profile is assumed. Such a model generates longitudinal forces but does not admit enveloping, although Davis (1974) has extended the model to allow for this. Torsional tyre distortion is not modelled.

2.7 Road Modelling

The representation of the road profile is vital for vehicle dynamic simulations because it is the main source of excitation. An accurate road model is as important as a good vehicle model. Sources of vehicle vibration include forces induced by road surface irregularities as well as aerodynamic forces and vibration that arise from the rotating mechanical parts of the vehicles (tyres, engine and transmission). However, the road surface elevation profile (identified by the co-ordinates z_{0i}, in the previous models) plays the major role. Road roughness includes any type of surface irregularities from bump and potholes to small deviations.

The reduction of forces transmitted to the road by moving vehicles (particularly for heavy vehicles) is also an important issue responsible for road damage. Heavy-vehicle suspensions should be designed accounting also for this constraint. The issue will be dealt in detail in a case study in Chapter 7.

Road inputs can be classified into three types: deterministic road (periodic and almost-periodic) inputs, random-type inputs and discrete events such as bumps and potholes.

As far as deterministic inputs are concerned, a variety of periodical waveforms can be used, such as sine waves, square waves or triangular waves. To a first approximation the road profile can be assumed to be sinusoidal. Although not realistic, it is useful in a preliminary study because it readily permits a comparison of the performance of different suspension designs both in the time and in the frequency domain (through transmissibility charts, plotted at different frequencies).

A multi-harmonic input which is closer to an actual road profile can be generated. A possible choice which approximates fairly well a real road profile is a so-called pseudo-random input (Sayers and Karamihas, 1998; Dukkipati, 2000) which results from summing several non-commensurately related sine waves (*i.e.*, the ratio of all possible pairs of frequencies is not a rational number) of decreasing amplitude, so as to provide a discrete approximation of a continuous spectrum of a random input. The trend can be proved to be non-periodic (sometimes referred to as almost periodic) in spite of being a sum of periodic waveforms (Bendat and

Piersol, 1971). To achieve a pseudo-random profile effect it is advisable to select spatial frequencies of the form

$$j\Delta\Omega + \text{trascendental term} \qquad j=1,\ldots,m. \tag{2.33}$$

where j is an integer, $\Delta\Omega$ is the separation between spatial frequencies and the added term could be $e/1000$ or $\pi/1000$ for example. The spatial frequency range is $(m\text{-}1)\ \Delta\Omega$. The RMS amplitude at each centre frequency is obtained from the power spectral law and multiplication by $\Delta\Omega$.

Another possible way to generate a realistic multi-harmonic input consists in making the ratio between frequencies constant and decreasing with amplitude, but using randomly generated phase angles (between 0 and 360 degrees) for each component. In this case the resulting waveform is periodical. Simulation results presented subsequently are based on the latter approach.

Figure 2.20 shows an example of road profile: 20 sine waves with random phases have been added together in order to create a pseudo-random profile. The amplitude of the profile is calculated to approximate a smooth highway by using the spatial frequency data suggested by the Society of Automotive Engineers (SAE).

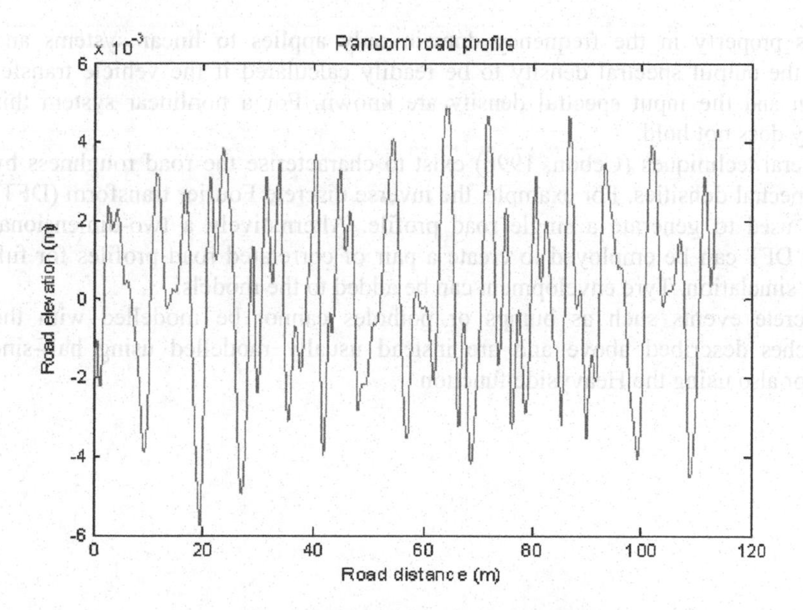

Fig. 2.20. Pseudo-random road profile

However, these road profiles are deterministic functions and as a consequence could not represent a real random pavement.

A stochastic model gives the more realistic representation of a road profile. The power spectral density (PSD) is the most common way to characterise the road roughness (British standard BS 7853, 1996). A road could well be approximated (Wong, 1993) by an ergodic process with spectral density expressed by:

$$S(\Omega) = C\Omega^{-n}, \tag{2.34}$$

where Ω is the spatial frequency, having units of cycles/m (*i.e.*, Ω is the inverse of wavelength), C is a coefficient dependent on the road roughness, f the frequency in Hz and n is a rational exponent. It follows that $\Omega = f/V$, where V is the forward speed of the car, so that:

$$S(f) = \left(\frac{C}{V^{-n}}\right) f^{-n}. \tag{2.35}$$

This approach involves an analysis in terms of power spectral densities. For a linear system the input and output spectral densities $Y_{in}(f)$ and $Y_{out}(f)$ are related through the transfer function of the system $G(f)$, from the equation:

$$Y_{out}(f) = |G(f)|^2 Y_{in}(f). \tag{2.36}$$

This property in the frequency domain only applies to linear systems and allows the output spectral density to be readily calculated if the vehicle transfer function and the input spectral density are known. For a nonlinear system this property does not hold.

Several techniques (Cebon, 1996) exist to characterise the road roughness by using spectral densities. For example, the inverse discrete Fourier transform (DFT) can be used to generate a single road profile. Alternatively a two-dimensional inverse DFT can be employed to create a pair of correlated road profiles for full vehicle simulation. Tyre envelopment can be added to the models.

Discrete events such as bumps or potholes cannot be modelled with the approaches described above and are instead usually modelled using half-sine waves or also using the Heavyside function.

3

Human Body Analysis

3.1 Introduction

The response of vehicle occupants to road inputs is an important factor in the design of any vehicle. A suspension ought to provide the greatest possible comfort.

On the other hand, on poor roads and in off-road operation, protection of the occupants from actual damage is most important. Tractor drivers in particular are at risk of lower-back injuries and possibly damage to viscera (liver, heart and brain).

In order to assess the merits of a suspension it is accordingly necessary to model the seat characteristics, particular those of the cushion. A model for the occupant(s) is also highly desirable, as will be described below.

The efficiency of isolation of 67 conventional seats and 33 suspension seats has been reported (Paddan and Griffin, 2002). The measure used was seat effective amplitude transmissibility (SEAT), which is the frequency-weighted root mean square acceleration experienced with the seat compared to that experienced with a rigid seat.

For 25 car seats the SEAT value varied between 57% and 122% with a median value of 78%. For the 16 trucks tested the SEAT range varied between 44% and 115%, with a mean of 87%. For seven tractors the SEAT value varied between 57% and 118%.

These results indicate the importance of good seat design. A key parameter is the characteristics of the foam material used for the seat.

Yu and Khameneh (1999) measured the transmissibility of three different types of foam formulations: two rather similar toluene diisocyanate (TDI) formulations, and one methylene diphenyl diisocyanate (MDI).

Using a shaped 50 kg load, displacement transmissibility for each foam was recorded over the range 2.5–6 Hz for inputs of 5 mm and 20 mm. At low-amplitude inputs (5 mm) the natural frequency was around 4 Hz. For all three foams, the natural frequency for 20 mm inputs was found to be about 0.5 Hz lower than that for 5 mm. The peak transmissibility was also reduced.

It is clear that the foams have a softening spring characteristic with damping ratio increasing with amplitude.

3.2 Human Body Response

Tests of 12 seated humans to vertical random acceleration in the range 0.2–20 Hz (Mansfield and Griffin, 2000) indicated that the human body also demonstrates a softening spring characteristic. The apparent mass resonance frequency fell from 5.4 Hz to 4.2 Hz as the magnitude of vibration increased from 0.25 m/s to 2.5 m/s. From these tests an equivalent mechanical system was developed (Wei and Griffin, 1998a and 1998b). While Wei and Griffin were careful to point out that no specific identification with parts of the body can be made, it appears plausible that this response is that of internal organs (viscera) within the skeleton. Excessive stretching of the intestinal attachment tissue (the mesentery) could lead to rupture and internal bleeding. Similarly the liver could be damaged, which implies serious implications for that vital organ. The visceral mass includes the brain, which *sloshes* within the skull. The issue here is that of pressure waves. Rebound of the brain causes reduced pressure which is thought to be the cause of concussion.

The resonant frequencies reported by Mansfield and Griffin for the human body are little greater than those found by Yu and Khameneh for mass foam cushions. The possibility of resonant interaction between cushion and occupant is quite possible. This may explain why some seats are significantly less comfortable than others.

Torsional chirp (swept sine) excitation of the wrist of subjects with forearm supported (Lakie *et al.*, 1984) indicated that the wrist resonant frequency fell markedly as the magnitude of the oscillatory input was increased. Here is evidence of the stiffness of tendons decreasing drastically with increasing force and hence displacement.

In view of the reported experimental work, it is plausible to adopt a spring force of the form $K(x-\varepsilon x^3)$, where K is the linear stiffness and $\varepsilon > 0$ for both foam and human body in the low-frequency range.

Hysteric damping is assumed and can be modelled via a complex stiffness $K(1+\mathrm{i}\beta)$, where β is the loss factor. This formulation guarantees non-linear damping which increases with amplitude, as will be shown below. The same form of damping in the human viscera is assumed as for foam.

3.3 Hysteretic Damping

For sinusoidal response (and the hysteretic model is really only valid for that condition) adopting the complex response:

$$x = A\,\mathrm{e}^{\mathrm{i}\omega t}, \tag{3.1}$$

$$\dot{x} = \mathrm{i}\omega x\,, \tag{3.2}$$

$$\dot{x}^3 = -\mathrm{i}\omega^3 x^3\,, \tag{3.3}$$

The damping term is hence obtained as:

$$K\beta\left(\frac{\dot{x}}{\omega}+\varepsilon\left(\frac{\dot{x}}{\omega}\right)^3\right), \tag{3.4}$$

The non-linear term produces increased damping when $\varepsilon > 0$.

An effective damping ratio can be obtained as:

$$\zeta_{\text{eff}} = \frac{\omega_n\,\beta\left[1+\varepsilon\left(\dfrac{\dot{x}}{\omega}\right)^2\right]}{2\,\omega}, \tag{3.5}$$

ω_n being the natural frequency of the linear system.

Assuming now that $x = C\,\sin(\omega t + \varphi)$, a mean damping of the form

$$\zeta_{\text{eff}} = \frac{\omega_n\,\beta\,(1+0.5\varepsilon\,C^2)}{2\,\omega} \tag{3.6}$$

can be obtained.

The frequency response of the system can be obtained and is useful in indicating the general behaviour of the system.

3.3.1 The Duffing Equation

A system governed by an equation of Duffing type (Hagerdon, 1998)

$$M\ddot{x} + B\dot{x} + K\left(x - \varepsilon x^3\right) = F\cos(\omega t + \varphi) \tag{3.7}$$

can exhibit jumps in amplitude if the damping is sufficiently low or the forcing F sufficiently large. Such jumps are undesirable, particularly so in the case of the visceral organs, where injury could result.

The type of behaviour is indicated in Figure 3.2 where the amplitude X of the response:

$$x = X\cos(\omega t) + \text{higher-order terms} \tag{3.8}$$

is plotted as a function of $p = \dfrac{\omega}{\omega_n}$, where $\omega_n = \left(\dfrac{K}{M}\right)^{0.5}$ is the natural frequency of the linear system.

For small F the response is virtually that of the linear system, but for larger values of F the response curves exhibit a buckled shape. For the top curve in Figure 3.1, as p is decreased slowly from a large value, when the point A is reached the response jumps to B. The critical condition is the vertical tangent. The solution at A is no longer real, while the formerly unreal solution at B becomes real.

As p is increased slowly again, when the point C is reached the response jumps to D (for a system in which the spring hardens with amplitude, the curves lean to the right).

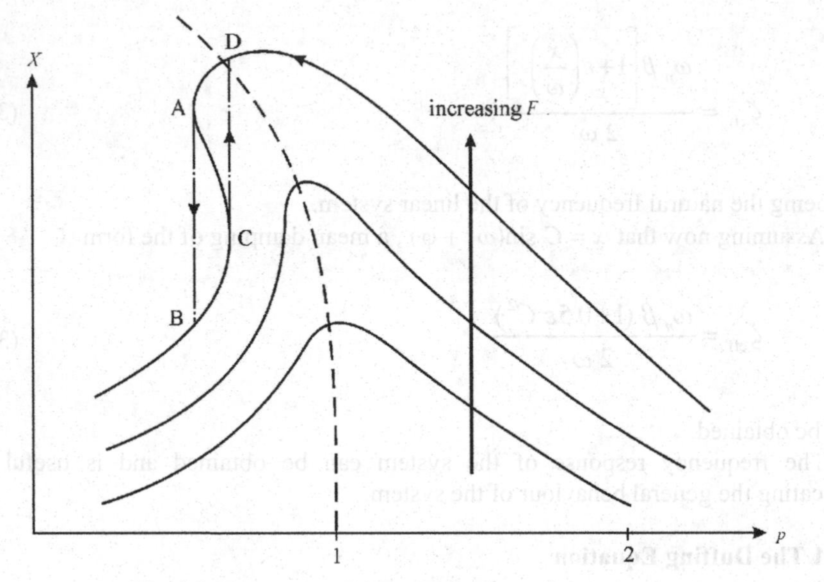

Fig. 3.1. Frequency response, softening spring; $p = \omega/\omega_n$

3.3.2 Suppression of Jumps

The loss factor β to prevent jumps can be obtained from the condition that no vertical tangent exists in the frequency response plot.

The equation of motion for a single-degree-of-freedom system has the form:

$$M\ddot{x} + K\beta\left(\frac{\dot{x}}{\omega} + \varepsilon\left(\frac{\dot{x}}{\omega}\right)^3\right) + K(x - \varepsilon x^3) = F_1\sin(\omega t) + F_2\cos(\omega t). \qquad (3.9)$$

It is convenient to non-dimensionalise the equation of motion by setting $\omega_n t = \tau$. The equation governing motion becomes

$$\ddot{x}+\beta\left(\frac{\dot{x}}{p}+\varepsilon\left(\frac{\dot{x}}{p}\right)^{3}\right)+\left(x-\varepsilon x^{3}\right)=\left(f_{1}\sin(p\tau)+f_{2}\cos(p\tau)\right), \tag{3.10}$$

where $f_j=F_j/M$, $p=\omega/\omega_n$ and differentiation is now with respect to τ.

Assuming a solution $x=C\sin(p\tau)$:

$$\left(1-p^{2}\right)C-\varepsilon C^{3}=f_{1}, \tag{3.11}$$

$$\beta C+\beta e C^{3}=f_{2}, \tag{3.12}$$

where

$$e=0.75\varepsilon. \tag{3.13}$$

Then

$$\left[\left(1-p^{2}\right)C-e\,C^{3}\right]^{2}+\left(\beta C+\beta e C^{3}\right)^{2}=f_{1}^{2}+f_{2}^{2}=f^{2}, \tag{3.14}$$

where f is a chosen input.

Jumps occur when $\dfrac{\partial C}{\partial p}$ is infinite, or more usefully, when $\dfrac{\partial p}{\partial C}=0$.

Differentiating (3.14) with respect to C, with the condition $\dfrac{\partial p}{\partial C}=0$

$$2\left[\left(1-p^{2}\right)C-eC^{3}\right]\left(1-p^{2}-3eC^{2}\right)+2\left(\beta C+\beta eC^{3}\right)\left(\beta+3\beta eC^{2}\right)=0. \tag{3.15}$$

Dividing by $2C$ and arranging as an equation in p

$$p^{4}-2p^{2}\left(1-2eC^{2}\right)+\left(1-3eC^{2}\right)\left(1-eC^{2}\right)+\beta^{2}\left(1+eC^{2}\right)\left(1+3eC^{2}\right)=0. \tag{3.16}$$

Jumps are impossible if this equation has no real roots for p^2, *i.e.*, if

$$\left(1-2\,e\,C^{2}\right)^{2}<\left(1-3\,e\,C^{2}\right)\left(1-e\,C^{2}\right)+\beta^{2}\left(1+eC^{2}\right)\left(1+3\,e\,C^{2}\right) \tag{3.17}$$

or

$$\beta^{2}>\frac{\left(1-2eC^{2}\right)^{2}-\left(1-3\,e\,C^{2}\right)\left(1-e\,C^{2}\right)}{\left(1+e\,C^{2}\right)\left(1+3\,e\,C^{2}\right)}, \tag{3.18}$$

i.e.,

$$\beta^2 > \frac{e^2\ C^4}{\left(1+e\ C^2\right)\left(1+3\ e\ C^2\right)}. \tag{3.19}$$

The value of β to suppress jumps is a function of eC^2. This is a measure of the magnitude of the nonlinear component of the spring force; see Equation 3.9. When $e\ C^2 << 1$, to prevent jumps it is necessary that

$$\beta > e\ C^2, \tag{3.20}$$

where $e = 0.75\ \varepsilon$.

For $e\ C^2 >> 1$, the required value of β to prevent all jumps is 0.58, greater than one would expect for human tissue. However, this is an extreme case.

Studies of the human visceral model with β around 0.3 (the level indicated by experimental work on cushions) suggest that jumps would occur for sinusoidal oscillations of the viscera in excess of 3 mm in magnitude. On the other hand, the value of β to prevent jumps of a foam cushion is less than 0.1.

3.4 Low-frequency Seated Human Model

Various detailed models of the human body exist. However, cushion and visceral natural frequencies are below 6 Hz. Moreover because the amplitude of road profile fluctuations falls with decreasing wavelength, road inputs experienced by vehicle occupants are predominantly at low frequency. For realistic vehicle speeds higher frequency inputs are not important until wheel-hop is experienced at around 12 Hz for cars and nearer 10 Hz for freight vehicles. When traversing rough ground, drivers instinctively slow, reducing the frequency of input. Hence a simple human model suitable for low-frequency inputs is adopted here.

The model adopted for the human body is one of those developed by Wei and Griffin (Wei and Griffin, 1998a). This is shown in the upper part of Figure 3.3. The non-linear spring K_v models stiffness effects with hysteretic effects providing damping.

The motion of the visceral mass M_v is denoted by x, and that of the remainder M_c of the body by y. This second mass could be that of the skeleton. The authors also produced a two-degree-of-freedom model in order to model response at frequencies greater than about 8 Hz. Vehicle simulations using this model indicated no significantly different response for realistic road inputs and vehicle speeds. Hence the single-degree-of-freedom model is adopted here.

The seat is modelled as a mass M_s and a (non-linear) spring K_s. As indicated above the damping terms are deduced from a hysteric model.

A seat control force F_c is considered. This would be provided by an actuator fixed to the vehicle floor beneath the seat.

The sprung mass M_s on a spring and damper suspension is modelled since it acts as a low-pass filter of road inputs. In the frequency range of interest, the unsprung mass is neglected, as is common with truck models.

The experimental work using vertical inputs indicates a natural frequency of the human body in the range 4.2–5.4 Hz, depending on the amplitude of input (Mansfield and Griffin, 2000). This is not far from the resonant frequencies reported for a loaded foam cushion (Yu and Khameneh, 1999).

3.4.1 Multi-frequency Input

Vehicle-occupant model for low-frequency vibration is depicted in Figure 3.2.

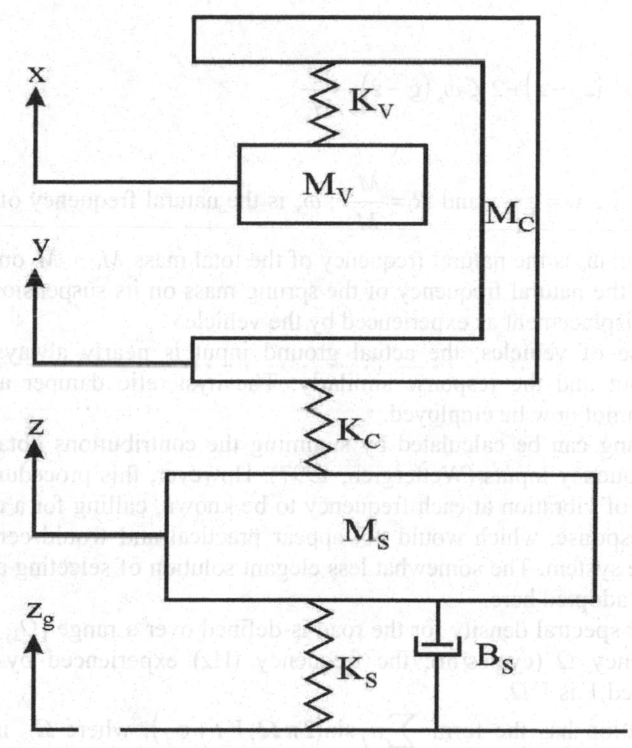

Fig. 3.2. Vehicle and occupant model (copyright Elsevier (1998), reproduced with minor modifications and with the addition of the bottom part of the figure from Wei and Griffin, The prediction of seat transmissibility from measures of seat impedance, J Sound Vib, Vol. 214, N 1, used by permission)

Ground input is indicated by z_g. The unsprung mass is not modelled as at the low frequencies considered (below 6 Hz) it follows the road.

M_s represents the sprung mass of the vehicle, restrained by a linear spring and viscous damper, as is commonly modelled.

The cushion is modelled as a complex spring K_c with loss factor β_c; M_c and M_v represent the two masses of the Wei and Griffin 1DOF model (Wei and Griffin, 1998b), with K_v a non-linear spring with loss factor β_v.

The equations of motion for an input z_g of frequency ω are:

$$\ddot{x} = \omega_v^2 u \left(1 - \varepsilon_v u^2\right) + \omega_v^2 \beta_v \dot{u} \frac{\left(1 + \varepsilon_v \frac{\dot{u}^2}{\omega^2}\right)}{\omega}, \tag{3.21}$$

$$\ddot{y} = -R\,\ddot{x} + (R+1)\omega_c^2 w \left(1 - \varepsilon_c w^2\right) + (R+1)\,\omega_c^2 \beta_c \dot{w} \frac{\left(1 + \varepsilon_c \frac{\dot{w}^2}{\omega^2}\right)}{\omega} + \frac{F_c}{M_c}, \tag{3.22}$$

$$\ddot{z} = \omega_s^2 \left(z_g - z\right) + 2\,\zeta\,\omega_s \left(\dot{g} - \dot{z}\right) - \frac{F_c}{M_s} \tag{3.23}$$

where $u = y - x$, $w = z - y$ and $R = \dfrac{M_v}{M_c}$; ω_v is the natural frequency of the linear visceral system, ω_c is the natural frequency of the total mass $M_v + M_c$ on the linear cushion, ω_s is the natural frequency of the sprung mass on its suspension, z_g is the road surface displacement as experienced by the vehicle.

In the case of vehicles, the actual ground input is nearly always a multi-frequency input and the response similarly. The hysteretic damper analysis of Section 3.3 cannot now be employed.

The damping can be calculated by summing the contributions obtained from individual frequency inputs (Wettergren, 1997). However, this procedure requires the amplitude of vibration at each frequency to be known, calling for a continuous FFT of the response, which would not appear practical and would certainly add expense to the system. The somewhat less elegant solution of selecting a typical ω is the strategy adopted here.

The power spectral density for the road is defined over a range $[\Omega_1, \Omega_2]$ of the spatial frequency Ω (cycles/m); the frequency (Hz) experienced by a vehicle moving at speed V is $V\,\Omega$.

The excitation has the form $\sum a_j \sin\left(2\pi \Omega_j V t + \varphi_j\right)$, where Ω_j is a spatial frequency (cycles/m), V the vehicle speed and φ_j a random phase angle; g is the time derivative $\sum a_j\, 2\,\pi\,\Omega_j V \cos\left(2\pi \Omega_j V t + \varphi_j\right)$ of the road surface displacement.

This model can be used to obtain the frequency response of the system, be it passive or controlled.

3.5 Semi-active Control

A semi-active device can only dissipate energy. Hence the damper can be on only when:

$$F_{\mathrm{c,\,des}}\,\dot{w} < 0 \tag{3.24}$$

otherwise $F_{c,\,des}$ should be set to zero. This can be achieved in a dry friction control system by separating the plates. For a viscous damper or for a magnetorheological damper zero force is not possible and the best that can be achieved is to demand the minimal setting.

The response F_{c} of the actuator is assumed to be governed by a first-order system of the form:

$$T_{\mathrm{const}}\,\dot{F}_{\mathrm{c}} + F_{\mathrm{c}} = F_{\mathrm{c,des}}\,, \tag{3.25}$$

where T_{const} is the time for the error in the response to a step demand to fall to 36%.

The error caused by the non-zero time constant of the actuator can be reduced by a process of gain compensation. The demanded force is multiplied by a factor $\gamma > 1$, which is found to be effective when the integration time step is less than 20% of the actuator time constant. For zero error after one time constant, $\gamma = \left(1 - e^{-1}\right)^{-1} = 1.58$.

As long as switching decisions are made several times a time constant, there should be no overshoot; the gain compensation procedure is found to reduce visceral accelerations by around 10% at vehicle speeds up to 20 m/s.

3.6 State Observer

To achieve comfort, it is necessary to reduce the acceleration $\ddot{x}$ of the viscera. This could be achieved if the relative internal displacement u and relative velocity $\dot{u}$ could be controlled. However, it is not possible in practice to measure these quantities.

It is therefore necessary to construct a state observer. The observer concept was first proposed by Luenberger (1964).

3.6.1 Luenberger State Observer

The most general case is the one in which no state variable are measured. This is termed the full-order observer.

Consider a system with state variables $\boldsymbol{x} = [x_1, x_2, \ldots, x_n]$ which generates an output $\boldsymbol{y} = \boldsymbol{C}\boldsymbol{x}$ which can be measured.

If the system dynamics are given by

$$\dot{\boldsymbol{x}} = \boldsymbol{A}\boldsymbol{x} + \boldsymbol{B}\boldsymbol{u}\,, \tag{3.26}$$

where $\boldsymbol{u}$ is a vector of control inputs, an estimate $\boldsymbol{z}$ of $\boldsymbol{x}$ is given by the equation

$$\dot{\boldsymbol{z}} = \boldsymbol{A}\boldsymbol{z} + \boldsymbol{B}\boldsymbol{u} + \boldsymbol{K}(\boldsymbol{y} - \boldsymbol{C}\boldsymbol{z}) \tag{3.27}$$

and $\boldsymbol{K}$ is the observer gain matrix, the error $\boldsymbol{e} = \boldsymbol{x} - \boldsymbol{z}$ is found to satisfy

$$\dot{\boldsymbol{e}} = (\boldsymbol{A} - \boldsymbol{K}\boldsymbol{C})\boldsymbol{e}\,. \tag{3.28}$$

The estimate $\boldsymbol{z}$ of $\boldsymbol{x}$ is governed by:

$$\dot{\boldsymbol{z}} = (\boldsymbol{A} - \boldsymbol{K}\boldsymbol{C})\boldsymbol{z} + \boldsymbol{B}\boldsymbol{u} + \boldsymbol{K}\boldsymbol{y}\,. \tag{3.29}$$

If the eigenvalues of the matrix $\boldsymbol{A} - \boldsymbol{K}\boldsymbol{C}$ are chosen suitably (by an appropriate choice of $\boldsymbol{K}$) the error $\boldsymbol{e} = \boldsymbol{x} - \boldsymbol{z}$ should decay rapidly.

The calculation of $\boldsymbol{K}$ involves, among other steps, the formation of a controllability matrix and an observability matrix, and is not simple. The reader is referred to Crossley and Porter (1979) or Burns (2001) for details.

In many cases, as in the application considered here, some of the variables (such as seat relative displacement and relative velocity) can be measured. However, what cannot be measured are the displacement and velocity of the viscera relative to the skeleton and the skeleton relative to the seat. In the case where only some of the state variables need to be estimated, the observer is known as a reduced-order state observer.

A similar mathematical path is required for the generation of $\boldsymbol{K}$ as in the full-order observer.

3.6.2 Simple State Observer

The method outlined here (Stammers and Sireteanu, 2004) requires no pre-processing, and is simple enough to be implemented in real time.

The relative acceleration $\ddot{u}$ can be expressed from (3.21) and (3.22) as

$$\begin{aligned}\ddot{u} = {} & -K_{\mathrm{v}}\left(\frac{1}{M_{\mathrm{v}}} + \frac{1}{M_{\mathrm{c}}}\right)u\left(1 - e_{\mathrm{v}}u^2\right) - \frac{\beta_{\mathrm{v}}K_{\mathrm{v}}\left(\frac{1}{M_{\mathrm{v}}} + \frac{1}{M_{\mathrm{c}}}\right)\dot{u}\left(1 + \varepsilon_{\mathrm{v}}\frac{\dot{u}^2}{\omega^2}\right)}{\omega} \\ & + \frac{K_{\mathrm{c}}\,w\left(1 - \varepsilon_{\mathrm{c}}w^2\right)}{M_{\mathrm{c}}} + K_{\mathrm{c}}\,\beta_{\mathrm{c}}\,\dot{w}\,\frac{\left(1 + \varepsilon_{\mathrm{c}}\frac{\dot{w}^2}{\omega^2}\right)}{\omega M_{\mathrm{c}}} + \frac{F_{\mathrm{c}}}{M_{\mathrm{c}}}.\end{aligned} \tag{3.30}$$

If F_{c} is chosen so that

$$K_c w \frac{\left(1-\varepsilon_c w^2\right)}{M_c} + \frac{K_c \beta_c \dot{w}\left(1+\varepsilon_c \frac{\dot{w}^2}{\omega^2}\right)}{\omega M_c} + \frac{F_c}{M_c} = 0 \tag{3.31}$$

whenever control is possible, u is the response of a damped oscillator and will decay toward zero. The relative displacement w and relative velocity $\dot{w}$ of the seat with respect to the sprung mass can be obtained by the use of an LVDT or a pull-wire transducer.

Alternatively the accelerations of the seat and the vehicle floor could be recorded and the difference integrated and low-pass filtered to obtain the relative velocity and displacement, although this system might be more expensive than that needed for the displacement method.

If the system were active and the time constant of the device low enough, the relative displacement and velocity of the viscera could be driven to zero and no discomfort would be experienced.

For the semi-active device to be on it is necessary that power be dissipated, namely that

$$F_c \dot{w} < 0 . \tag{3.32}$$

Experience with a semi-active control for a random input shows that the damper can only be on about half of the time.

3.6.3 Ideal Control

The performance of the proposed observer can be assessed by a comparison with the ideal situation in which the relative displacement u and its derivative are known.

In this case reference to Equation (3.30) shows that F_c should be chosen to satisfy

$$-K_v\left(\frac{1}{M_v}+\frac{1}{M_c}\right) u \left(1-\varepsilon_v u^2\right) - \frac{\beta_v K_v\left(\frac{1}{M_v}+\frac{1}{M_c}\right)\dot{u}\left(1+\varepsilon_v \frac{\dot{u}^2}{\omega^2}\right)}{\omega} + \tag{3.33}$$

$$+\frac{K_c w\left(1-\varepsilon_c w^2\right)}{M_c} + \frac{K_c \beta_c \dot{w}\left(1+\varepsilon_c \frac{\dot{w}^2}{\omega^2}\right)}{\omega M_c} + \frac{F_c}{M_c} = 0.$$

3.7 Results

The value of ε_c was estimated from the transmission ratios found experimentally by Yu and Khameneh (1999). Their work indicated that ε_c is approximately 400 m^{-2}. The results of Yu and Khameneh were also used to obtain values of f_c and β_c. Values of 4.35 Hz and 0.375 Hz, respectively, were deduced. The work of Mansfield and Griffin (2000) suggests ε_v is approximately 5000 m^{-2}.

Figure 3.3 shows the loss factor β required to prevent jumps in the cushion and viscera as a function of amplitude of response C.

The required value of β for the cushion is sufficiently low that even with oscillations of 20 mm amplitude, jumps will not in practice occur.

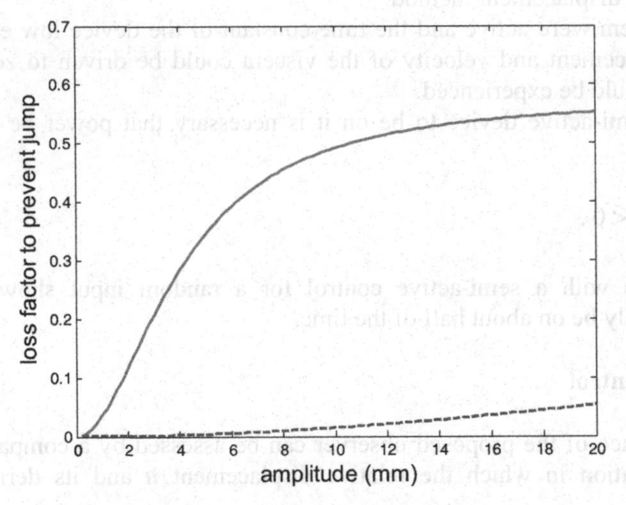

Fig. 3.3. Loss factor β to prevent jumps as a function of amplitude: viscera (solid line), cushion (dashed line)

The corresponding results for viscera indicate that, if visceral input oscillations are kept below 2 mm, a value of β_v of only 0.1 or greater is required. If on the other hand oscillations of up to 4 mm in amplitude occured, β_v would need to be greater than 0.3 (of similar magnitude to that estimated for cushion material) to prevent jumps. Jumps in the viscera would be physically damaging quite apart from the effect of the oscillation itself. The value of β_v likely to exist is not known but such predictions appear quite credible.

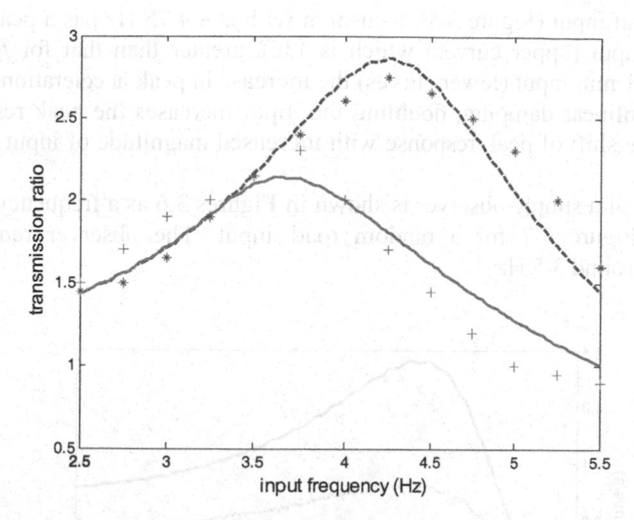

Fig. 3.4. Comparison of predicted (solid curves) and measured cushion transmissibility, * 5 mm input, + 20 mm input (measured data reproduced from Yu and Khameneh, Automotive seating foam: subjective dynamic comfort study. Reprinted with permission from SAE Paper # 1999-01-0588 © 1999 SAE International)

In Figure 3.4 the quoted transmissibility for cushion C (acc. out/acc. in) is compared with the predicted value for sinusoidal inputs of amplitude (a) 5 mm and (b) 20 mm; $\beta_c = 0.3$, $\varepsilon_c = 100$ N/m^3 and $f_n = 4.25$ Hz.

Agreement is quite good for the 5 mm inputs, and fair for the 20 mm inputs. Higher-order stiffness effects could be introduced if required.

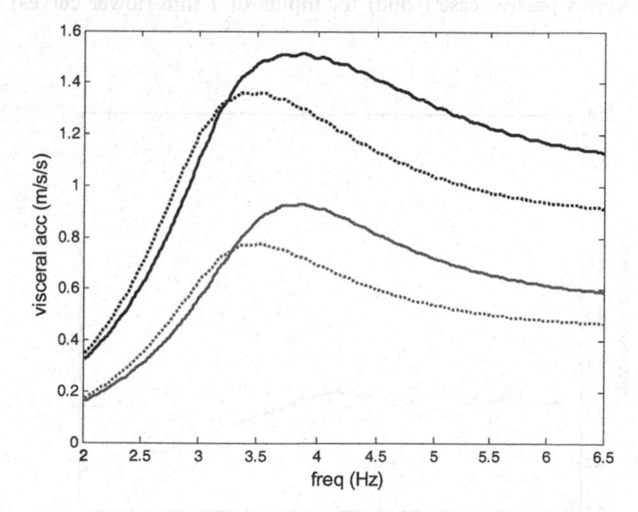

Fig. 3.5. $\varepsilon_v = 5000$ m^{-2}, $\varepsilon_c = 100$ m^{-2}, $\beta = 0.3$ for both viscera and cushion; viscera $f_v = 5$ Hz. Solid lines: $f_c = 4.75$ Hz, dashed lines: $f_c = 4$ Hz. Lower curves: 1 mm input, upper curves: 2 mm input

With 2 mm seat input (Figure 3.5), a cushion with f_c = 4.75 Hz has a peak visceral acceleration input (upper curves) which is 11% greater than that for f_c = 4 Hz. However, for 1 mm input (lower curves) the increase in peak acceleration is 20%.

Due to nonlinear damping, doubling the input increases the peak response by only 67%. The shift of peak response with increased magnitude of input is modest but detectable.

The effct of a simple observer is shown in Figures 3.6 as a frequency response plot and in Figure 3.7 for a random road input. The observer removes the resonance at around 3.5 Hz.

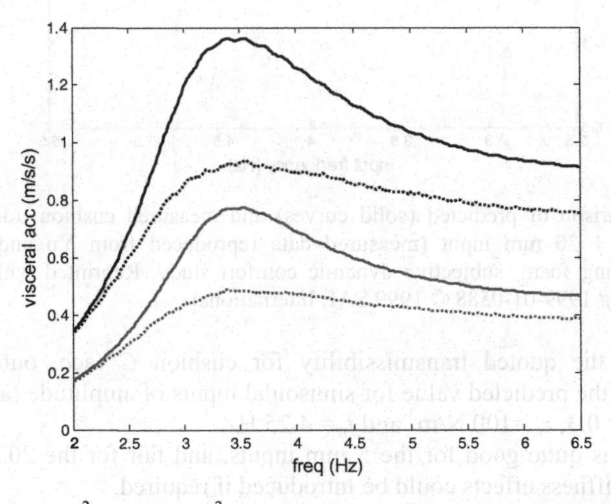

Fig. 3.6. $\varepsilon_v = 5000\ \text{m}^{-2}$, $\varepsilon_c = 100\ \text{m}^{-2}$, $\beta_c = \beta_v = 0.4$, $f_c = 4$ Hz, $f_v = 5$ Hz. Observer-based seat control (dotted) versus passive case (solid) for inputs of 1 mm (lower curves) and 2 mm (upper curves)

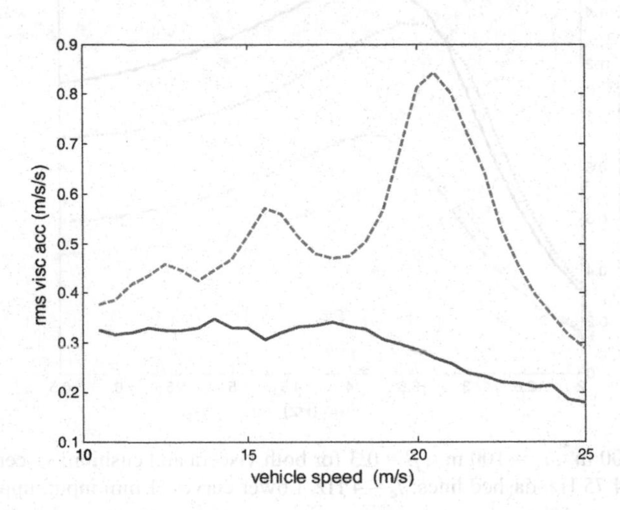

Fig. 3.7. Visceral response passive (dashed), simple observer (solid line); data as above

3.8 Seated Human with Head-and-Neck Complex

The seated human model has four subsystems: seat, cushion, driver body and the head-and-neck complex (HNC). The latter is represented as an inverted double pendulum.

The entire seated human model, together with the seat mass and under seat semi-active damper, is presented in Figure 3.8.

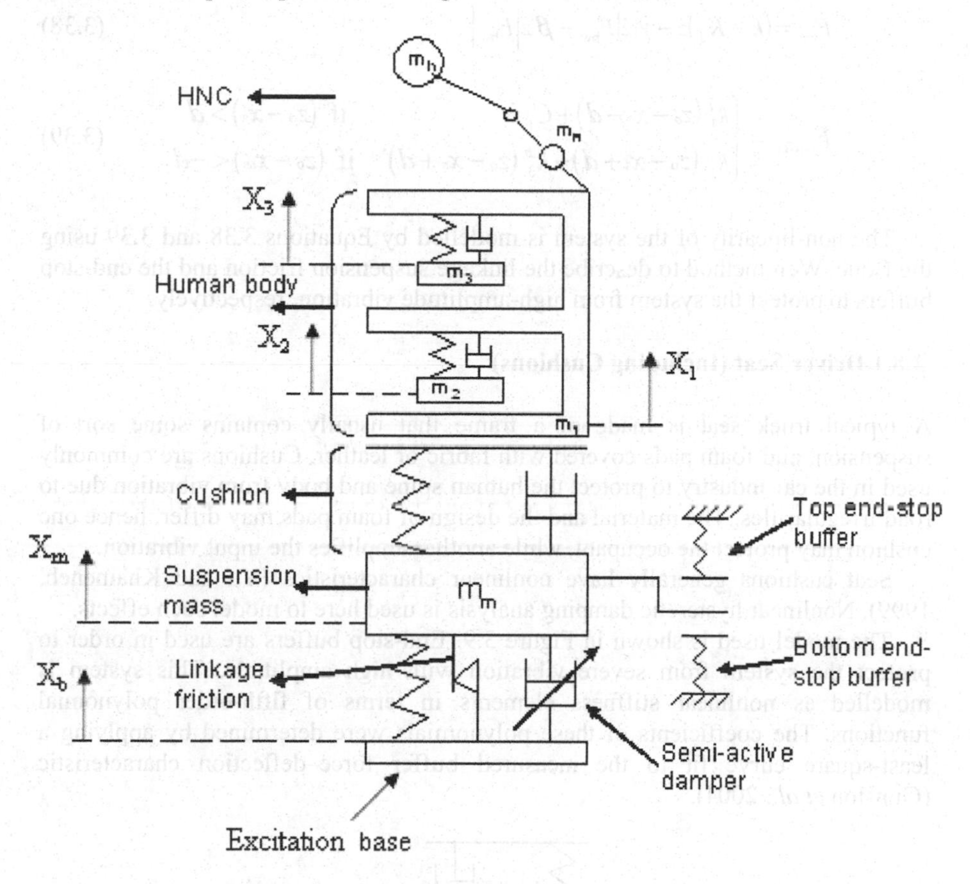

Fig. 3.8. Schematic diagram of the seated human model with HNC

The equations of motion for the body and seat are:

$$m_3\ddot{x}_3 = C_3\left(\dot{x}_1 - \dot{x}_3\right) + K_3\left(x_1 - x_3\right) = f_3(t), \tag{3.34}$$

$$m_2\ddot{x}_2 = C_2\left(\dot{x}_1 - \dot{x}_2\right) + K_2\left(x_1 - x_2\right) = f_2(t), \tag{3.35}$$

$$m_1 \ddot{x}_1 = C_c \left(\dot{x}_s - \dot{x}_1\right) + K_c \left(x_s - x_1\right) - f_2(t) - f_3(t), \tag{3.36}$$

$$m_m \ddot{x}_s = F_c(t) + K_s \left(z_0 - x_s\right) - C_c \left(\dot{x}_s - \dot{x}_1\right) - \left(x_s - x_1\right) + F_{bw}(t) + F_{buffer}(t). \tag{3.37}$$

$$\dot{F}_{bw} = \left(k - K_s\right)\dot{z} - \gamma\left|\dot{z}\right|F_{bw} - \beta\,\dot{z}\left|\dot{F}_{bw}\right|, \tag{3.38}$$

$$F_{buffer} = \begin{cases} k_1^t \left(z_o - x_s - d\right) + C_1\,\dot{z} & \text{if } \left(z_o - x_s\right) > d \\ k_1^b \left(z_o - x_s + d\right) + k_3^b \left(z_o - x_s + d\right)^3 & \text{if } \left(z_o - x_{ss}\right) < -d \end{cases} \tag{3.39}$$

The non-linearity of the system is modelled by Equations 3.38 and 3.39 using the Bouc–Wen method to describe the linkage suspension friction and the end-stop buffers to protect the system from high-amplitude vibration, respectively.

3.8.1 Driver Seat (Including Cushions)

A typical truck seat is made of a frame that usually contains some sort of suspension, and foam pads covered with fabric or leather. Cushions are commonly used in the car industry to protect the human spine and body from vibration due to road irregularities. The material and the design of foam pads may differ, hence one cushion may protect the occupant, while another amplifies the input vibration.

Seat cushions generally have nonlinear characteristics (Yu and Khameneh, 1999). Nonlinear hysteretic damping analysis is used here to model such effects.

The model used is shown in Figure 3.9. End-stop buffers are used in order to protect the system from severe vibration with high amplitude. This system is modelled as nonlinear stiffness elements in terms of fifth-order polynomial functions. The coefficients of these polynomials were determined by applying a least-square curve fit to the measured buffer force–deflection characteristic (Gunston *et al.*, 2004).

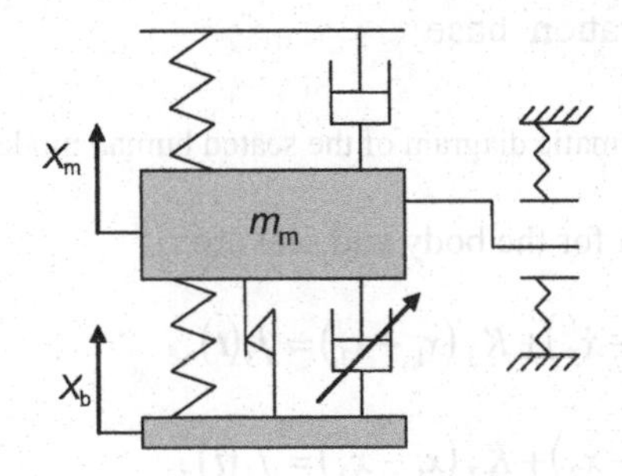

Fig. 3.9. Non-linear driver seat model

The input to the driver seat is the motion of the vehicle chassis which is combined heave and pitch. Only the vertical component is considered in the body response analysis.

3.8.2 Driver Body

The driver model (Figure 3.10) is based on experimental work of Wei and Griffin and consists of a light frame (m_1) and two suspended masses (m_2 and m_3) each with a linear spring and damper. The three masses do not represent actual human organs but are chosen so that the model reproduces the force response of vibrated subjects.

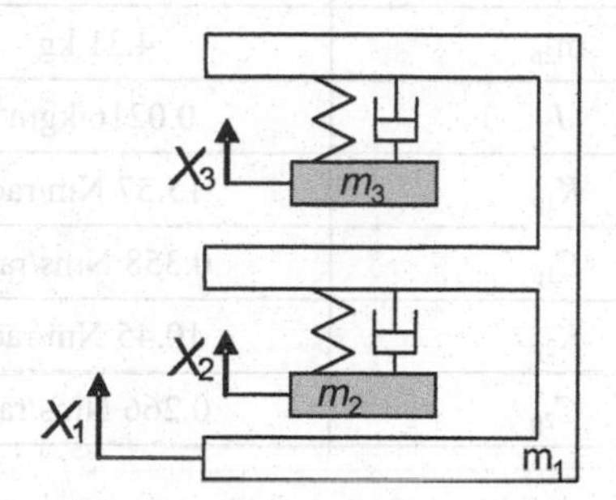

Fig. 3.10. Driver body model [copyright Elsevier (1998), reproduced with modifications from Wei and Griffin, The prediction of seat transmissibility from measures of seat impedance, J Sound Vib,Vol 214, N 1, pp 121–137, used by permission]

3.8.3 Head-and-Neck Complex (HNC)

A two-degree-of-freedom model (Figure 3.8) is used to describe the head-and-neck system (Fard *et al.*, 2003); a linearised model of the double inverted pendulum is used to emulate the motion of the head-and-neck complex. The first centre of rotation is assumed to be very close to the centre of the neck O_2, while the second centre of rotation is situated at O_1.

The determination of the viscoelastic parameters is a very difficult task that requires much experimental work and data with human volunteers. Based on published work of Fard *et al.* (2003) these parameters are summarised in Table 3.1. The head-and-neck complex is attached to the driver body in order to represent the human seated model including the driver body and the head-and-neck motion. For vehicle applications the driver body is assumed to vibrate in the vertical direction only, while the HNC is able to rotate in three dimensions in response to driver body vertical motion and the vertical, pitch and lateral motion of the vehicle chassis. The HNC system for 3D analysis is described using the Gibbs–Appel method.

Table 3.1. Head-and-neck complex parameters (copyright Elsevier, reproduced from Tsampardoukas G, Stammers CW and Guglielmino E, Hybrid balance control of a magnetorheological truck suspension, accepted for publication in J Sound Vib, used by permission)

Parameter	Value
L_{1p}	0.042 m
m_{1p}	1.07 kg
J_1	0.0012 kgm^2
L_{2p}	0.071 m
m_{2p}	4.31 kg
J_2	0.0216 kgm^2
K_{1p}	15.57 Nm/rad
C_{1p}	0.358 Nms/rad
K_{2p}	10.45 Nm/rad
C_{2p}	0.266 Nms/rad

3.8.4 Analysis of the Head-and-Neck System

The motion of the head-and-neck complex (HNC) due to vertical, pitch and roll motions of the vehicle chassis is presented using the Gibbs–Appel method (Blundell and Harty, 2004), an alternative method to Lagrange. With the Gibbs–Appel method the kinetic energy of the Lagrange method is replaced with the "energy" of acceleration. The potential energy of the system (Equation 3.46) is used just as with the Lagrange method. In complicated systems, Gibbs–Appel can be a simpler tool than Lagrange for the derivation of the equations of motion.

The Appel function A of the system (the acceleration "energy") in three dimensions is presented in Equation 3.47. Equations of motion are obtained via derivation of the total acceleration of each direction as presented by Equations 3.48–3.51.

The external accelerations (Figures 3.11 and 3.12) applied both to the head and neck due to the motion of the vehicle chassis are given by Equations 3.40–3.42, taking into account that the pitch and roll chassis accelerations can be analysed into two components, one vertical and one horizontal. The component in lateral direction is not illustrated in Figures 3.11 and 3.12.

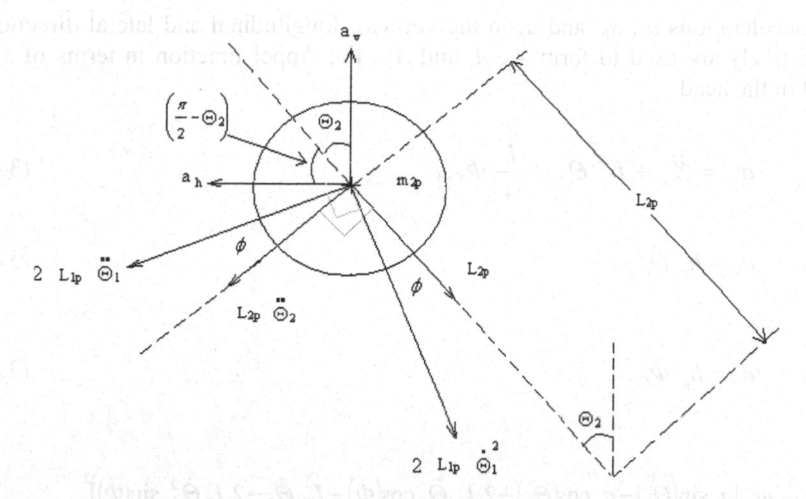

Fig. 3.11. Head accelerations [copyright IMechE (2008), reproduced from Tsampardoukas G, Stammers CW and Guglielmino E, Semi-active control of a passenger vehicle for improved ride and handling, accepted for publication in Proceedings of the Institution of Mechanical Engineers, Part D: Journal of Automobile Engineering, Publisher: Professional Engineering Publishing, ISSN 0954/4070, Vol 222, D3/2008, pp 325–352, used by permission]

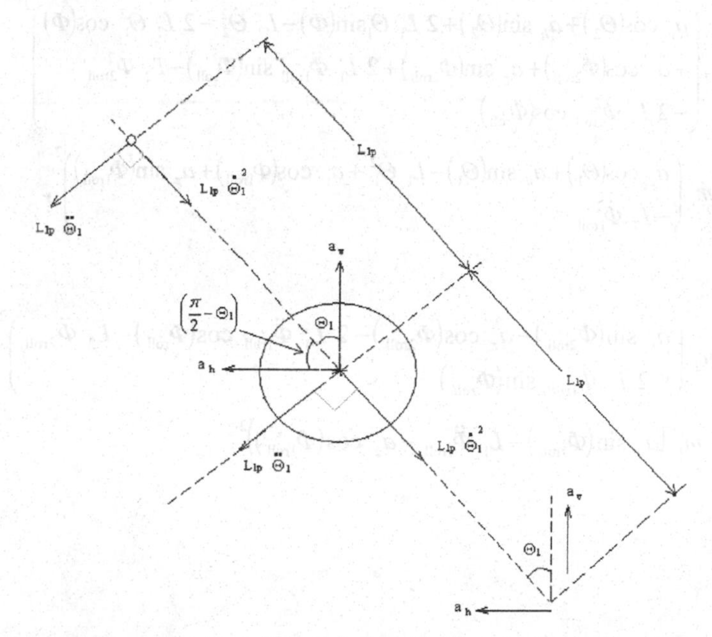

Fig. 3.12. Neck accelerations [copyright IMechE (2008), reproduced from Tsampardoukas G, Stammers CW and Guglielmino E, Semi-active control of a passenger vehicle for improved ride and handling, accepted for publication in Proceedings of the Institution of Mechanical Engineers, Part D: Journal of Automobile Engineering, Publisher: Professional Engineering Publishing, ISSN 0954/4070, Vol 222, D3/2008, pp 325–352, used by permission]

The accelerations a_{v}, a_{h} and a_{z} in the vertical, longitudinal and lateral directions, respectively are used to form A_x, A_y and A_z, the Appel function in terms of axes fixed in the head.

$$a_{\mathrm{v}} = \ddot{X}_{\mathrm{c}} + L_3\ \ddot{\Theta}_{\mathrm{C}} + \frac{L}{4}\ \ddot{\Phi}_{\mathrm{C}}, \tag{3.40}$$

$$a_{\mathrm{h}} = h_{\mathrm{p}}\ \ddot{\Theta}_{\mathrm{C}}, \tag{3.41}$$

$$a_{\mathrm{z}} = h_{\mathrm{p}}\ \ddot{\Phi}_{\mathrm{C}}, \tag{3.42}$$

$$\begin{aligned} A_x = & \frac{1}{2}\, m_2\, \left(a_v \sin(\Theta_2) - a_h \cos(\Theta_2) - 2\, L_1\, \ddot{\Theta}_1 \cos(\Phi) - L_2\, \ddot{\Theta}_2 - 2\, L_1\, \dot{\Theta}_1^2 \sin(\Phi)\right)^2 \\ & + \frac{1}{2}\, m_1\, \left(a_v \sin(\Theta_1) - a_h \cos(\Theta_2) - L_1\, \ddot{\Theta}_1\right)^2, \end{aligned} \tag{3.43}$$

$$\begin{aligned} A_y = & \frac{1}{2}\, m_2 \begin{pmatrix} a_v \cos(\Theta_2) + a_h \sin(\Theta_2) + 2\, L_1\, \ddot{\Theta}_1 \sin(\Phi) - L_2\, \dot{\Theta}_2^2 - 2\, L_1\, \dot{\Theta}_1^2 \cos(\Phi) \\ + a_v \cos(\Phi_{2\mathrm{roll}}) + a_z \sin(\Phi_{2\mathrm{roll}}) + 2\, L_1\, \ddot{\Phi}_{1\mathrm{roll}} \sin(\Phi_{\mathrm{roll}}) - L_2\, \dot{\Phi}_{2\mathrm{roll}}^2 \\ - 2\, L_1\, \dot{\Phi}_{1\mathrm{roll}}^2 \cos(\Phi_{\mathrm{roll}}) \end{pmatrix}^2 \\ & + \frac{1}{2}\, m_1 \begin{pmatrix} a_v \cos(\Theta_1) + a_h \sin(\Theta_1) - L_1\, \dot{\Theta}_1^2 + a_v \cos(\Phi_{1\mathrm{roll}}) + a_z \sin(\Phi_{1\mathrm{roll}}) \\ - L_1\, \dot{\Phi}_{1\mathrm{roll}}^2 \end{pmatrix}^2, \end{aligned} \tag{3.44}$$

$$\begin{aligned} A_z = & \frac{1}{2}\, m_2 \begin{pmatrix} a_v \sin(\Phi_{2\mathrm{roll}}) - a_z \cos(\Phi_{2\mathrm{roll}}) - 2\, L_1\, \ddot{\Phi}_{1\mathrm{roll}} \cos(\Phi_{\mathrm{roll}}) - L_2\, \ddot{\Phi}_{2\mathrm{roll}} \\ - 2\, L_1\, \dot{\Phi}_{1roll}^2 \sin(\Phi_{roll}) \end{pmatrix}^2 \\ & + \frac{1}{2}\, m_1\, \left(a_v \sin(\Phi_{1\mathrm{roll}}) - L_1\, \ddot{\Phi}_{1\mathrm{roll}} - a_z \cos(\Phi_{1\mathrm{roll}})\right)^2, \end{aligned} \tag{3.45}$$

$$PE = m_1 L_1 g \cos(\Theta_1) + m_2 g (2 L_1 \cos(\Theta_1) + L_2 \cos(\Theta_2)) + \frac{1}{2} K_2 (\Theta_2 - \Theta_1)^2$$
$$+ \frac{1}{2} C_2 \left(\dot{\Theta}_2 - \dot{\Theta}_1\right)^2 + \frac{1}{2} K_1 \Theta_1^2 + \frac{1}{2} C_1 \dot{\Theta}_1^2 + m_1 L_1 g \cos(\Phi_{1roll})$$
$$+ m_2 g \left(2 L_1 \cos(\Phi_{1roll}) + L_2 \cos(\Phi_{2roll})\right) + \frac{1}{2} K_2 (\Phi_{2roll} - \Phi_{1roll})^2 \quad (3.46)$$
$$+ \frac{1}{2} C_2 \left(\dot{\Phi}_{2roll} - \dot{\Phi}_{1roll}\right)^2 + \frac{1}{2} K_1 \Phi_1^2 + \frac{1}{2} C_1 \dot{\Phi}_1^2 ,$$

$$A = A_x + A_y + A_z , \quad (3.47)$$

The equations of motion are:

$$\frac{\partial A}{\partial \ddot{\Theta}_1} + \frac{\partial PE}{\partial \dot{\Theta}_1} + \frac{\partial PE}{\partial \Theta_1} = 0 \Leftrightarrow \frac{\partial A_x}{\partial \ddot{\Theta}_1} + \frac{\partial A_y}{\partial \ddot{\Theta}_1} + \frac{\partial A_z}{\partial \ddot{\Theta}_1} + \frac{\partial PE}{\partial \dot{\Theta}_1} + \frac{\partial PE}{\partial \Theta_1} = 0 , \quad (3.48)$$

$$\frac{\partial A}{\partial \ddot{\Theta}_2} + \frac{\partial PE}{\partial \dot{\Theta}_2} + \frac{\partial PE}{\partial \Theta_2} = 0 \Leftrightarrow \frac{\partial A_x}{\partial \ddot{\Theta}_2} + \frac{\partial A_y}{\partial \ddot{\Theta}_2} + \frac{\partial A_z}{\partial \ddot{\Theta}_2} + \frac{\partial PE}{\partial \dot{\Theta}_2} + \frac{\partial PE}{\partial \Theta_2} = 0 , \quad (3.49)$$

$$\frac{\partial A}{\partial \ddot{\Phi}_{1roll}} + \frac{\partial PE}{\partial \dot{\Phi}_{1roll}} + \frac{\partial PE}{\partial \Phi_{1roll}} = 0 \Leftrightarrow \frac{\partial A_x}{\partial \ddot{\Phi}_{1roll}} + \frac{\partial A_y}{\partial \ddot{\Phi}_{1roll}} + \frac{\partial A_z}{\partial \ddot{\Phi}_{1roll}} + \frac{\partial PE}{\partial \dot{\Phi}_{1roll}} + \frac{\partial PE}{\partial \Phi_{1roll}} = 0, \quad (3.50)$$

$$\frac{\partial A}{\partial \ddot{\Phi}_{2roll}} + \frac{\partial PE}{\partial \dot{\Phi}_{2roll}} + \frac{\partial PE}{\partial \Phi_{2roll}} = 0 \Leftrightarrow \frac{\partial A_x}{\partial \ddot{\Phi}_{2roll}} + \frac{\partial A_y}{\partial \ddot{\Phi}_{2roll}} + \frac{\partial A_z}{\partial \ddot{\Phi}_{2roll}} + \frac{\partial PE}{\partial \dot{\Phi}_{2roll}} + \frac{\partial PE}{\partial \Phi_{2roll}} = 0, \quad (3.51)$$

The equations of motion for the head-and-neck complex in three dimensions are obtained. In matrix form:

$$\boldsymbol{M}(\boldsymbol{q})\ddot{\boldsymbol{q}} + \boldsymbol{B}(\boldsymbol{q})\,\dot{\boldsymbol{q}}^2 + \boldsymbol{C}\,\dot{\boldsymbol{q}} + \boldsymbol{D}\,\boldsymbol{q} + \boldsymbol{E}(\boldsymbol{q})\sin(\boldsymbol{q}) = \boldsymbol{P}(\boldsymbol{q})\,\boldsymbol{U} , \quad (3.52)$$

where

$$\boldsymbol{q} = \begin{bmatrix} \Theta_1 \\ \Theta_2 \\ \phi_{1roll} \\ \phi_{2roll} \end{bmatrix}, \quad \boldsymbol{U} = \begin{bmatrix} U_V \\ U_H \\ U_Z \end{bmatrix}$$

These equations can be then incorporated with those for the seat and body.

3.8.5 Head Accelerations During Avoidance Manoeuvre

The acceleration of the driver's head during an avoidance manoeuvre (modelled as a rapid lane change) are shown in Figure 3.13 as a function of vehicle speed. The performance of three different algorithms for the semi-active control (discussed subsequently in Chapters 4 and 7) of the suspension, namely skyhook, balance control cancelling (BCC) and balance control additive (BCA).

Skyhook was designed to achieve improved ride, and the benefit of skyhook control compared with the passive case is evident in all three directions — longitudinal, vertical and lateral. The BCC algorithm was designed to limit dynamic tyre loads and hence reduce road damage in the case of heavy vehicles. It is of no help to the driver in terms of comfort, but is very helpful in improving handling and thus achieving the intended rapid lane change. The BCA algorithm is employed to reduce vehicle roll. This is important for a laden freight vehicle which will have a higher centre of gravity, but is of secondary importance for a passenger vehicle. The switching between different algorithms depending on the driving conditions could be implemented automatically on the basis of some feedback indicators. This could be made in a variety of ways. A possible approach could be using a rapid steering input at high speeds to select the BCC algorithm (to achieve improved safety). Conversely at lower speeds, skyhook could be selected for driver comfort.

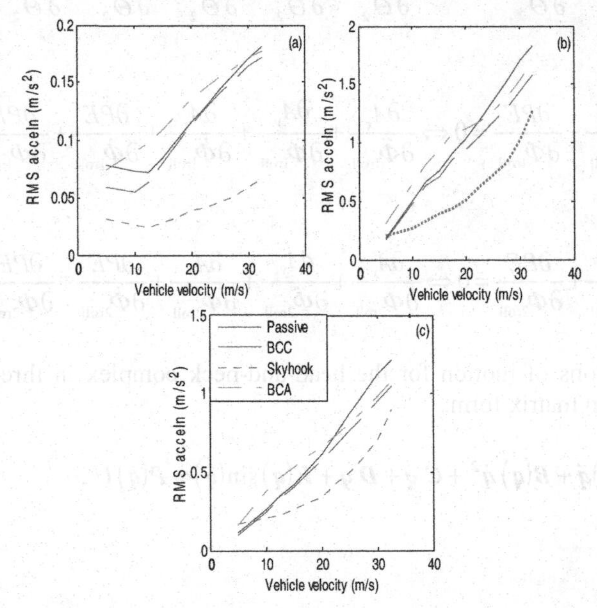

Fig. 3.13. RMS acceleration of the driver head-and-neck complex: (a) longitudinal direction, (b) vertical direction, (c) lateral direction [copyright IMechE (2008), reproduced from Tsampardoukas G, Stammers CW and Guglielmino E, Semi-active control of a passenger vehicle for improved ride and handling, accepted for publication in Proceedings of the Institution of Mechanical Engineers, Part D: Journal of Automobile Engineering, Publisher: Professional Engineering Publishing, ISSN 0954/4070, Vol 222, D3/2008, used by permission]

4

Semi-active Control Algorithms

4.1 Introduction

Feedback control radically alters the dynamics of a system: it affects its natural frequencies, its transient response as well as its stability. These aspects must be carefully studied whenever a closed-loop control system is designed. This is the aim of this chapter which focusses on the "brain" of the semi-active suspension, the control algorithm.

A brief and succinct overview of control fundamentals (PID, adaptive and robust control) is firstly provided, then robust control algorithms will be analysed, with a particular focus on variable structure control. Subsequently the classical semi-active damper control algorithms will be reviewed with emphasis on balance logic (which is a spring force cancellation strategy).

A sound mathematical analysis of the balance logic algorithm will be presented. The theory is discussed based on a friction damper (FD) but the results can be applied straightforwardly to a magnetorheological damper (MRD) as there are deep analogies between the FD and MRD mathematical models.

In Chapter 2 it was stated that the simplest suspension model has one degree of freedom. Such a system can be regarded as a non-linear oscillator constituted by a spring–mass–damper system whose governing equation is second order

$$m_1\ddot{x}_1 + c(\dot{x}_1 - \dot{y}) + k(x_1 - y) + F_d(x_1 - y, \dot{x}_1 - \dot{y}, \ddot{x}_1) = 0 \,. \tag{4.1}$$

Figure 4.1 depicts the system described by Equation 4.1: k and c are, respectively, the elastic and viscous coefficients, m_1 is the sprung mass, y and $\dot{y}$ represent the external inputs, x_1 the mass displacement, $\dot{x}_1$ its velocity and $\ddot{x}_1$ its acceleration. The function $F_d(\cdot)$ is the controlled damping force generated by the variable damper in the figure. The coefficient c can be regarded as a residual passive viscous damping).

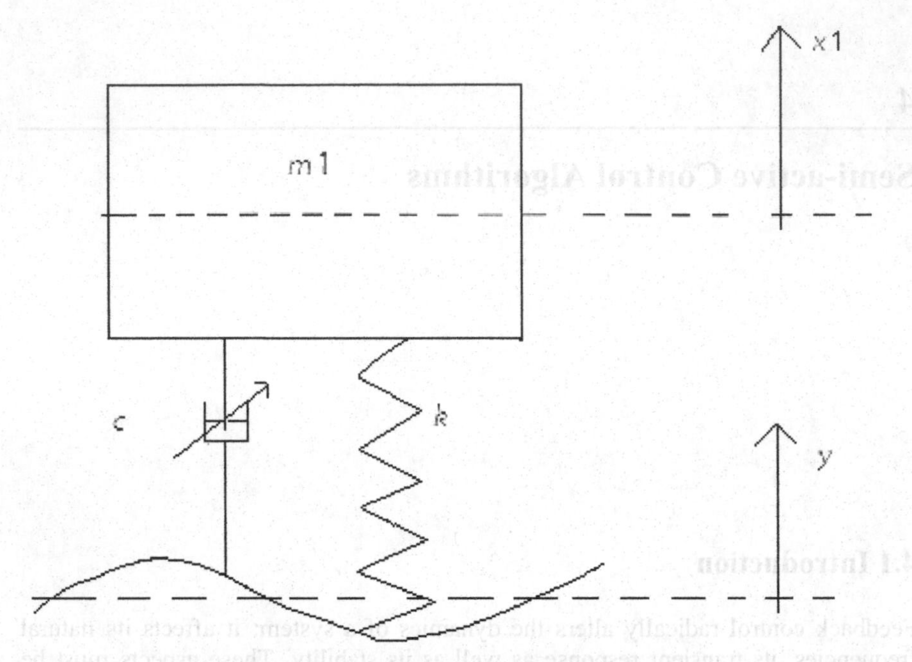

Fig. 4.1. 1DOF spring–mass–damper system model

From a sheer mathematical viewpoint the suspension control problem can be regarded as finding a piecewise continuous function $F_\mathrm{d}(x_1 - y, \dot{x}_1 - \dot{y}, \ddot{x}_1)$ that forces the solutions of the equation (chassis position) and its derivatives (*i.e.*, velocity and acceleration) to behave in a predefined manner, according to a set of design specifications. Ideally these quantities should all be minimised.

The function $F_\mathrm{d}(\cdot)$ is physically a controlled damping force that adds up to the spring, inertial and residual passive viscous damping forces of the system. The specification of this function (in terms of ride and road holding) is the target of the control. Equation 4.1 is in general nonlinear and its coefficients may not be known precisely (parameter uncertainties), hence the system parameters cannot be identified precisely and may deviate significantly from their actual values, thus hampering the controller performance.

Control theory offers a number of tools which can be employed to cope with this class of systems. They can be grouped into three families:

- PID controllers
- adaptive controllers
- robust controllers

The following sections will provide some qualitative insights into each.

4.2 PID Controllers

Proportional–integral–derivative (PID) controllers are used in a wide range of applications. These controllers are frequently designed in the frequency domain, and on the hypothesis of linearity. Therefore they work well if this hypothesis is close to the actual behaviour of the system. If this is not the case, performance is likely to be poor. A linear system is defined in the frequency domain by its transfer function $G(i\omega)$, which is often the result of a linearisation of a mathematical model. The actual response of a PID-controlled system depends on how close the linear model represents the actual behaviour of the real system. The major sources of uncertainty are due to parameter changes, unmodelled dynamics, time delays, changes in the operating point, sensor noise and unpredicted disturbance inputs. Although a PID controller is not supposed to perform well in conditions different from those for which it has been designed, its performance can be optimised using a robust design approach or by including an additional adaptive loop (see Section 4.3). In this way, it can be made to exhibit the desired performance over a larger range.

A controller is said to be robust (Dorf and Bishop, 1995) if it has low sensitivities, is stable and continues to meet its nominal specification over a typical range of parameter variations. A robust control design is one that satisfactorily meets its control specification, even in the presence of parameter uncertainties and other modelling errors. A measure of the robustness to small parameter variations is the sensitivity, defined as:

$$S_\alpha = \frac{\partial G / G}{\partial \alpha / \alpha} \tag{4.2}$$

with α being a parameter of the system. The design of a PID controller entails the choice of the proportional, integral and derivative gains. More generally the design of the controller transfer function entails defining its static gain, and its poles and zeroes. Classical linear control system theory offers a number of tools for controller design and stability analysis of the closed-loop response, such as the root locus, Bode and Nyquist diagrams *etc.*, as well as several design criteria.

The desirable features of a robust design in the frequency domain are the largest possible bandwidth and the largest possible loop gain, attained primarily in the controller and in the forward path transfer functions (in this way the disturbance rejection is increased). However these specifications are in conflict and trade-offs must be pursued in order to guarantee both a swift response and system stability. Several design and tuning methods have been developed over the years for PID controllers. The most classical tuning method involves the so-called Ziegler and Nichols rules (Ziegler and Nichols, 1942); they permit the tuning of the PID parameters, based on the partial knowledge of the transfer function, which can be obtained by simple tests on the system; another classical method is based on the Cohen and Coon rules (Cohen and Coon, 1953); in this method the PID parameters are tuned by fixing a damping ratio of 0.25 in response to a disturbance input. In the subsequent years several methods have been developed, such as the Smith predictor (Smith, 1957), the Haalman method (Haalman, 1965) and the Dahlin regulator (Dahlin, 1968). Other methods exist that minimise a performance index of the error (ISE, IAE *etc.*). A review of all these and other methods can be found in Åström *et al.* (1993). Research is still active on PID controller tuning and new methods are being proposed due to the widespread industrial use of these controllers worldwide.

As opposed to linear frequency-domain-based methods, typically employed in PID controller design, in the past 40 years non-linear control methods, namely adaptive and robust algorithms (Slotine and Li, 1992), have been developed as a consequence of the derivation (circa 1960–1980) of various mathematical methods (state space and phase plane methods, Lyapunov stability theory *etc.*). Other control methods have been developed in those years including optimal control (Berkovitz, 1974; Hocking, 1991; Kirk, 2004) and neuro-fuzzy control (Passino and Yurkovich, 1998; Abdi *et al.*, 1999).

4.3 Adaptive Control

A possible approach to deal with nonlinear systems is adaptive control. This control philosophy is opposite to robust control. Although this work is focussed on the use of the latter type of control, it is worthwhile describing the basic principles of adaptive control in the context of this chapter. Adaptive controllers are used if the parameters of the process are time varying and a fixed-parameter controller would not yield acceptable results.

Adaptive control can be thought as control with on line estimation of uncertain parameters (Slotine, 1984). The parameters, estimated on the basis of the measured plant signals, are then used in the control action. Systems with constant or slowly varying uncertain parameters are the most likely to have an improvement in their responses with this type of control. In such systems the on line estimation requires a small amount of computational power and therefore the controller learns more quickly (the more the adaptation goes on, the more an adaptive controller improves its performance). The target of the adaptive control is to try to keep consistent performance in systems with unknown variations in its parameters. A controller with fixed parameters (such as a PID) can become inaccurate or even unstable for large parameter variations; however an adaptive loop can be added to a PID controller, enhancing its performance.

Typical applications where adaptive algorithms could be successfully employed can be found in robotics (where the payload can vary), electrical power systems (where the power demand varies slowly over the day) and process control (*e.g.*, chemical or metallurgical processes). Two main approaches exist to design adaptive controllers. The first one is referred to as model-reference adaptive control; a reference model is used to set the ideal response of the system, with the error between the ideal and the actual response being the input to the adaptation law. The second approach is self-tuning control where the estimation of the parameters is obtained by measuring the inputs and the outputs to the plant. These are then used for the on line tuning of the controller, which is continually updated as parameters change with operating conditions. There is an enormous amount of literature on the topic. For a comprehensive overview on adaptive and other non-linear control methods the reader can refer to the work by Slotine and Li (1992).

4.4 Robust Control

A robust controller is meant to provide a reasonable level of performance in systems with uncertain parameters, no matter how fast they vary, but it usually requires the knowledge (or an estimate) of the bounds of the uncertainty. The typical structure of a robust control law is composed of a nominal part (such as a state feedback or an inverse model control law), plus a term which deals with model uncertainty.

A small degree of robustness can be achieved with PID controllers, as discussed before. However robust control is often associated with H_∞ control and variable structure control (VSC). In this book only the latter control strategy is considered. Details on H_∞ control can be found in Chen (2000).

Variable structure controllers are a very large class of robust controllers (Gao and Hung, 1993; Hung *et al.*, 1993; De Carlo *et al.*, 1998). The distinctive feature of VSC is that the structure of the system is intentionally changed according to an assigned law. This can be obtained by switching on or cutting off feedback loops, scheduling gains and so forth (Itkis, 1976). By using VSC, it is possible to take the best out of several different systems (more precisely structures), by switching from one to the other. The control law defines various regions in the phase space and the controller switches between a structure and another at the boundary between two different regions according to the control law. Therefore the designer is no longer forced to trade off between static and dynamic requirements, as is the case in linear control systems design. It is possible to synthesize a wide range of trajectories by switching between two or more systems according to a predefined law, improving therefore the transient and steady-state responses of the new system with respect to the responses of the original systems.

In principle the laws governing the change in the structure can be very different; however the law that is given greatest attention is that producing the so-called sliding regime. The reason why it has been so widely studied lies in the fact that, when the sliding motion occurs, the system becomes insensitive (ideally invariant) to parametric and external disturbances of the plant. This particular type of VSC is known as sliding mode control (Utkin, 1992). It must be stressed that a VSC system can be devised without a sliding mode, but in this general case the robustness properties are not guaranteed.

The principles of adaptive and robust control have been now outlined. There is no general rule to make a decision on which control is more suitable to a particular type of system to be controlled; however some rules of thumb can be given. In very broad terms adaptive control performs better with constant or slowly varying parameters. An adaptive controller improves its performance as adaptation proceeds, whereas a robust controller only attempts to keep good performance. Robust control however can cope better with quickly varying parameters and unmodelled dynamics. Some advanced controllers exist which combine adaptive and robust features; they are known as robust adaptive controllers. Furthermore, the implementation of an adaptive controller requires more computational effort to perform the on line identification and this require more computational power, whilst robust control algorithms are usually simpler from a computational viewpoint.

4.5 Balance, Skyhook and Groundhook

After having overviewed control fundamentals, attention is now centred on semi-active suspension control algorithms. In particular the focus is on balance, skyhook, groundhook logic and their numerous variants. They can all be categorised as VSC algorithms.

4.5.1 Balance Logic

This logic first was introduced by Rakheja and Sankar (1985) and developed by the authors (Stammers and Sireteanu, 1997). With reference to Figure 1.3 (the case of a friction damper) balance logic aims at reducing chassis acceleration. If the relative displacement x and relative velocity $\dot{x}$ across the suspension are measured then, according to the balance logic, the damping force $F_{\mathrm{d}}(x,\dot{x})$ must be modulated sequentially such that:

$$F_{\mathrm{d}} = F_{\mathrm{balance}} = \begin{cases} k_{\mathrm{s}}|x|\operatorname{sgn}\dot{x} & x\dot{x} \le 0 \\ 0 & x\dot{x} > 0 \end{cases} \tag{4.3}$$

4.5.2 Skyhook Logic

This logic, originally devised by Karnopp *et al.* (1974) is called skyhook, as the damper is regarded as being hooked to a fixed point in the sky. As this is obviously not possible, the approximation to an ideal skyhook is given by the following control logic:

$$F_{\mathrm{d}} = F_{\mathrm{skyhook}} = \begin{cases} c_{\mathrm{sky}}\dot{x}_1 & \dot{x}\dot{x}_1 > 0 \\ 0 & \dot{x}\dot{x}_1 \le 0 \end{cases} \tag{4.4}$$

where $\dot{x}$ is the relative velocity and $\dot{x}_1$ the absolute chassis velocity.

4.5.3 Groundhook Logic

This logic (Novak and Valasek, 1996) aims at reducing dynamic tyre force, thus improving handling and at the same time reducing road damage (particularly useful in the case of heavy vehicles). Like skyhook, the groundhook damper is supposed to be hooked to a fixed point, in this case the ground. The algorithm is:

$$F_d = F_{\text{groundhook}} = \begin{cases} c_{\text{gnd}}\dot{x}_2 & -\dot{x}\dot{x}_2 > 0 \\ 0 & -\dot{x}\dot{x}_2 \leq 0 \end{cases} \tag{4.5}$$

where $\dot{x}$ is the wheel absolute displacement and $\dot{x}_2$ the relative displacement.

4.5.4 Displacement-based On–Off Groundhook Logic

This is an interesting algorithm based on an on–off switched damping logic: it is a form of groundhook blended with balance. The switching condition is given by the product of a displacement and a velocity rather than two velocities (as in the pure groundhook). The algorithm was originally proposed by Koo *et al.* (2004). The algorithm is the following:

$$F_d = c_{\text{controllable}}\dot{x}, \tag{4.6}$$

where

$$c_{\text{controllable}} = \begin{cases} c_{\text{on}} & x_2\dot{x} > 0 \\ c_{\text{off}} & x_1\dot{x} \leq 0 \end{cases} \tag{4.7}$$

4.5.5 Hybrid Skyhook–Groundhook Logic

This logic is aimed at reducing both body acceleration and dynamic tyre force and is obtained by combining skyhook and groundhook:

$$F_{\text{hybrid}} = G\left[\mu F_{\text{skyhook}} + (1-\mu)F_{\text{groundhook}}\right], \tag{4.8}$$

with the following four-state switching condition:

$$\begin{cases} F_{\text{skyhook}} = \dot{x}_1 & \dot{x}\dot{x}_1 > 0 \\ F_{\text{skyhook}} = 0 & \dot{x}\dot{x}_1 \leq 0 \\ F_{\text{groundhook}} = \dot{x}_2 & -\dot{x}\dot{x}_2 > 0 \\ F_{\text{groundhook}} = 0 & -\dot{x}\dot{x}_2 \leq 0 \end{cases} \tag{4.9}$$

If μ is set to 1, the hybrid control policy is switched to pure skyhook. On the other hand, if μ is set to 0, the hybrid control is switched to pure groundhook.

An interesting problem is to choose μ so as to minimise a performance index which takes into account, with different weight terms proportional to dynamic tyre force, suspension working space and chassis acceleration (coefficients α, β and γ in Equation 4.10) such as

$$J(\mu) = \int_0^T \{[\alpha(x_2 - z_0)]^2 + [\beta(x_1 - x_2)]^2 + (\gamma\, \ddot{x}_1)^2\} \mathrm{d}t \ , \qquad (4.10)$$

If the integral is solved numerically over an integer number of periods or a sufficiently long time period, the functional $J(\mu)$ is a function of μ only, and hence it can be plotted, and the minimum can be easily found. The main drawback of this approach is that the optimal set depends heavily on the type of road (and clearly on the type of performance index too). Hence, in order to implement it, it would be necessary to add a further adaptive loop which chooses in real time the most appropriate set of values (previously calculated for various road conditions and stored in a look-up table in the microcontroller memory) by the knowledge of the type of road, inferred via some type of observer. This is complex and of questionable benefit.

4.6 Balance Logic Analysis

The previous section has outlined three classical damper semi-active algorithms and two variants of these logics. The focus is now placed on the balance logic. The subsequent theory will be developed considering a friction damper (using a Coulomb friction model and not a Bouc–Wen model for the mathematical developments to simplify the notation) but the results can readily be extended also to an MRD or any damper which can be modelled via a Bouc–Wen model.

The basic idea of this semi-active control strategy is to balance the elastic force by means of the damping force (as long as these forces act in opposite directions) and to set the damping force to a minimum value (possibly zero) otherwise. Therefore, the force transmitted through the isolation system is significantly reduced or even cancelled during the motion sequences in which the damper is active and is only slightly higher (ideally equal) to the elastic force otherwise. As firstly introduced by Federspiel (1976), this type of damping modulation is also known as sequential damping. Analytical models for sequential damping have been proposed and analysed (Stammers and Sireteanu, 1997; Guglielmino, 2001), showing better vibration isolation properties than the optimal passive damping, for both deterministic and random excitations.

The effectiveness of using balance logic in the semi-active control of vibration can be intuitively illustrated for the 1DOF models of a vehicle suspension and a machine foundation shown in Figure 4.2.

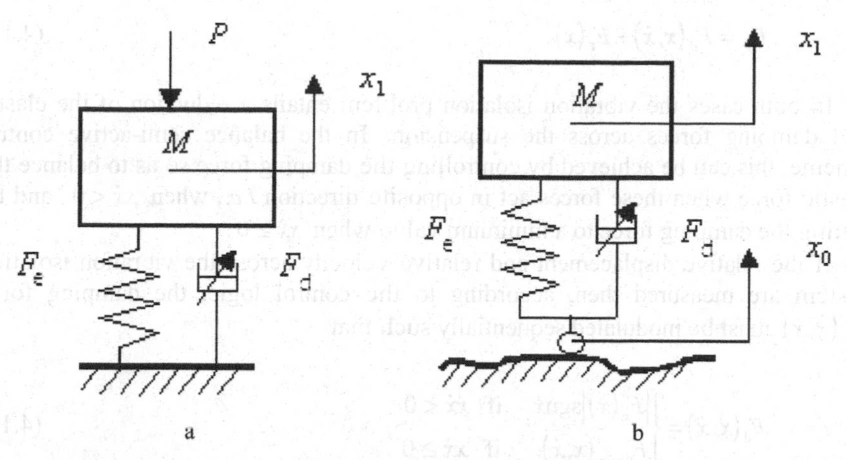

Fig. 4.2. a. 1DOF vehicle suspension model; b. 1DOF machine foundation model (copyright Publishing House of the Romanian Academy (2003), reproduced from Topics in Applied Mechanics, Vol I, Ch 12, edited by Sireteanu T and Vladareanu L, used by permission)

The motion of the sprung mass M is described by:

$$M\ddot{x}_1 + F_d(x,\dot{x}) + F_e(x) = P(t), \tag{4.11}$$

where x_1 and x are the absolute and relative displacements of the sprung mass, $F_e(x)$ is the passive elastic force, $F_d(x,\dot{x})$ is the semi-active damping force and $P(t)$ is the exciting force.

For a vehicle suspension model:

$$x_1(t) = x(t) + x_0(t), \quad P(t) \equiv 0, \tag{4.12}$$

where $x_0(t)$ is the imposed displacement of the wheel centre, assumed to follow the road surface (without losing contact with it).

In the case of a machine foundation model

$$x_1(t) = x(t), \quad P(t) = m_e e\omega^2 \sin\omega t, \tag{4.13}$$

where the perturbation of the sprung mass (machine and foundation block) is produced by a rotor unbalance, modelled via an eccentric mass m_e spinning with angular frequency ω and placed at a distance e from the machine rotation axis.

The absolute acceleration of the vehicle body $\ddot{x}_1$ and the transmitted force F_T to the foundation base are given by:

$$\ddot{x}_1 = -\frac{1}{M}\left[F_d(x,\dot{x}) + F_e(x)\right], \tag{4.14}$$

$$F_{\mathrm{T}} = F_{\mathrm{d}}(x,\dot{x}) + F_{\mathrm{e}}(x). \tag{4.15}$$

In both cases the vibration isolation problem entails a reduction of the elastic and damping forces across the suspension. In the balance semi-active control scheme, this can be achieved by controlling the damping force so as to balance the elastic force when these forces act in opposite direction *i.e.*, when $x\dot{x} < 0$, and by setting the damping force to a minimum value when $x\dot{x} \geq 0$.

If the relative displacement and relative velocity across the vibration isolation system are measured then, according to the control logic, the damping force $F_{\mathrm{d}}(x,\dot{x})$ must be modulated sequentially such that

$$F_{\mathrm{d}}(x,\dot{x}) = \begin{cases} |F_{\mathrm{e}}(x)|\,\mathrm{sgn}\dot{x} & \text{if } x\dot{x} < 0 \\ F_{\mathrm{d\,min}}(x,\dot{x}) & \text{if } x\dot{x} \geq 0 \end{cases}, \tag{4.16}$$

where $F_{\mathrm{e}}(x)$ is the elastic force and $F_{\mathrm{d\,min}}$ is the minimum possible setting of the damping force. From an intuitive point of view, Equation 4.16 implies that the damping force is zero or very small as long as the sprung mass is moving away from its static equilibrium position. This force suddenly increases when the stroke reverses and then gradually decreases as the system returns to its static equilibrium position.

The semi-active closed-loop control law defined by (4.16) can be practically implemented by using a controllable friction damper or an MR damper.

For a friction coefficient μ (assuming a Coulomb friction model), the normal force $F_{\mathrm{n}}(x,\dot{x})$ applied to the friction plates of the device must be controlled by an actuator such that:

$$F_{\mathrm{n}}(x,\dot{x}) = \begin{cases} \left(\dfrac{2\alpha}{\mu}\right)|F_{\mathrm{e}}(x)| & \text{if } x\dot{x} < 0 \\ F_{\mathrm{d\,min}} & \text{if } x\dot{x} \geq 0 \end{cases}, \tag{4.17}$$

where α is a dimensionless gain factor. Theoretically, complete (ideal) balance is achieved for $\alpha = 0.5$ and $F_{\mathrm{d\,min}} = 0$. However, in order to avoid damper lock-up (no motion across the damper), which could occur when the relative displacement shifts from bound to rebound stroke (or conversely), it is sufficient to ensure that the friction force produced is always less than the spring force, *i.e.*, $\alpha < 0.5$.

Assuming $F_{\mathrm{e}}(0) = 0$, the switching logic (4.17) can be rewritten in the form:

$$F_{\mathrm{n}}(x,\dot{x}) = \mathrm{sw}(x,\dot{x})\left(\frac{2\alpha}{\mu}\right)|F_{\mathrm{e}}(x)|, \tag{4.18}$$

where sw(·) is the two-valued condition function which takes on the value 1 when the friction damper is active and 0 otherwise

$$\mathrm{sw}(x,\dot{x}) = \frac{1}{2}[\mathrm{sgn}(x\dot{x}) - 1]\mathrm{sgn}(x\dot{x}). \tag{4.19}$$

Equation 4.17 implies an instantaneous switch between the on and off settings of the force. A more realistic model must take the switching time into account. The gain factor α has to be replaced by a demanded value α_{dem} since the force achieved is not that demanded by the logic. The variation of the damping force is assumed to be described by the following first-order differential equations:

$$\begin{aligned} T_{\mathrm{c}}\dot{F}_{\mathrm{d}} + F_{\mathrm{d}} &= -\alpha_{\mathrm{dem}} F_{\mathrm{e}}(x) & \text{if } x\dot{x} < 0 \\ T_{\mathrm{c}}\dot{F}_{\mathrm{d}} + F_{\mathrm{d}} &= F_{\mathrm{dmin}} & \text{if } x\dot{x} \geq 0 \end{aligned}, \tag{4.20}$$

where T_{c} is the switching time constant.

The control law (4.17) is discontinuous and consequently the controller prone to chatter. Chattering occurs in the neighbourhood of a switch point (*e.g.*, when either x or $\dot{x}$ is zero). The smaller the switching time constant T_{c} the more likely chatter is to occur. On the other hand, increasing the switching time constant T_{c} too much will inevitably lessen the effectiveness of the semi-active controller because the elastic force will be only partially balanced by the damping force in the "on" sequences. The drawbacks of the switchable semi-active system are all related with the rate of change between the two values of the condition function. Unfortunately, a condition function as that involved in the balance logic, *i.e.*, $\mathrm{sw}(x\dot{x})$ has a large rate of change between 0 and 1, mainly due to the derivative $\dot{x}$, especially when the input x_0 is a filtered white noise.

4.7 Chattering Reduction Strategies

The control law outlined above is of the switched type and consequently prone to induce chatter. Furthermore fast switching produces acceleration and jerk peaks which overall negatively affect ride quality. Chattering occurs in the neighbourhood of a switch point (*i.e.*, when either x or $\dot{x}$ is zero). The faster the drive dynamics the more likely chatter is to occur. On the other hand, sluggish drive dynamics will inevitably lessen damper performance. Therefore appropriate anti-chatter strategies need to be devised. Amongst the practical anti-chatter methods applied to smooth the variation of the condition function, the following can be mentioned (Sireteanu *et al.*, 1997; Guglielmino *et al.*, 2005):

- filtering of feedback signals
- introduction of a dead band around the switching points
- introduction of a fuzzy semi-active controller

The first method consists of either low-pass filtering both the feedback signals $x(t)$ and $\dot{x}(t)$ or only the relative velocity which, in most cases, has a broader frequency spectrum than the displacement. It should be pointed out that any signal filtering would result in a lag between the filtered input and output, and therefore in an increase of the switching time constant T_c. This method, although the most straightforward, penalises the performance of the dampers, as the effective working bandwidth is reduced.

The second method allows for relative displacement and velocity dead bands $-x_\varepsilon \le x \le x_\varepsilon$, $-\dot{x}_\varepsilon \le \dot{x} \le \dot{x}_\varepsilon$, within which the condition function is given only the value 0, irrespective of $\mathrm{sgn}(x\dot{x})$. Then the new condition function is:

$$\mathrm{sw}(x,\dot{x},x_\varepsilon)=\begin{cases}0 & \text{if } |x|\le x_\varepsilon,\ |\dot{x}|\le \dot{x}_\varepsilon \\ \mathrm{sw}(x,\dot{x}) & \text{if } |x|> x_\varepsilon,\ |\dot{x}|> \dot{x}_\varepsilon\end{cases}. \tag{4.21}$$

In most cases it is sufficient to introduce only a displacement dead band since the elastic force is not very high for small relative displacements and therefore system performance is not significantly affected by the spring force not being balanced. This approach is somewhat similar to the introduction of a boundary layer in sliding mode controllers (which are themselves particular variable structure algorithms; Utkin, 1992).

The third method is the synthesis of a hybrid fuzzy-variable structure controller (Guglielmino *et al.*, 2005). The variable structure algorithm can be fuzzified by choosing as fuzzy variables the relative displacement and velocity. This makes it possible to soften the fast switching action of the VSC "crisp" controller, without low-pass filtering which would result in a system bandwidth reduction. An application of this method is presented in Chapter 7.

In order to illustrate qualitatively how the first two anti-chatter policies work, let us consider the time histories of the relative displacement and of the unfiltered and filtered relative velocity across a vehicle suspension, depicted in Figures 4.3, 4.4 and 4.5. Since in this example the semi-active control is implemented to mitigate body absolute acceleration within the low-frequency range, the controller dynamic response can be assumed to be instantaneous.

The effects of the anti-chatter policies are readily observed in Figures 4.3, 4.4 and 4.5, showing the time trends of displacement and velocity (both filtered and unfiltered) and in Figures 4.6, 4.7 and 4.8 plotting the associated condition functions. As can be evinced from Figure 4.8 the number of on–off switches could be reduced by a factor of four (compared with Figure 4.6) in the same time interval if a combination of the first two anti-chattering strategies is applied.

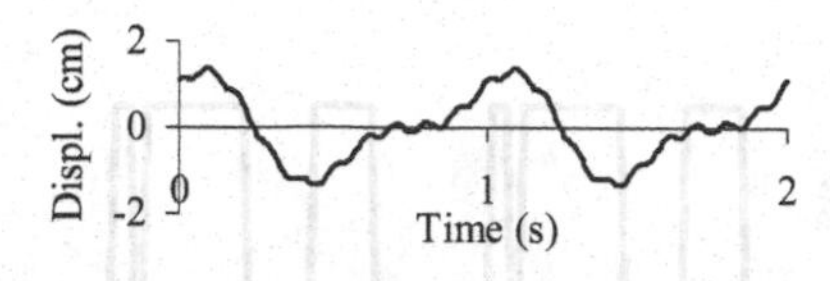

Fig. 4.3. Relative displacement versus time (copyright Publishing House of the Romanian Academy (2003), reproduced from Topics in Applied Mechanics Vol I, Ch 12, edited by Sireteanu T and Vladareanu L, used by permission)

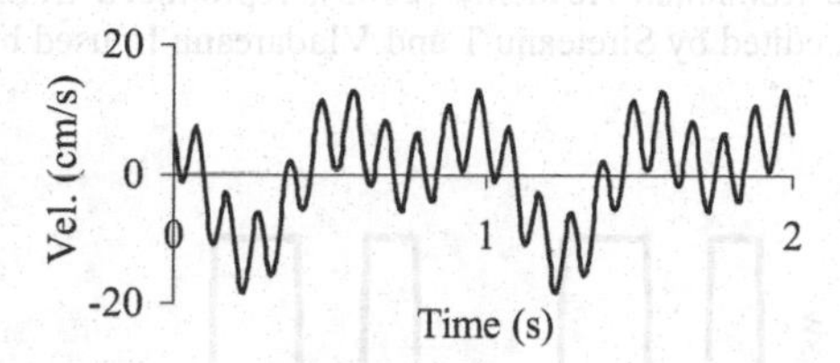

Fig. 4.4. Unfiltered relative velocity versus time (copyright Publishing House of the Romanian Academy (2003), reproduced from Topics in Applied Mechanics Vol I, Ch 12, edited by Sireteanu T and Vladareanu L, used by permission)

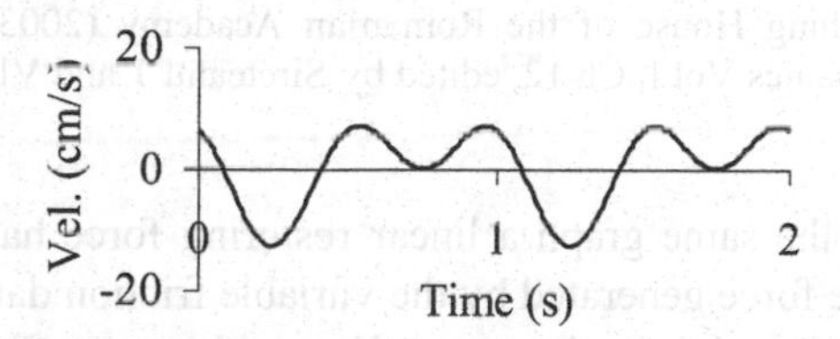

Fig. 4.5. Filtered relative velocity versus time (copyright Publishing House of the Romanian Academy (2003), reproduced from Topics in Applied Mechanics Vol I, Ch 12, edited by Sireteanu T and Vladareanu L, used by permission)

Fig. 4.6. Condition function versus time without anti-chatter logic (copyright Publishing House of the Romanian Academy (2003), reproduced from Topics in Applied Mechanics Vol I, Ch 12, edited by Sireteanu T and Vladareanu L, used by permission)

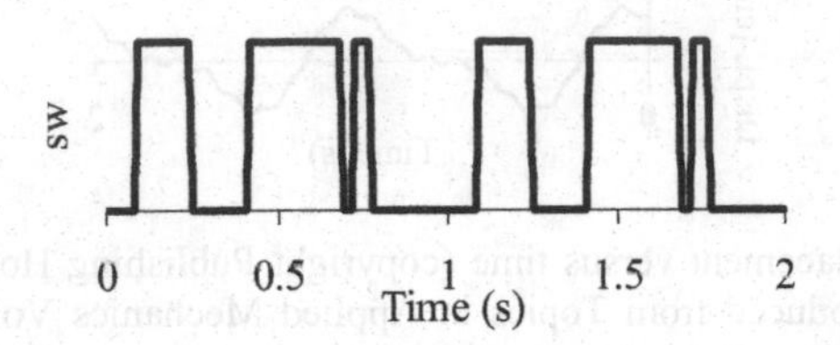

Fig. 4.7. Condition function versus time with dead band anti-chatter logic (copyright Publishing House of the Romanian Academy (2003), reproduced from Topics in Applied Mechanics Vol I, Ch 12, edited by Sireteanu T and Vladareanu L, used by permission)

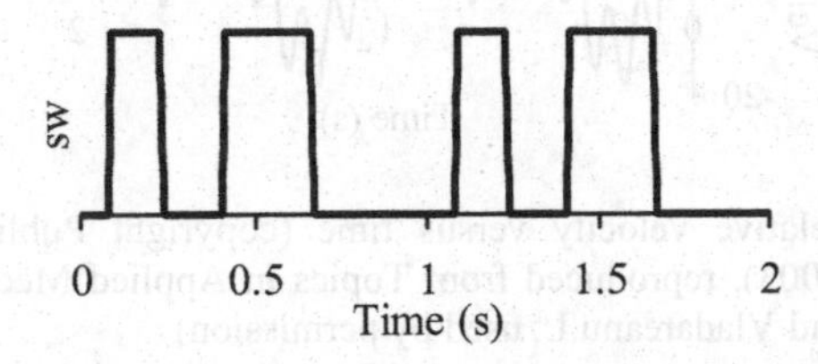

Fig. 4.8. Condition function versus time with combined filtering and dead band anti-chatter logic (copyright Publishing House of the Romanian Academy (2003), reproduced from Topics in Applied Mechanics Vol I, Ch 12, edited by Sireteanu T and Vladareanu L, used by permission)

Figure 4.9 shows on the same graph a linear restoring force having a spring rate $k = 120$ kN/m and the force generated by the variable friction damper. The balance logic control is applied applying a low pass filter with a cut-off frequency of 8 Hz and a displacement dead band $x_e = 0.1|x|_{max}$.

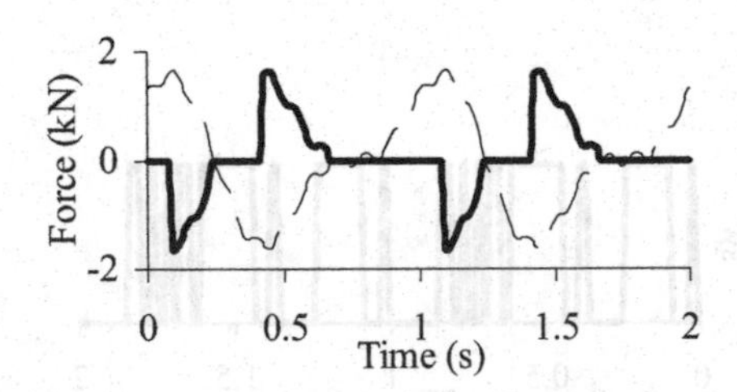

Fig. 4.9. Elastic force (thin line) and damping force (thick line) in balance logic (copyright Publishing House of the Romanian Academy (2003), reproduced from Topics in Applied Mechanics Vol I, Ch 12, edited by Sireteanu T and Vladareanu L, used by permission)

4.8 SA Vibration Control of a 1DOF System with Sequential Dry Friction

The sequential semi-active (SA) damping force $F_{\mathrm{d}}(x,\dot{x})$ of the 1DOF systems pictured in Figure 4.2 will be considered as a function of the relative displacement $x(t)$ and the relative velocity $\dot{x}(t)$, having the general form:

$$F_{\mathrm{d}}(x,\dot{x}) = c\dot{x} + \frac{1}{2}[1-\mathrm{sgn}(x\dot{x})]f_1(x)f_2(\dot{x}),\ c \geq 0, \tag{4.22}$$

where $f_1(x) \geq 0$, $\dot{x}f_2(\dot{x}) \geq 0$ and $f_1(x) = f_2(\dot{x}) = 0$ if and only if $x = \dot{x} = 0$. Moreover, $f_1(x)$ and $f_2(\dot{x})$ are continuous functions on $\Re$ (with possible exception of $\dot{x} = 0$) and monotonic on each x and $\dot{x}$ semi-axis.

The motion of the sequentially damped 1DOF oscillator having linear spring force is described by

$$M\ddot{x} + F_{\mathrm{d}}(x,\dot{x}) + kx = P_{\mathrm{ext}}(t), \tag{4.23}$$

where the exciting force $P_{\mathrm{ext}}(t)$ is the force $P(t)$ acting directly on the sprung mass of the machine foundation model (Figure 4.2b). In the case of the vehicle suspension model shown in Figure 4.2a the exciting force is given by

$$P_{\mathrm{ext}}(t) = -M\ddot{x}_0(t). \tag{4.24}$$

From Equations 4.22 and 4.23 it is obvious that perfect semi-active balance of the spring force by the damping force is achieved if:

$$F_{\mathrm{d}}(x,\dot{x}) = c\dot{x} + f_1(x)f_2(\dot{x}) = -kx \quad \text{for} \quad x\dot{x} < 0. \tag{4.25}$$

The simplest physically implementable solution of the functional equation (4.25) is obtained for:

$$c = 0,\ f_2(\dot{x}) = \mathrm{sgn}\,\dot{x},\ f_1(x) = k|x| \quad \text{for} \quad x\dot{x} < 0, \tag{4.26}$$

i.e., when no viscous damping is present, the energy dissipation takes place only when the relative displacement and velocity are of opposite sign, and this can be achieved by a controllable friction damper. If a low level of viscous damping $c_{\min}$ is added, Equation 4.22 can be written in the form:

$$\begin{aligned} &M\ddot{x} + c_{\min}\dot{x} + \alpha_{\mathrm{dem}}k|x|\mathrm{sgn}\,\dot{x} + (1-\alpha_{\mathrm{dem}})kx = P_{\mathrm{ext}}(t) && \text{if } x\dot{x} < 0 \\ &M\ddot{x} + c_{\min}\dot{x} + kx = P_{\mathrm{ext}}(t) && \text{if } x\dot{x} \geq 0 \end{aligned}, \tag{4.27}$$

where α_{dem} is the demanded gain factor for the normal force applied to the friction plates. It should be chosen so as to obtain as much cancellation as possible of the damping and elastic forces in the "on" sequences.

It should be noticed that viscous damping can be also emulated by a suitable control of the normal force applied to the friction plates:

$$F_n(\dot{x}) = \frac{c_{min}}{\mu}|\dot{x}|. \tag{4.28}$$

Such damping should be more appropriately called pseudo-viscous damping, as it is obtained by velocity-controlling a friction damper.

Equation 4.27 can be regarded as describing the motion of an oscillator having in the "on" sequences (*i.e.*, when $x\dot{x} < 0$) the dissipative characteristic $c_{min}\dot{x} + \alpha_{dem}\, k|x|\text{sgn}\,\dot{x}$ and the elastic characteristic $(1-\alpha_{dem})kx$. In the "off" sequences (*i.e.*, when $x\dot{x} \geq 0$) these characteristics are $c_{min}\dot{x}$ and kx, respectively. The above form of the equation of motion shows a dynamic weakening of the spring stiffness in the "on" sequences caused by partial balance. This results in a reduction of the resonant frequency of the system, without any modification of the sprung mass static deflection. This is a very beneficial effect if a low tuned vibration isolation system is envisaged. Equation 4.27 shows that $\alpha_{dem} < 1$ is required in order to have a positive elastic coefficient. In fact α_{dem} should be chosen to be close to 0.5 if the minimisation of sprung mass acceleration is the main control objective.

For sake of generality the subsequent analysis will be carried out using dimensionless equations, using the following notation:

$$\omega = \sqrt{k/M},\ \varsigma_{min} = c_{min}/2\sqrt{kM},\ \tau = \omega t,$$
$$y(\tau) = \frac{x(\tau/\omega)}{a},\ y'(\tau) = \frac{dy}{d\tau},\ z(\tau) = \frac{P_{ext}(\tau/\omega)}{aM\omega}, \tag{4.29}$$

where a is the unit of length. The equation of motion can be written in the following dimensionless form

$$y'' + \delta(y, y') + y = z(\tau)\text{-} \tag{4.30}$$

The dimensionless sequential damping characteristic is:

$$\delta(y, y') = \begin{cases} 2\varsigma_{min} y' + 2\alpha_{dem}|y|\text{sgn}\, y' & \text{if } yy' < 0 \\ 2\varsigma_{min} y' & \text{if } yy' \geq 0 \end{cases} \tag{4.31}$$

For ease of notation in the following developments the dimensionless parameters will be referred to by the corresponding physical parameters (*e.g.*, τ

time, y relative displacement or elastic force, $\delta(y, y')$ damping force, $y_1'' = y'' - z$ absolute acceleration or transmitted force *etc.*).

4.8.1 Sequential Damping Characteristics

The behaviour of the damping force $\delta(y, y')$ as the relative displacement y varies sinusoidally is important not only from a theoretical but also from a practical point of view since most testing machines can generate such a relative motion between the mounting ends of the shock absorber.

For an imposed cyclic sinusoidal motion:

$$y(\tau) = Y_0 \sin \nu\tau \tag{4.32}$$

the damping force $\delta(y, y')$ varies as shown in Figures 4.10a and 4.10b in terms of the relative displacement y and the relative velocity y' respectively, for $\varsigma_{\min} = 0.25$ and $\alpha_{\text{dem}} = 0.45$

The energy loss per cycle is:

$$\Delta E = \int_0^{2\pi/\nu} \delta(y, y') y' \mathrm{d}\tau = 2Y_0 (\pi\nu\varsigma_{\min} + \alpha_{\text{dem}}). \tag{4.33}$$

From Equation 4.33 it can be deduced that the dissipative effect of the additional semi-active dry friction could be significant at low frequency but becomes less important when the frequency of the imposed motion increases.

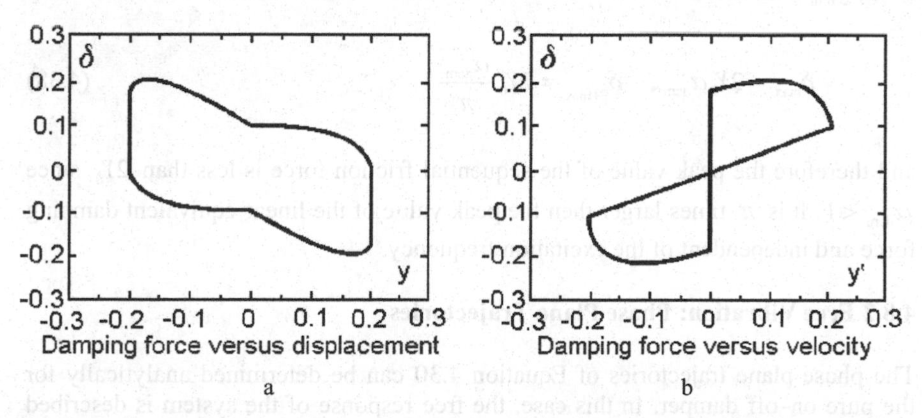

Fig. 4.10. Sequential damping characteristics (a) damping force versus displacement; (b) damping force versus velocity [copyright Publishing House of the Romanian Academy (2003), reproduced from Sireteanu T, Stoia N, Damping optimization of passive and semi-active vehicle suspension by numerical simulation, Proc Romanian Academy, Series A, Vol 4, N 2, used by permission]

The damping coefficient ς_{eq} of the equivalent linear damper, which provides the same energy loss per cycle as the sequential damper, is given by

$$\varsigma_{eq} = \varsigma_{min} + \frac{\alpha_{dem}}{\pi\nu} \tag{4.34}$$

for a pure on–off damping characteristic ($\varsigma_{min} = 0$) $\varsigma_{eq} < 1/\pi\nu$, since $\alpha_{dem} < 1$.

The peak value of the sequential damping force is reached at the time interval

$$\Delta\tau = \frac{1}{\nu}\left[\frac{\pi}{2} - \tan^{-1}\left(\frac{\alpha_{dem}}{\nu\varsigma_{min}}\right)\right] \tag{4.35}$$

from the instantaneous switch of the friction force from zero to its demanded value, which is given by:

$$\delta_{max} = 2Y_0\nu\sqrt{\varsigma_{min}^2 + \frac{\alpha_{dem}^2}{\nu^2}} \tag{4.36}$$

The peak value of the equivalent linear damping force is:

$$\delta_{eq\,max} = 2Y_0\nu\varsigma_{eq} = 2Y_0\nu\left(\varsigma_{min} + \frac{\alpha_{dem}}{\pi\nu}\right). \tag{4.37}$$

If $\varsigma_{min} = 0$, then

$$\delta_{max} = 2Y_0\alpha_{dem}, \quad \delta_{eq\,max} = 2Y_0\frac{\alpha_{dem}}{\pi}, \tag{4.38}$$

and therefore the peak value of the sequential friction force is less than $2Y_0$ since $\alpha_{dem} < 1$. It is π times larger then the peak value of the linear equivalent damping force and independent of the excitation frequency.

4.8.2 Free Vibration: Phase Plane Trajectories

The phase plane trajectories of Equation 4.30 can be determined analytically for the pure on–off damper. In this case, the free response of the system is described by the Cauchy problem:

$$\begin{aligned} y_1' &= y_2 & y_1(0) &= y_1^0 \\ y_2' &= -\delta(y_1, y_2) - y_1 & y_2(0) &= y_2^0. \end{aligned} \tag{4.39}$$

The analytical expressions of the phase trajectories are:

$$\begin{aligned} &y_1^2 + y_2^2 = \left(y_1^0\right)^2 + \left(y_2^0\right)^2 && \text{if } y_1^0 y_2^0 \geq 0,\ y_2^0 \neq 0 \\ &(1-\alpha_{\text{dem}})y_1^2 + y_2^2 = (1-\alpha_{\text{dem}})\left(y_1^0\right)^2 + \left(y_2^0\right)^2 && \text{if } y_1^0 y_2^0 \leq 0,\ y_1^0 \neq 0. \end{aligned} \tag{4.40}$$

These trajectories are piecewise circular and elliptic curves as shown in Figure 4.11 for different initial conditions. The ratio between two successive maxima or minima of the relative displacement $y(\tau)$ is constant and given by:

$$\rho = \frac{1}{1-\alpha_{\text{dem}}}. \tag{4.41}$$

Therefore the free motion amplitude decreases linearly. This behaviour is also encountered in the free response of a passive damping system with dry friction. The main difference is that the semi-active system with sequential dry friction will always return to its equilibrium position since there is no friction force when $y = 0$ (unlike a pure dry friction oscillator).

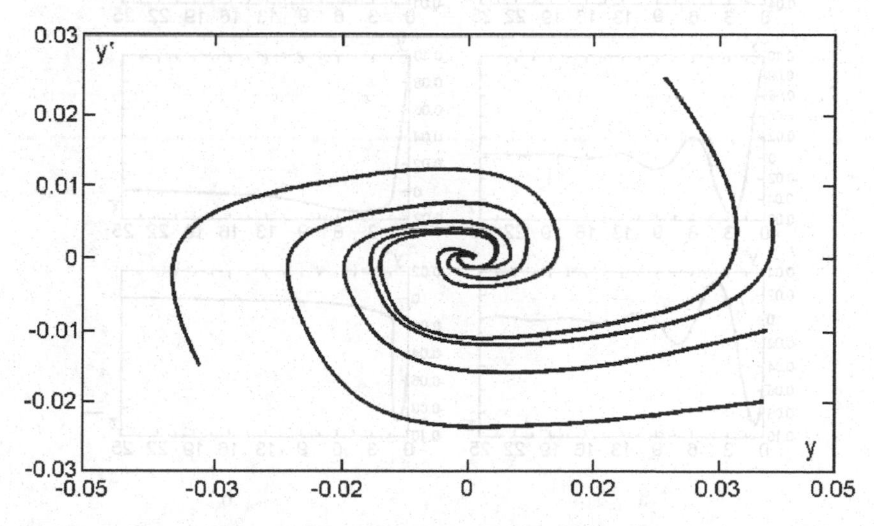

Fig. 4.11. Phase plane trajectories for several different initial conditions [copyright Publishing House of the Romanian Academy (2003), reproduced from Sireteanu T, Stoia N, Damping optimization of passive and semi-active vehicle suspension by numerical simulation, Proc Romanian Academy, Series A, Vol 4, N 2, used by permission]

4.8.3 Free Vibration: Shock Absorbing Properties

In this section the shock absorbing properties of the 1DOF suspension model with sequential dry friction will be compared to those of an optimally damped linear system. This aspect is very important as it is desirable to reduce not only the peak value of a transmitted impulsive force but also the number of after-shock free oscillations. It can be shown that the optimum value of the relative damping coefficient ς, which minimises the peak value $\hat{y}''$ of the transmitted force $y_1''(\tau) = y''(\tau)$ in the case of a passive linear vibration isolation system excited by a Dirac impulse force $\delta(\tau)$ applied to the sprung mass, is $\varsigma_0 = 0.25$ (Sireteanu and Balas, 1991). Since the first sequence of the motion described by Equation 4.30 for $z(\tau) = \delta(\tau)$ and initial conditions $y(0) = y'(0) = 0$ is always governed by a linear equation, the same optimal value $\hat{y}''_{opt}$ can be obtained by taking $\varsigma_{min} = \varsigma_0$ in the sequential damping characteristic (4.31). The effect of the additional sequential dry friction is then observed by comparing the evolution of the motion after the first peak value of the transmitted force is reached.

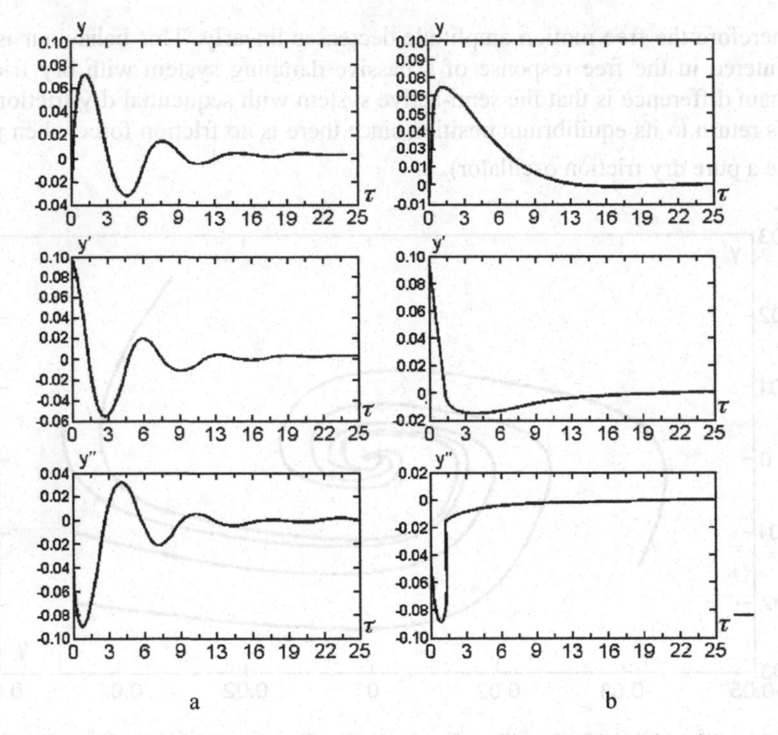

Fig. 4.12. Free vibration versus time for (a) passive case (linear); (b) semi-active case (non-linear) [copyright Publishing House of the Romanian Academy (2003), reproduced from Sireteanu T, Stoia N, Damping optimization of passive and semi-active vehicle suspension by numerical simulation, Proc of the Romanian Academy, Series A, Vol 4, N 2, used by permission]

Figure 4.12 depicts the optimal passive system free vibration time histories with $\varsigma = 0.25$ and those of the semi-active system with same viscous linear damping, to which a sequential dry friction with $\alpha_{dem} = 0.45$ is added. As shown in Figure 4.12b, the free motion of the semi-active system with sequential dry friction resembles that of an almost critically damped passive system, but the peak value of the transmitted force in the semi-active isolation system is significantly lower. For example, in the case of the passive system with $\varsigma = 0.7$, the peak value of the transmitted force is $\hat{y}'' = 1.4$, *i.e.*, 1.7 times larger than the optimal value $\hat{y}''_{\text{opt}} = 0.82$.

4.8.4 Harmonically Excited Vibration

4.8.4.1 Time Histories

The steady-state solution of Equation 4.30 with harmonic excitation

$$z(\tau) = Z \sin \nu\tau \tag{4.42}$$

has been determined by numerical integration using Newmark's method. Figure 4.13 depicts the time histories of the transmitted force $y''_1(\tau)$ and of the damping force $\delta(y, y')$ in the case of the linear system with $\varsigma = 0.25,\ \alpha_{\text{dem}} = 0.25$, for $Z = 0.2$ and $\nu = 1$ (*i.e.*, when the excitation frequency is equal to the undamped natural frequency of the system). It can be noticed that the additional sequential dry friction leads to a reduction of the transmitted force (46% for the peak value and 54% for the RMS value) for only a 3% increase in the damping force peak value.

4.8.4.2 Amplitude–Frequency Characteristics

Since the interest lies in the reduction of the loads transmitted through the vibration isolation system from the system base (road) to the sprung mass (in the case of a suspension) or conversely (in the case of a rotating machinery), the force transmissibility factor is the key indicator. For a linear system the force amplitude ratio $T(\nu)$ is given by:

$$T(\nu) = \hat{y}''_1 / Z = \sqrt{2}\,\tilde{y}''_1 / Z\,, \tag{4.43}$$

where $\hat{y}''_1$ and $\tilde{y}''_1$ are the peak and RMS values of $y''_1(\tau)$, respectively.

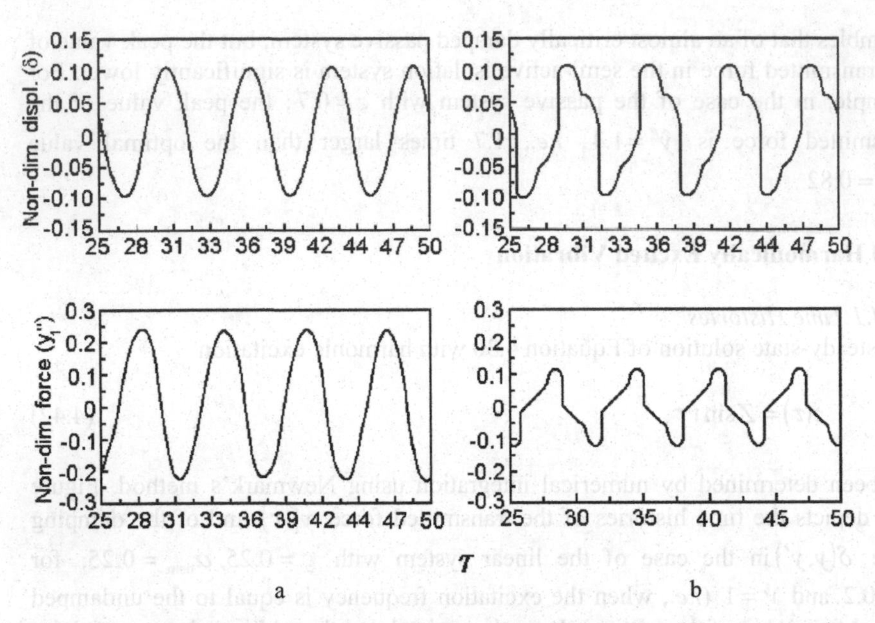

Fig. 4.13. (a) Response to harmonic excitation of passive system; (b) response to harmonic excitation of semi-active system (copyright Publishing House of the Romanian Academy (2003), reproduced from Sireteanu T, Stoia N, Damping optimization of passive and semi-active vehicle suspension by numerical simulation, Proc Romanian Academy, Series A, Vol 4, N 2, used by permission)

In the case of a semi-active system similar amplitude–frequency characteristics can be defined which, in general, depend on both input amplitude and frequency. However since the equation of motion (4.30) is piecewise linear, these functions are independent of the input amplitude Z, but they are not equal as in the case of a linear system:

$$\hat{T}(\nu) = \hat{y}_1''/Z, \qquad \tilde{T}(\nu) = \sqrt{2}\tilde{y}_1''/Z. \tag{4.44}$$

In Figure 4.14 the force transmissibility factors $T(\nu)$ are plotted for $\varsigma = 0.25$ and $\tilde{T}(\nu)$ for $\varsigma_{\min} = 0.25, \alpha_{\text{dem}} = 0.45$.

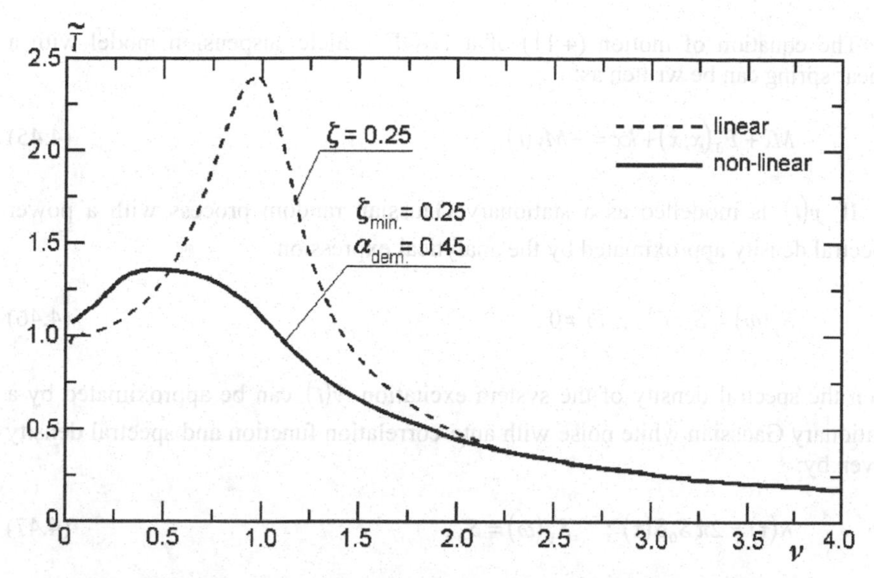

Fig. 4.14. Force transmissibility factors for passive (dashed line) and semi-active (solid line) systems (copyright copyright Publishing House of the Romanian Academy (2003), reproduced from Topics in Applied Mechanics Vol I, Ch 12, edited by Sireteanu T and Vladareanu L, used by permission)

Figure 4.14 indicates a noteworthy result: when the sequential dry friction damping is added to the passive damping a resonance shifting towards the lower-frequency range occurs. This effect is a consequence of the dynamic weakening shown by Equation 4.30 and is a very advantageous effect for both vehicle suspensions and machine foundations. In this manner, the system resonant frequency can be reduced without an increase in the static deflection for a given value of the sprung mass. Another important feature of semi-active vibration isolation system is the remarkable reduction of the RMS force amplification factor in the resonance range of the initial passive system. For $\nu > 2$ and same viscous damping the transmissibility factor of both passive and semi-active systems are virtually equal. This aspect is important, since for low values of ς, passive suspensions are very effective vibration isolators in the higher frequency range.

4.8.5 Random Vibration

In this section the mean square response to a stationary Gaussian white noise excitation for the 1DOF passive and semi-active suspension models is evaluted (Sireteanu and Stoia, 2003), using Newmark's method and Monte Carlo simulation.

The aim of this numerical analysis is to optimise the suspension damping with respect to the classical ride comfort criterion (minimum RMS body acceleration). It will be shown that the semi-active suspension with sequential dry friction can achieve a significant comfort enhancement in comparison with the optimum settings of linear or non-linear passive suspensions.

The equation of motion (4.11) of a 1DOF vehicle suspension model with a linear spring can be written as:

$$M\ddot{x} + F_d(x,\dot{x}) + kx = -M\ddot{y}(t). \tag{4.45}$$

If $y(t)$ is modelled as a stationary Gaussian random process with a power spectral density approximated by the analytical expression

$$S_y(\omega) = S_0\omega^{-4} \quad , \quad \omega \neq 0 , \tag{4.46}$$

then the spectral density of the system excitation $\ddot{y}(t)$ can be approximated by a stationary Gaussian white noise with auto-correlation function and spectral density given by:

$$R(\tau) = 2\pi S_0 \delta(\tau) \ ; \quad S_{\ddot{y}}(\omega) = S_0 , \tag{4.47}$$

where S_0 is a constant dependent upon the road roughness and the vehicle speed.

The damping characteristics of the passive benchmark suspensions considered are linear and quadratic:

$$F_d(\dot{x}) = c\dot{x} \tag{4.48}$$

$$F_d(\dot{x}) = q\dot{x}|\dot{x}| . \tag{4.49}$$

Defining:

$$\begin{gathered} f(x,\dot{x}) = \frac{1}{M} F_d(x,\dot{x}), \ \ z(t) = -\ddot{y}(t), \\ \omega = \sqrt{\frac{k}{M}}, \ \varsigma = \frac{c}{2M\omega}, \ \beta = \frac{q}{M\omega^2} a, \end{gathered} \tag{4.50}$$

where a is the acceleration unit (1 m/s^2), Equation 4.45 can be rewritten as:

$$\ddot{x} + 2\omega\varsigma\,\dot{x} + \omega^2 x = z(t) \tag{4.51}$$

for linear damping,

$$\ddot{x} + \beta\, a^{-1}\, \omega^2\, \dot{x}|\dot{x}| + \omega^2 x = z(t) \tag{4.52}$$

for quadratic damping, and

$$\begin{aligned} &\ddot{x} + \alpha\omega^2|x|\operatorname{sgn}\dot{x} + (1-\alpha)\omega^2 x = z(t) \quad \text{if} \quad x\dot{x} < 0 \\ &\ddot{x} + x = z(t) \quad \text{if} \quad x\dot{x} \geq 0 \end{aligned} \tag{4.53}$$

for sequential damping.

The damping optimisation procedure consists in determining the parameters ς, β and α such that the RMS absolute acceleration of the sprung mass is minimum. This quantity is given by:

$$\sigma_{\ddot{x}_1}^{\ 2} = E\left[(\ddot{x}-z)^2\right] = E\left[(f(x,\dot{x}) + \omega^2 x)^2\right], \tag{4.54}$$

where $E[\ ]$ is the mathematical expectation operator.

In the analysis of real systems, random processes describing physical phenomena represent families of individual realisations. Each realisation (sample function) of the system input leads to a unique solution trajectory if the problem is well posed. The collection of these solution trajectories is an output random process. If the output random process is a second-order process than this is the solution of the stochastic equation of motion in the mean square sense (Soong, 1973).

The numerical simulation methods used to evaluate the statistical properties of the output random process usually imply a numerical integration of the deterministic differential equation of motion for numerically simulated trajectories of the system random excitation. The statistical properties of the solution process are then determined by standard estimation procedures (Bendat and Piersol, 1980).

In this section, the Newmark's method and a pseudo-random number generation algorithm are used for the numerical integration of Equations 4.51, 4.52 and 4.53 and the simulation of the discrete-time trajectories of the white noise excitation *z(t)*.

In order to verify the accuracy of the mean square response evaluation, the approximate solution is compared with the exact solution (Dinca and Teodosiu, 1973), known for the system (4.51):

$$\sigma_x^{\ 2} = \frac{\pi S_0}{2\omega^3\varsigma}, \ \sigma_{\dot{x}}^{\ 2} = \frac{\pi S_0}{2\omega\varsigma}, \ \sigma_{\ddot{x}_1}^{\ 2} = \frac{\pi S_0\,\omega\left(1+4\varsigma^2\right)}{2\varsigma}. \tag{4.55}$$

The minimum value of mean square absolute acceleration is obtained for $\varsigma = 0.5$ and the corresponding mean square response is:

$$\sigma_x^{\ 2}(0.5) = \frac{\pi S_0}{\omega^3}, \quad \sigma_{\dot{x}}^{\ 2}(0.5) = \frac{\pi S_0}{\omega}, \quad \sigma_{\ddot{x}_1}^{\ 2}(0.5) = 2\pi\omega\, S_0. \tag{4.56}$$

4.8.5.1 Simulation of White Noise Sample Functions

A discrete-time history (sample function) of the stationary Gaussian white noise excitation

$$z_{n+1} = z(n\Delta t),\ n = 1,...,N \tag{4.57}$$

can be derived approximately from a sequence of pseudo-random numbers U_n, $n = 1,...,N$, uniformly distributed on the unit interval [0,1] (Abramowitz and Stegun, 1970). The sequence can be obtained using a linear congruential pseudo-random number generator (Monte Carlo simulation). This has the recursive form

$$X_{n+1} = aX_n + b \pmod{c}, \tag{4.58}$$

where a and c are positive integers and b a non-negative integer. For an integer initial value or seed X_0, the algorithm (4.58) generates a sequence taking integer values from 0 to $c-1$ (the remainders when the $aX_n + b$ are divided by c). When the coefficients a, b and c are chosen appropriately, the numbers

$$U_n = X_n / c \tag{4.59}$$

seem to be uniformly distributed on the unit interval [0,1]. Since the number sequences are finite, the modulus c should be chosen as large as possible. To prevent cycling with a period smaller than c, the multiplier a should be also taken to be relatively prime to c.

According to the Box–Muller method, if U_1 and U_2 are two independent uniformly-distributed random variables on [0,1], then N_1 and N_2 defined by:

$$\begin{aligned} N_1 &= \sqrt{-2\ln U_1}\cos 2\pi U_2 \\ N_2 &= \sqrt{-2\ln U_1}\sin 2\pi U_2 \end{aligned} \tag{4.60}$$

are two independent standard Gaussian random variables. Therefore:

$$\begin{aligned} z_{2k-1} &= \sqrt{2\pi S_0 \Delta t}\ \sqrt{-2\ln U_{2k-1}}\ \cos 2\pi U_{2k} \\ z_{2k} &= \sqrt{2\pi\ S_0 \Delta t}\ \sqrt{-2\ln U_{2k-1}}\ \sin 2\pi U_{2k} \end{aligned} \tag{4.61}$$

are independent Gaussian random variables with:

$$E[z_n] = 0\ ,\ E[z_m z_n] = 2\pi\ S_0 \Delta t\, \delta_{mn}\,. \tag{4.62}$$

As can be seen from (4.58) and (4.59), the discrete random process defined by

$$z^{\Delta}(t) = z_n \ \text{for}\ (n-1)\Delta t \le t \le n\Delta t \tag{4.63}$$

is mean square convergent to the white noise process $z(t)$ when $\Delta t \to 0$.

4.8.5.2 Numerical Solution of the Equation of Motion

Newmark's discrete-time method in five steps is now applied in order to obtain the approximate solution of the equation

$$\ddot{x} + f(x,\dot{x}) + kx = z(t) \tag{4.64}$$

with the initial conditions

$$x(0) = x_0, \quad \dot{x}(0) = \dot{x}_0 \,. \tag{4.65}$$

At the initial time $t_0 = 0$, the initial value of the acceleration $\ddot{x}(0) = \ddot{x}_0$ is evaluated from:

$$\ddot{x}_0 = z(t_0) - f(x_0, \dot{x}_0) - kx_0 \,. \tag{4.66}$$

The principle of the method consists in the approximation of the discrete values $x_{n+1}, \dot{x}_{n+1}, \ddot{x}_{n+1}$ by using the values obtained at the time step t_n. The steps of the method are:

- Initialisation of $\ddot{x}_{n+1}$ with an arbitrary value $\ddot{x}_{n+1,i}$
- Evaluation of $\dot{x}_{n+1}$ from

$$\dot{x}_{n+1} = \dot{x}_n + (\ddot{x}_n + \ddot{x}_{n+1,i})\frac{\Delta t}{2} \tag{4.67}$$

- Approximation of x_{n+1} by

$$x_{n+1} = x_n + \dot{x}_n \Delta t + (\ddot{x}_n + \ddot{x}_{n+1,i})\frac{(\Delta t)^2}{4} \tag{4.68}$$

- With x_{n+1} and $\dot{x}_{n+1}$ from (4.67) and (4.68) the following value of acceleration is found:

$$\ddot{x}_{n+1,c} = z(t_{n+1}) - f(x_{n+1}, \dot{x}_{n+1}) - kx_{n+1} \tag{4.69}$$

- The values $\ddot{x}_{n+1,c}$ and $\ddot{x}_{n+1,i}$ are then compared. If the difference is not sufficiently small, $\ddot{x}_{n+1,i}$ is replaced by $\ddot{x}_{n+1,c}$ and the algorithm is repeated from (4.67), otherwise, a new iteration is initiated. Usually, for the initialisation of the unknown acceleration value, the value at the previous step is used.

4.8.5.3 Numerical Results

The aim of the numerical analysis is to compare the mean square response of the suspension analytical models (4.51), (4.52) and (4.53) for optimal linear, quadratic and sequential damping characteristics. The optimal values of the parameters ς, β and α are determined so as to minimise the RMS absolute acceleration for different values of the excitation intensity S_0. In order to determine a realistic variation range of the excitation intensity S_0, appropriate RMS values of the sprung mass absolute acceleration $\ddot{x}_1$ must be considered. The measured values of $\sigma_{\ddot{x}_1}$ for a medium-sized saloon car were obtained within the range 0.9–2.15 m/s^2. Assuming that the undamped natural frequency of the car suspension is 1 Hz, then $\omega = 2\pi$. Therefore, considering Equation 4.57, values of S_0 between 0.01 m^2/s^3 and 0.1 m^2/s^3 seem reasonable.

The optimal values of the damping coefficients ς, β and α determined by using the numerical solutions of Equations 4.51–4.53, are almost constant for all values of S_0 within the range $\varsigma_0 = 0.5$, $\beta_0 = 0.9$, $\alpha_0 = 0.8$.

The mean square response values for the three optimal damping characteristics (**L** – linear, **Q** – quadratic and **S** – sequential) are given in Table 4.1.

Table 4.1. RMS response of passive and semi-active systems (copyright Publishing House of the Romanian Academy (2003), reproduced from Sireteanu T, Stoia N, Damping optimization of passive and semi-active vehicle suspension by numerical simulation, Proc Romanian Academy, Series A, Vol 4, N 2, used by permission)

	S_0 [m/s^2]	0.01	0.02	0.03	0.04	0.05	0.06	0.07	0.08	0.09	0.10
L	σ_x [mm]	11	16	19	22	24	27	29	31	33	36
	$\sigma_{\dot{x}}$ [m/s]	0.07	0.11	0.12	0.14	0.16	0.17	0.19	0.20	0.21	0.23
	$\sigma_{\ddot{x}_1}$ [m/s^2]	0.61	0.89	1.06	1.24	1.38	1.52	1.66	1.77	1.87	2.05
Q	σ_x [mm]	19	24	26	28	30	31	33	35	38	39
	$\sigma_{\dot{x}}$ [m/s]	0.12	0.15	0.16	0.17	0.18	0.19	0.20	0.22	0.23	0.24
	$\sigma_{\ddot{x}_1}$ [m/s^2]	0.83	1.06	1.21	1.36	1.49	1.63	1.76	1.87	1.98	2.17
S	σ_x [mm]	44	58	67	74	80	86	92	97	101	108
	$\sigma_{\dot{x}}$ [m/s]	0.13	0.17	0.19	0.21	0.23	0.25	0.26	0.27	0.29	0.30
	$\sigma_{\ddot{x}_1}$ [m/s^2]	0.45	0.65	0.78	0.92	1.03	1.14	1.26	1.34	1.43	1.58

Figure 4.15 shows a comparison of the minimum RMS absolute acceleration versus excitation intensity S_0 for the three methods.

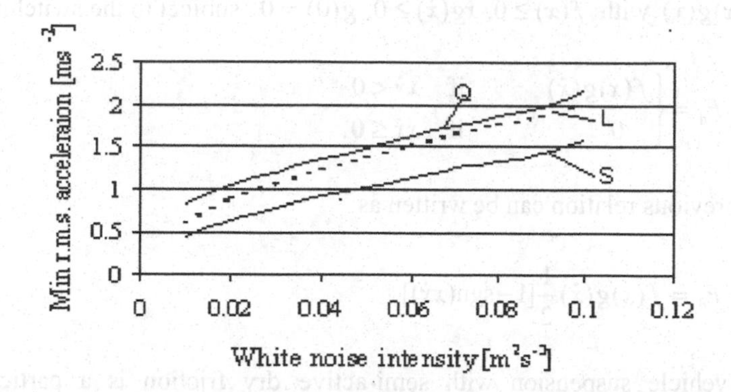

Fig. 4.15. RMS body acceleration for optimum passive (L), quadratic (Q) and semi-active (S) damping (copyright Publishing House of the Romanian Academy (2003), reproduced from Sireteanu T, Stoia N, Damping optimization of passive and semi-active vehicle suspension by numerical simulation, Proc Romanian Academy, Series A, Vol 4, N 2, used by permission)

As can be observed from Figure 4.15, the semi-active suspension can provide a reduction of approximately 25–30% of the RMS sprung mass acceleration over the whole excitation intensity range, when compared with both linear and non-linear passive suspensions. The improvement in terms of comfort is clear. It should be mentioned that this improvement is obtained at the expense of a significant increase in the suspension relative displacement. Therefore, a certain compromise between comfort and working space requirements should be considered in the optimisation cost function.

The results obtained by the numerical simulation presented in this section show that the acceleration experienced by a system controlled by a semi-active control strategy can be significantly lower than that of a passive system (either linear or non-linear).

4.9 Stability of SA Control with Sequential Dry Friction

In this paragraph the stability properties obtained for the semi-active control strategy with sequential dry friction is discussed. The stability is studied employing the direct Lyapunov method and it is shown that the closed-loop system is asymptotically stable, even for a more general class of damping forces, which includes sequential dry friction as a particular case. Consider the system described by

$$M\ddot{x} + c\dot{x} + kx + F_{\mathrm{d}} = 0 \,. \tag{4.70}$$

where F_d is assumed to be a generic nonlinear controlled damping force of the type $F_\mathrm{d} = f(x)g(\dot{x})$ with $f(x) \geq 0,\ \dot{x}g(\dot{x}) \geq 0,\ g(0) = 0$, subject to the switching law

$$F_\mathrm{d} = \begin{cases} f(x)g(\dot{x}) & \text{if} \quad x\dot{x} < 0 \\ 0 & \text{if} \quad x\dot{x} \leq 0. \end{cases} \tag{4.71}$$

The previous relation can be written as:

$$F_\mathrm{d} = f(x)g(\dot{x})\frac{1}{2}[1 - \operatorname{sgn}(x\dot{x})]. \tag{4.72}$$

The vehicle suspension with semi-active dry friction is a particular case obtained for

$$f(x) = 2\alpha k|x|, \qquad g(\dot{x}) = \operatorname{sgn}\dot{x}. \tag{4.73}$$

However the sgn (signum) function is a non-smooth discontinuous function and this makes it very difficult to use classical Lyapunov theory as the equation is piecewise linear. Hence for the purpose of the proof the sgn function is replaced by a saturation-shaped function which is smooth and continuous over its entire domain. Such an approximation is often used also in numerical simulation in order to make system equations less numerically stiff.[1]

If the saturation function is defined as:

$$\operatorname{sat}(\cdot) = \frac{2\tan^{-1}(\cdot)}{\pi} \tag{4.74}$$

the discontinuous sgn functions can be replaced by the correspondent sat functions.

Using the expression (4.72), Equation 4.70 can be rewritten as follows:

$$M\ddot{x} + c\dot{x} + kx + f(x)g(\dot{x})\frac{1 - \operatorname{sat}(x\dot{x})}{2} = 0. \tag{4.75}$$

---

1. A system of ordinary differential equations $\dot{x} = f(x,t)$ is said to be stiff (according to Lambert, 1973) if the eigenvalues λ_i of the Jacobian matrix $J = \dfrac{\partial f}{\partial x}$ satisfy $\mathrm{Re}(\lambda_i) < 1$ and $\dfrac{\max[-\mathrm{Re}(\lambda_i)]}{\min[-\mathrm{Re}(\lambda_i)]} << 1$. In more operative terms (Richards *et al.*, 1990) the stiffness can be measured via the dimensionless quantity $\dfrac{\textit{Integration range}}{\textit{Smallest time constant}}$.

By introducing the state variables $x_1 = x$ and $x_2 = \dot{x}$, Equation 4.75 becomes:

$$\dot{x}_1 = x_2,$$
$$\dot{x}_2 = -\frac{k}{M}x_1 - \frac{c}{M}x_2 - \frac{f(x_1)g(x_2)}{2M}[1-\mathrm{sat}(x_1x_2)]. \tag{4.76}$$

Consider the Lyapunov function associated with Equation 4.76:

$$V(x_1, x_2) = \frac{1}{2}(\frac{k}{M}x_1^2 + x_2^2). \tag{4.77}$$

From the definition it follows directly that V is a continuous and positive-definite function. The proof of stability is based on the following theorem (Barbashin–Krasowsky).

If $\dot{V} \le 0$ (negative semi-definite) and if the set of the points of the state space for which $\dot{V} = 0$ does not include any complete trajectory (except the origin), then the system is asymptotically stable.

The first derivative of V is:

$$\dot{V} = \frac{\partial V}{\partial x_1}\dot{x}_1 + \frac{\partial V}{\partial x_2}\dot{x}_2 = -\frac{c}{M}x_2^2 - x_2\frac{f(x_1)g(x_2)}{2M}[1-\mathrm{sat}(x_1x_2)]. \tag{4.78}$$

Using the conditions $f(x) \ge 0,\ \dot{x}g(\dot{x}) \ge 0, g(0) = 0$ it is obvious from (4.78) that $\dot{V} \le 0$ and the set of points for which $\dot{V} = 0$ is $A = \{(x_1, x_2) \mid x_2 = 0\}$. Next, according to the previous theorem, in order to prove the asymptotic stability, it is sufficient to show that the trajectories of system defined by (4.76) are not contained within the abscissa axis. This can be readily verified considering the tangent vector to the trajectory for $x_2 = 0$. Its components are $(0, -\frac{k}{M}x_1)$, and hence it is orthogonal to the abscissa axis. Therefore the hypotheses of the above stability theorem are fulfilled and hence the equilibrium O (0, 0) is asymptotically stable.

4.10 Quarter Car Response with Sequential Dry Friction

The theory developed for a 1DOF system can be readily extended to the classical quarter car model with viscous damping and sequential dry friction damping. As shown in Figure 4.16, the equations of motion are:

$$M\ddot{x} + F_{\mathrm{d}} + k(x - x_1) = 0,$$
$$M_1\ddot{x}_1 - F_{\mathrm{d}} - k(x - x_1) + k_1(x_1 - r) = 0, \tag{4.79}$$

where

$$F_{\mathrm{d}} = \begin{cases} c_{\min}(\dot{x}-\dot{x}_1) + 2\alpha_{\mathrm{dem}}|x-x_1|\operatorname{sgn}(\dot{x}-\dot{x}_1) & \text{if } (x-x_1)(\dot{x}-\dot{x}_1) < 0 \\ c_{\min}(\dot{x}-\dot{x}_1) & \text{if } (x-x_1)(\dot{x}-\dot{x}_1) \geq 0 \end{cases} \tag{4.80}$$

and $r(t)$ is the excitation induced by the road profile for a constant vehicle speed V.

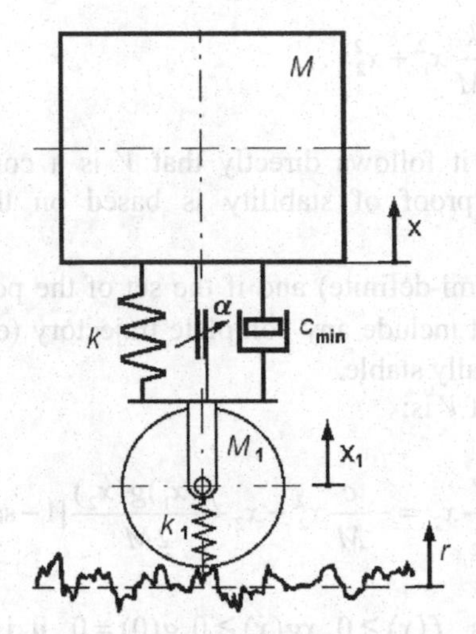

Fig. 4.16. Quarter car model with controlled friction damper (copyright Publishing House of the Romanian Academy (2003), reproduced from Topics in Applied Mechanics Vol I, Ch 12, edited by Sireteanu T and Vladareanu L, used by permission)

By using the notation

$$\begin{aligned} &\omega = \sqrt{k/M},\ \varsigma_{\min} = c_{\min}/2\sqrt{kM}, \\ &\tau = \omega t,\ \gamma = \frac{M}{M_1},\ \chi = \frac{k_1}{k}, \\ &y(\tau) = \frac{x(\tau/\omega)}{a},\ y_1(\tau) = \frac{x_1(\tau/\omega)}{a}, \\ &z(\tau) = \frac{r(\tau/\omega)}{a}, \end{aligned} \tag{4.81}$$

Equation 4.79 can be written in the dimensionless form:

$$\begin{aligned} &y'' + \delta + y - y_1 = 0, \\ &y_1' - \gamma\delta - \gamma(y - y_1) + \gamma\chi(y_1 - z) = 0, \end{aligned} \tag{4.82}$$

where

$$\delta = \begin{cases} 2\varsigma_{\min}(y' - y_1') + 2\alpha_{dem}|y - y_1|\operatorname{sgn}(y' - y_1') & \text{if } (y - y_1)(y' - y_1') < 0 \\ 2\varsigma_{\min}(y' - y_1') & \text{if } (y - y_1)(y' - y_1') \geq 0. \end{cases} \tag{4.83}$$

Equation 4.83 assumes instantaneous switching of the friction damper force from zero to its demanded value $-2\alpha_{\text{dem}}(y - y_1)$ and conversely. If switching dynamics is modelled (via a first-order lag) the equations become:

$$\tau_c\delta' + \delta = -2\alpha_{\text{dem}}(y - y_1) \quad \text{if} \quad (y - y_1)(y - y') < 0 \tag{4.84a}$$

$$\tau_c\delta' + \delta = 0 \quad \text{if} \quad (y - y_1)(y - y') \geq 0, \tag{4.84b}$$

where $\tau_c = \omega T_c$ is the dimensionless switching time constant.

5

Friction Dampers

5.1 Introduction

This chapter deals with the design of a controlled automotive friction damper. A damper of this type is indeed non-conventional in the world of semi-active suspensions (as opposed to the magnetorheological damper, the use of which is more widespread). Controlling friction involves revisiting an early-day damping technology and enhancing it in the light of the latest mechatronic advances.

An overview of the phenomenology and modelling of the friction force is firstly given. Subsequently its modelling and design are illustrated. The focus is on the methodological approach required when designing a novel damping system: from its modelling and experimental concept proofing to building a working prototype, including designing an appropriate electrohydraulic driving system and eventually enhancing its design and optimising damper performance.

5.2 Friction Force Modelling

Friction arises as a consequence of the dissipative microscopic phenomena taking place between the microscopical asperities of two nominally smooth surfaces. The contact between two surfaces can be of two types: conformal or punctual. The former occurs, for instance, in a slider; in this case the macroscopic contact area is proportional to the dimensions of the object. The second type of contact is defined as punctual, *e.g.*, the (ideal) contact between tyre and road or between two gear teeth, where the actual contact area is proportional to load and material strength. The latter is usually referred to as a Hertzian contact (Hertz, 1881).

5.2.1 Static Friction Models

A large number of friction models have been developed over the years to describe the frictional interaction between two materials. In broad terms essentially two main friction regimes exist: the pre-sliding regime and the sliding regime. In the pre-sliding regime there is no true sliding; this motion arises with surface deformation: the asperity junctions deform elastoplastically behaving as springs and friction force appears to be a function of displacement rather than velocity. As the displacement increases, more and more junctions will break resulting eventually in sliding. In the sliding regime all the asperity junctions are broken and the friction force becomes also a function of the velocity (Lampaert *et al.*, 2004).

In this brief review the approach and categorisation proposed by Armstrong-Helouvry *et al.* (1994) is followed. In a static model the common approach is to describe friction level in terms of a velocity-dependent friction coefficient. Friction is predominantly a function of velocity (although it does depend on other factors, such as pressure, temperature, wear, type of lubrication and so forth) because the physical process of shear at the junction changes with velocity. This can be mathematically expressed through an expression of the type

$$F_{\mathrm{d}}(\dot{x}) = -\mu(\dot{x}) F_{\mathrm{n}} \operatorname{sgn} \dot{x} \tag{5.1}$$

F_{d} being the friction force, F_{n} the normal force and $\dot{x}$ the relative sliding velocity between the two surfaces. The minus sign indicates that friction always opposes the motion.

Dynamic effects occur at the breakaway, in response to change in velocity and normal force.

The classical Coulomb friction model was originally envisaged by Leonardo da Vinci (1519), revisited by Amontons in 1699 and subsequently rediscovered and formalised by Coulomb (1785). In this model the function $\mu(\dot{x})$ is taken to be constant. The Coulomb model, although very approximate, is still largely used because it is easy to handle and provides a sufficient level of approximation in many engineering problems. In numerical simulations the discontinuity expressed by the sgn(·) function may create problems, therefore it is sometimes replaced by an equivalent continuous function around the origin (for instance a saturation-shaped function as mentioned in Chapter 4).

The stiction, or zero-velocity friction, was introduced by Morin (1833). Subsequently Reynolds (1886) took viscous effects into account and Stribeck (1902) further improved the model. His model describes the friction regimes between two lubricated surfaces with grease or oil and is usable in a variety of applications. The Stribeck model accounts for a negative slope of the frictional characteristics and can be described by:

$$F_{\mathrm{d}}(\dot{x}) = -\mu(\dot{x}) F_{\mathrm{n}} \operatorname{sgn} \dot{x} - (F_{\mathrm{s}} - F_{\mathrm{c}}) \mathrm{e}^{(-\frac{\dot{x}}{v_{\mathrm{s}}})^{\delta}} \operatorname{sgn} \dot{x} - k_{\mathrm{v}} \dot{x} \,, \tag{5.2}$$

where F_s and F_c are the stiction and the minimum level of friction, k_v is a velocity-dependent coefficient, v_s and δ are model parameters; v_s is known as the Stribeck velocity, which corresponds to the minimum frictional force.

Following the approach of Armstrong-Helouvry, Dupont and Canudas De Wit (1994) the frictional interaction between two lubricated surfaces as a function of velocity can be described by four regimes: pre-sliding or static friction, boundary lubrication, partial fluid lubrication and full fluid lubrication.

Pre-sliding or static friction: in this regime the asperities of the two materials deform elastically producing pre-sliding micro-displacements and plastically causing rising static friction (stiction). A junction in the static friction regime behaves in a spring-like manner and hence friction force is proportional to micro-slip (order of microns):

$$F_d(\dot{x}) = -k_{tan} x , \tag{5.3}$$

k_{tan} being the tangential stiffness of the contact, which depends upon asperity, geometry, material elasticity and normal load (Johnson, 1987). If tangential force increases, there exists a breakaway force which causes the actual sliding to begin (hence friction passes from static to sliding friction). In this situation the undesired stick–slip effect may occur.

From a control viewpoint the effect of pre-sliding can be significant in position and pointing servos where leverage systems can amplify this micro-slip to an actual displacement of the order of millimetre (Canudas de Wit *et al.*, 1993), which can trigger the feedback loop. Dynamic effects occur as well in the pre-sliding to actual sliding transition, which are described in Subsection 5.2.2.

Boundary lubrication: this regime arises at very low velocities, when velocity is not large enough to maintain a fluid film between the two materials. A high stiction force occurs when lubricants which provide small boundary lubrication are employed. If lubricants having low-stiction additives (Wills, 1980) are chosen, stick–slip can be reduced or even eliminated. Stick–slip can also be reduced by increasing the stiffness of the system.

Partial fluid lubrication: this is the most complicated regime to model (Sadeghi and Sui, 1989). In this regime a dynamic effect occurs in response to change of velocity in the form of a delay between a change of velocity and a change of friction force. This phenomenon is known as frictional memory (Rice and Ruina, 1983).

Full fluid lubrication: in this regime the fluid film between the two materials is fully developed and dictates the trend of the friction characteristic, which is of viscous type. The transition between the partial and full fluid lubrication regime may produce a negative slope in the friction characteristic (the aforementioned Stribeck effect), which can result in a destabilising effect because under such conditions the system has a negative damping term. Hence if the working velocity of the system is in that range, the system may become unstable and self-sustained oscillations and limit cycles arise.

5.2.2 Dynamic Friction Models

Dynamic phenomena occur in the transition from static to Coulomb friction and in response to changes in velocity. These phenomena are known as relaxation oscillations and were first investigated by Rabinowicz (1958). Dynamic phenomena also exist in response to changes in the normal force.

The dynamics associated with the breakaway transition have been modelled by a number of researchers using, for instance, exponential (Kato *et al.*, 1972) or linear (Armstrong-Helouvry, 1990) models. The rising time is known as the dwell time.

The dynamic effect in response to changes in velocity is the so-called frictional memory: it results in a lag between a change of velocity and the subsequent change of the friction force to a new steady value; this delay can range from milliseconds to seconds depending upon the materials and appears to be independent of the input frequency (Hess and Soom, 1990).

From a control viewpoint frictional memory can help reduce the destabilising effect of the negative slope of the Stribeck effect. Rabinowicz (1965) verified that, if the time constants of the system are short compared to the frictional memory (*i.e.*, if the system is stiff), the resulting limit cycle is unstable.

Dynamics associated with changes in the normal force also exist. Anderson and Ferri (1990) studied this effect. However Dupont and co-workers (1997) verified that friction dynamics associated with variations in normal force are fast in comparison with the dynamics associated with relaxation oscillations; therefore this effect is negligible in many engineering applications, including vehicle suspensions.

5.2.3 Seven-parameter Friction Model

Taking into account both static and dynamic effects a very comprehensive model known as the seven-parameter friction model has been proposed (Armstrong-Helouvry *et al.*, 1994). This model is summed up by the following set of equations.

Pre-sliding displacement:

$$F_{\mathrm{d}}(\dot{x}) = -k_{\tan} x \tag{5.4}$$

Sliding (Coulomb, viscous, Stribeck effects and frictional memory):

$$F_{\mathrm{d}}(\dot{x}, t) = -[F_{\mathrm{d}} + k_{\mathrm{v}}|\dot{x}| - F_{\mathrm{s}}(\gamma, t_2)\frac{1}{1+(\frac{\dot{x}(t-T_{\mathrm{L}})}{\dot{x}_s})^2}]\ \mathrm{sgn}\,\dot{x} \tag{5.5}$$

Rising static friction (friction levels at breakaway):

$$F_{\mathrm{s}}(\gamma, t_2) = F_{\mathrm{s,a}} + (F_{\mathrm{s},\infty} - F_{\mathrm{s,a}})\frac{t_2}{t_2+\gamma} \tag{5.6}$$

Parameters are defined in Table 5.1. In this model a polynomial model has replaced the exponential model used to represent the Stribeck effect.

Table 5.1. Seven-parameter friction model coefficients

F_s	Friction force at the breakaway [N]
$F_{s,a}$	Breakaway friction force at the end of the previous sliding period [N]
$F_{s,\infty}$	Friction force at the breakaway after a long time at rest [N]
k_v	Viscous coefficient in Stribeck friction model [Ns/m]
T_L	Frictional memory [ms]
t_2	Dwell time [s]
γ	Friction model coefficient [–]

In recent decades several other friction models (both static and dynamic models) have been developed to meet different application needs, here briefly summarised. The interested reader can refer to the specialised literature cited in the references. Several other friction models exist; amongst the static ones, Karnopp (1985) developed a model to deal with the zero-speed issue in computer simulations. If hysteresis is present in the static friction characteristic the Bouc–Wen model (Wen, 1976) described in Chapter 2 can be usefully used. Dahl (Dahl, 1968 and 1976) developed a dynamic model based on the stress–strain curve of materials. Another dynamic friction model is the LuGre model (so called because it was developed at the universities of Lund and Grenoble; Canudas de Wit, Olsson *et al.*, 1995) which captures internal frictional dynamics as well as friction velocity hysteresis, spring-like response at stiction and varying breakaway force. Another dynamic model is the Leuven model (Swevers *et al.*, 2000) based on the experimental findings that the friction force in the pre-sliding regime is a hysteresis function of the position, with non-local memory. The Leuven model attempts to fit this specific behaviour into the LuGre model. Other comprehensive models are the generalised Maxwell slip friction model (Lampaert *et al.*, 2003) and the elasto-plastic model (2000) which renders both pre-sliding displacement and stiction.

Several other friction models exist, such as the Bliman–Sorine model (1991 and 1995). It is also finally worth mentioning other models recently developed for haptic interface applications. Haessig and Friedland (1991) proposed two models known as the bristle model and reset-integrator model; another model was proposed in 1998 by Hollerbach and two years later Hayward (2000) developed a computationally efficient model for the same type of applications.

This brief survey of friction models has outlined the main phenomena occuring in the sliding friction between two surfaces and their modelling. Depending upon the application the designer can adopt a more or less advanced friction model. However, whatever the model adopted, a major problem is the identification of the correct numerical values for the coefficients of the model. An effective method is to carry out experimental tests on the materials under the working conditions of the system. This approach has been undertaken in order to identify the most suitable friction model for an automotive friction damper.

5.3 The Damper Electrohydraulic Drive

In order to implement a proportional-type semi-active control logic (balance, skyhook *etc.*) friction force must be controlled in a proportional fashion, which can be achieved with an electrohydraulic drive. Hence the force control problem translates into a pressure control problem, *i.e.*, the modulation of pressure in a constant-volume chamber. In a pressure control system the flow and hence the required power is negligible. This results in smaller components (pump, pipes, valves). In principle a force-controlled electrical drive could be an alternative solution but this would need a large solenoid to generate the forces required for the application. A pneumatic actuation would not be effective because of the slow dynamic response of pneumatic systems due to the compressibility of air.

Various hydraulic circuits can be envisaged to perform the task of controlling pressure. A typical solution is based on a pressure control valve mounted in parallel with a pump (Figure 5.1). The main drawback of this configuration is that, if several independent pressures are to be controlled, an equivalent number of pumps and valves would be necessary. This solution is therefore convenient only when a single FD is to be controlled.

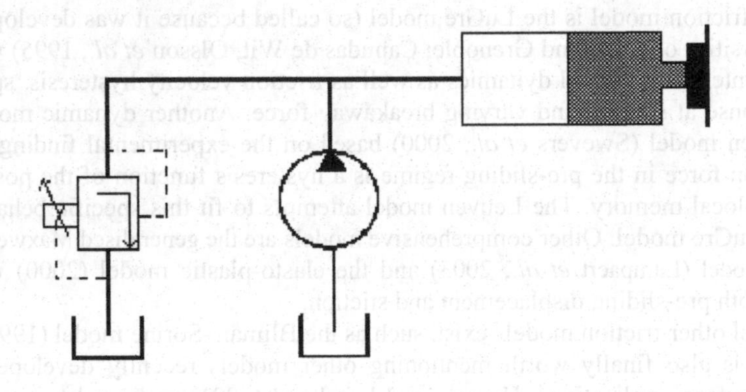

Fig. 5.1. Pressure control valve system

An alternative configuration employs two two-way valves (Figure 5.2): one for the loading phase (to control the pressure rise) and the other for the unloading phase (to control the pressure decrease). In a simplified version the unloading valve can be replaced by a fixed orifice. Such a circuit configuration is common in braking systems.

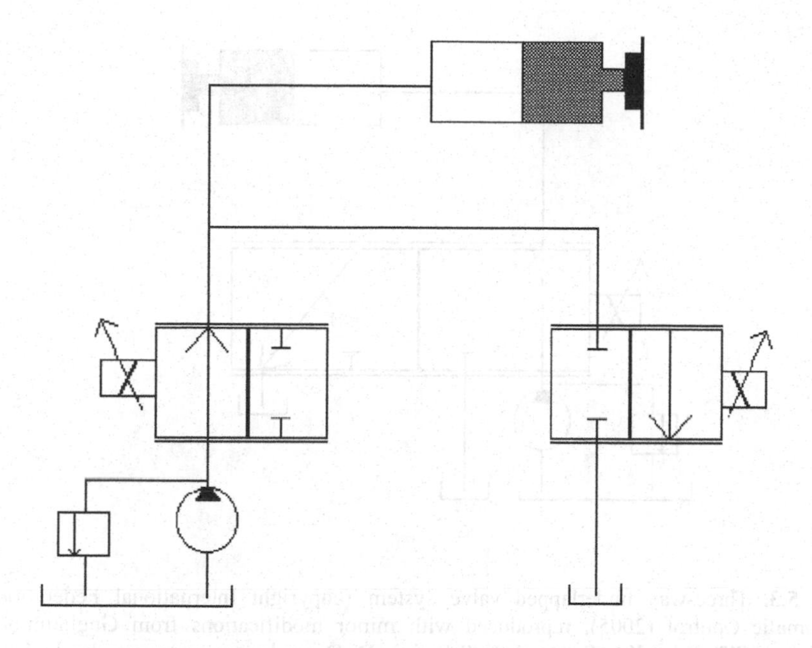

Fig. 5.2. Two two-way valves system

A third type of drive can be devised which minimises the number of components if several pressures (*i.e.*, several FDs) are to be controlled independently. This solution is based on a three-way proportional flow control underlapped valve used in pressure control mode driving a single-chamber actuator (Figure 5.3). With such a configuration it is possible to control several independent pressures using only one pump and a number of valves equivalent to the number of pressures to be controlled, as Figure 5.4 depicts.

The control valve behaves in a manner analogous to a resistive potential divider, *i.e.*, the actuator chamber pressure is modulated by metering both the in-flow and the out-flow. Pressure versus spool current demand characteristics (pressure gain) have a saturation-shaped trend with saturation limits at supply and return pressures and a gradient dependent upon the leakage flows (an ideal leak-free valve would have an on–off characteristic). Such a characteristic allows pressure to be controlled in an almost proportional fashion for small demand signals, whereas for large signals pressure can be switched in a relay-like fashion. This allows the implementation of both proportional and bang–bang type control algorithms (Guglielmino and Edge, 2000 and 2001). Detailed modelling of this hydraulic drive is presented in the following section.

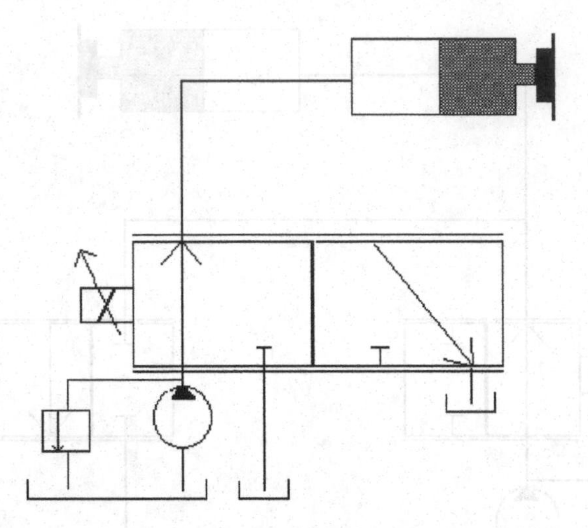

Fig. 5.3. Three-way underlapped valve system (copyright International Federation of Automatic Control (2005), reproduced with minor modifications from Guglielmino E, Stammers CW, Edge KA, Sireteanu T, Stancioiu D, Damp-by-wire: magnetorheological vs. friction dampers, used by permission)

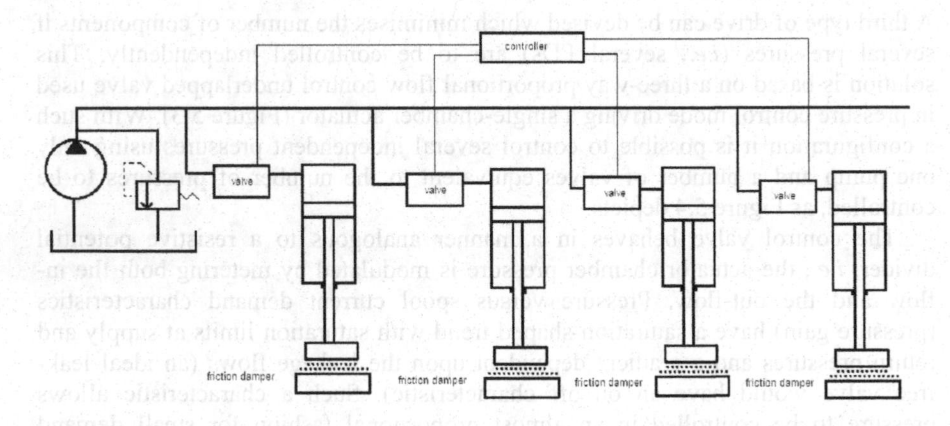

Fig. 5.4. Hydraulic circuit for the independent control of four friction dampers

From the viewpoint of controlling the pressure the three configurations are virtually equivalent, provided valves with sufficient bandwidth are employed.

When designing a hydraulic circuit a major limiting factor is the valve dynamic response which could be negatively affected by external factors independent of the component chosen such as the presence of air in the hydraulic oil. However costly and sophisticated the system, the presence of free air in the circuit results in a reduction of the hydraulic oil bulk modulus and consequently adversely affects the system response. All these issues will be dealt with in detail in Section 5.9.

5.4 Friction Damper Hydraulic Drive Modelling

The mathematical model of the three-way valve-based hydraulic drive is now presented. The model here described has been implemented in the Simulink® software.

In this design an electrohydraulic proportional underlapped valve, supplied by a pump in parallel with a relief valve, drives a single-chamber actuator. The hydraulic circuit behaves as a resistive potential divider (Figure 5.5).

From consideration of Figure 5.5, the continuity equation can be written as:

$$Q_p = Q_1 + Q_{rv} + Q_c \qquad \text{for} \quad P_s > P_c, \tag{5.7}$$

P_c being the relief valve cracking pressure and Q_c the compressibility flow in the line connecting the outlet of the pump with the inlet of the relief valve and with the supply port of the control valve (also $Q_p = Q_1 + Q_c$ for $P_s < P_c$, however in the model the pump is supposed to supply enough flow so as to keep the relief valve always open).

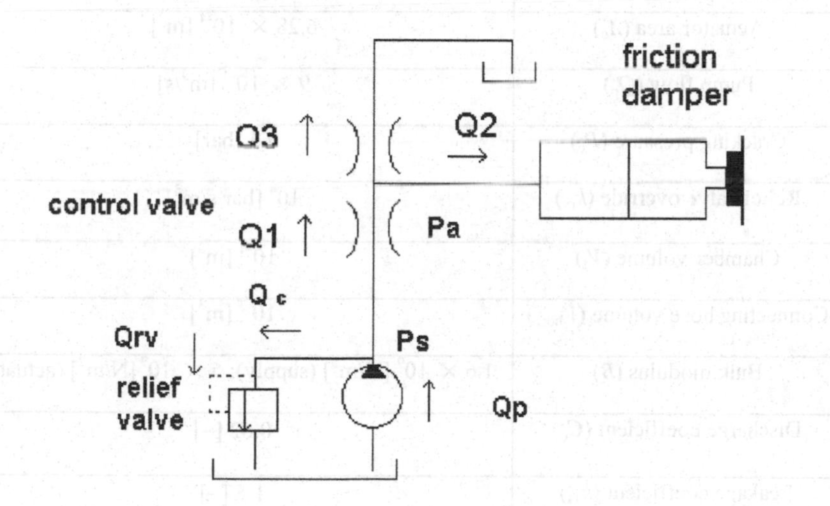

Fig. 5.5. Equivalent hydraulic circuit [copyright Elsevier (2003), reproduced from Guglielmino E, Edge KA, Controlled friction damper for vehicle applications, Control Engineering Practice, Vol 12, N 4, pp 431–443, used by permission]

The governing equation of the relief valve is (refer to Table 5.2 for notation):

$$P_s = P_c + k_{rv}(Q_p - Q_1 - Q_c), \tag{5.8}$$

where Q_p is the rated flow of the pump and k_{rv} is the relief valve override coefficient.

Relief valve dynamics have been not included because they are typically very fast (around 200 Hz). The compressibility flow in the connecting hose is:

$$Q_c = \frac{V_{hose}}{B} \frac{dP_s}{dt}. \tag{5.9}$$

Applying the continuity equation at the second node of the circuit of Figure 5.5, yields

$$Q_1 = Q_2 + Q_3 \tag{5.10}$$

Table 5.2. Key parameters used in the hydraulic drive simulation [copyright Elsevier (2003), reproduced from Guglielmino E, Edge KA Controlled friction damper for vehicle applications, Control Engineering Practice, Vol 12, N 4, pp 431–443, used by permission]

Parameter	Value
Underlap (u)	0.1 [m]
Actuator area (A_c)	6.28×10^{-4} [m^2]
Pump flow (Q_p)	9×10^{-5} [m^3/s]
Cracking pressure (P_c)	64 [bar]
Relief valve override (k_{rv})	10^4 [bar·s/m^3]
Chamber volume (V_t)	10^{-4} [m^3]
Connecting hose volume (V_{hose})	10^{-3} [m^3]
Bulk modulus (B)	1.6×10^9 [N/m^2] (supply); 5×10^7 [N/m^2] (actuator)
Discharge coefficient (C_q)	0.62 [–]
Leakage coefficient (k_{1s})	1.5 [–]
Valve spool damping ratio	0.6 [–]
Valve spool resonant frequency	105 [Hz]
Hydraulic oil density	870 [kg/m^3]

A sufficiently accurate model of the flow past the valve is crucial for capturing the behaviour of the valve. When it works in pressure control mode and the spool moves around the central position with small displacements, leakage flows play an important role. Leakage flow could be considered to a first approximation as being laminar and therefore expressed by a linear relationship between flow and pressure drop. However depending upon the length of the leakage flow path the regime can be either laminar, turbulent or transitional. The model described here was proposed by Eryilmaz and Wilson (2000) and uses a turbulent model and an empirical correlation to model the valve opening. Based on this model, the governing equations are (see Figure 5.6):

$$Q_1 = C_q \pi D(u+z)\sqrt{\frac{2(P_S - P_A)}{\rho}} \quad \text{if } z \geq 0 \tag{5.11a}$$

$$Q_1 = C_q \pi D\sqrt{\frac{2(P_S - P_A)}{\rho}\frac{u^2}{(u-k_{1s}z)}} \quad \text{if } z < 0 \tag{5.11b}$$

$$Q_2 = \frac{V_t}{B}\frac{dP_A}{dt} \tag{5.12}$$

$$Q_3 = C_q \pi D\sqrt{\frac{2(P_A - P_T)}{\rho}\frac{u^2}{(u+k_{1s}z)}} \quad \text{if } z \geq 0 \tag{5.13a}$$

$$Q_3 = C_q \pi D(u-z)\sqrt{\frac{2(P_A - P_T)}{\rho}} \quad \text{if } z < 0 \tag{5.13b}$$

with $-u \leq z \leq u$, where

$$k_{1s} = \frac{1}{2}\sqrt{\frac{P_S + P_A(u) - P_T}{P_S - P_A(u) - P_T}} - 1. \tag{5.14}$$

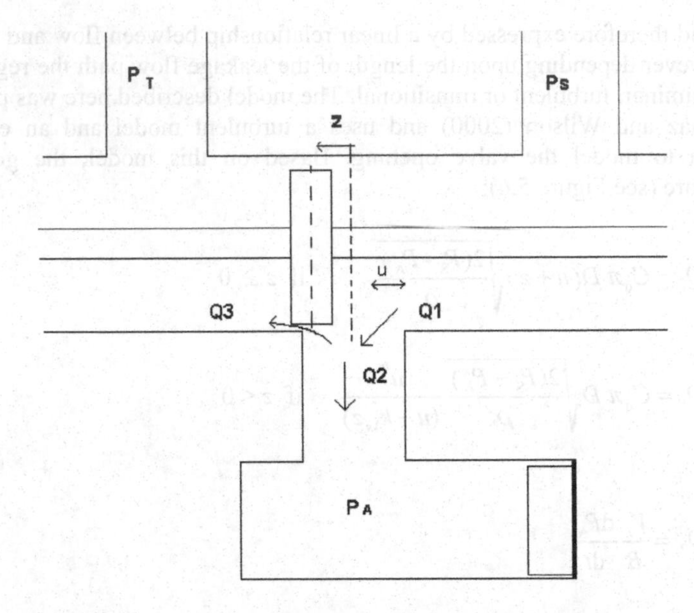

Fig. 5.6. Three-way control valve internal geometry [copyright ASME (2001), reproduced from Guglielmino E, Edge KA, Modelling of an electrohydraulically-activated friction damper in a vehicle application, Proc ASME IMECE 2001, New York, used by permission]

Spool–solenoid electromechanical dynamics can be expressed through a second-order linear model with transfer function

$$\frac{z}{i}(s)=\frac{k_z}{\frac{s^2}{\omega_n^{\ 2}}+2\frac{\xi_v}{\omega_n}s+1} \tag{5.15}$$

z being the spool displacement and i the solenoid current. The valve amplifier voltage–current dynamics are fast enough to be neglected.

The transfer function relating pressure to current is obtained to show which parameters (valve lap, volume, bulk modulus) affect the dynamic response of the drive. A linearised model cannot be used for detailed performance assessment because the pressure variations are too large to allow a linearisation procedure.

Neglecting leakage flows past the valve, linearising (see Equation 5.10) around the equilibrium position yields (lower-case letters denote small variations):

$$q_1 = q_2 + q_3\,, \tag{5.16}$$

where

$$q_1 = \left.\frac{\partial Q_1}{\partial z}\right|_{P_A} z + \left.\frac{\partial Q_1}{\partial P_A}\right|_{z} p_A = k_{c1} z + k_{q1} p_A \tag{5.17}$$

and

$$q_3 = \frac{\partial Q_3}{\partial z}\bigg|_{P_A} z + \frac{\partial Q_3}{\partial P_A}\bigg|_z p_A = k_{c3} z + k_{q3} p_A \tag{5.18}$$

hence

$$\frac{p_A}{z}(s) = \frac{\frac{k_{q1} - k_{q3}}{k_{c3} - k_{c1}}}{1 + \frac{V_t}{B(k_{c3} - k_{c1})} s} . \tag{5.19}$$

Therefore, the overall transfer function between the solenoid current and pressure is third order:

$$\frac{p_A}{i}(s) = \frac{\frac{k_{q1} - k_{q3}}{k_{c3} - k_{c1}}}{1 + \frac{V_t}{B(k_{c3} - k_{c1})} s} \frac{k_z}{(\frac{s^2}{\omega_n^2} + 2\frac{\xi_v}{\omega_n} s + 1)} , \tag{5.20}$$

where

$$k_{q1} = C_q \pi D \sqrt{\frac{2(P_S - P_A)}{\rho}} , \tag{5.21}$$

$$k_{c1} = -\frac{C_q \, \pi D(u+z)}{\sqrt{2\rho(P_S - P_A)}} , \tag{5.22}$$

$$k_{q3} = -C_q \pi D \sqrt{\frac{2(P_A - P_T)}{\rho}} , \tag{5.23}$$

$$k_{c3} = \frac{C_q \pi D(u-z)}{\sqrt{2\rho(P_A - P_T)}} . \tag{5.24}$$

Linearising around $(z = 0; \; P_A = \frac{P_S}{2})$ yields

$$k_{q1} = C_q \pi D \sqrt{\frac{P_S}{\rho}} , \tag{5.25}$$

$$k_{c1} = -\frac{C_q \pi D u}{\sqrt{\rho P_s}}, \tag{5.26}$$

$$k_{q3} = -C_q \pi D \sqrt{\frac{P_s}{\rho}}, \tag{5.27}$$

$$k_{c3} = \frac{C_q \pi D u}{\sqrt{\rho P_s}}. \tag{5.28}$$

The pressure-to-current transfer function (5.20) indicates that the dynamic response is dependent on the lap size and on the volume of the actuator chamber and connecting hose characteristics. A block diagram of the whole system, made with the Simulink® software, is depicted in Figure 5.7.

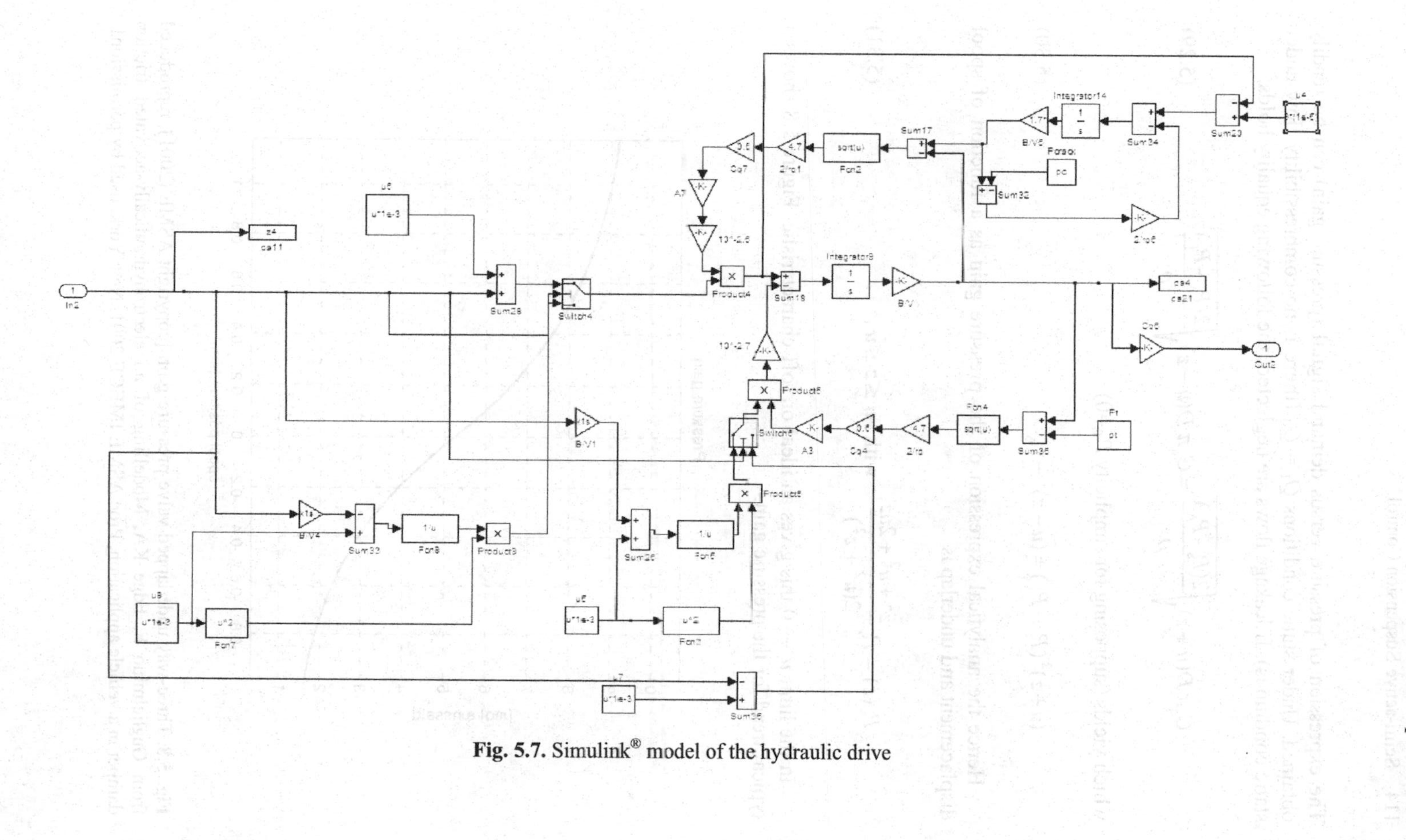

Fig. 5.7. Simulink® model of the hydraulic drive

The expression of pressure versus demand signal (pressure gain) can be readily obtained. Under static conditions $Q_1 = Q_3$ (there is no compressibility flow under static conditions). If leakage flows are neglected, the following equality holds:

$$C_q \pi D(u+z)\sqrt{\frac{2(P_S - P_A)}{\rho}} = C_q \pi D(u-z)\sqrt{\frac{2(P_A - P_T)}{\rho}}, \tag{5.29}$$

which yields (supposing for simplicity P_T=0)

$$(u+z)^2 (P_S - P_A) = (u-z)^2 - P_A . \tag{5.30}$$

Hence the analytical expression of the pressure gain as a function of spool displacement and underlap is

$$P_A(z) = P_S \frac{z^2 + u^2 + 2uz}{2(u^2 + z^2)} \quad \text{with } -u \le z \le u . \tag{5.31}$$

In the limit $u \rightarrow 0$ this gives an ideal on–off characteristic. Figure 5.8 shows a typical trend for the pressure gain.

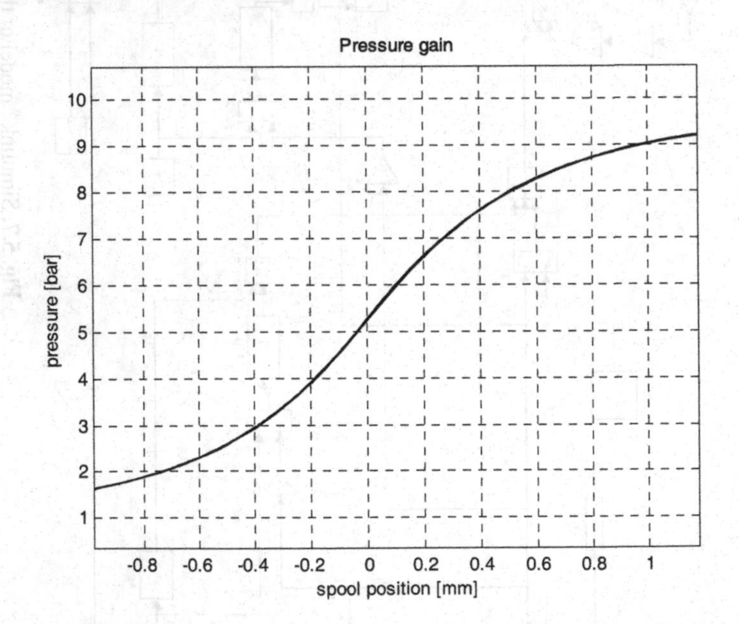

Fig. 5.8. Three-way underlapped valve pressure gain [copyright ASME (2001), reproduced from Guglielmino E, Edge KA, Modelling of an electrohydraulically-activated friction damper in a vehicle application, Proc ASME IMECE 2001, New York, used by permission]

If leakage flows defined by (5.11b) and (5.13a) are included, it is still possible to obtain an analytical expression for the pressure gain but it is far more involved (Eryilmaz and Wilson, 2000).

It is interesting to note that the pressure versus opening characteristic of an underlapped valve inherently constitutes a sliding-mode controller with boundary layer proportional to valve lap; an ideal zero-lapped valve would behave as an on–off controller. If piloted with small demand signal it behaves as an almost proportional controller. Hysteresis, if present, only affects the motion inside the boundary layer and can contribute to reduction of chattering (Gerdes and Hedrick, 1999). However from the viewpoint of minimising chattering (to improve comfort), a strategy based on a switched proportional controller is of greater interest. Balance logic, skyhook and almost all the other control strategies described in Chapter 4 can be classified as proportional-type variable structure controllers, or in more classical control terms as switched state feedback controllers.

5.4.1 Power Consumption

A major benefit of the pressure-controlled semi-active suspension is the low power required compared not only to an active suspension but also to other semi-active suspensions whose control involves throttling flows past orifices. The hydraulic circuit described controls only pressure. Flow is negligible in the force control hydraulic drive described, and therefore the power required is smaller than that in a more conventional flow system.

It is possible to roughly estimate the power consumption. The power required is given by:

$$W = \frac{P_S Q_P}{\eta}, \tag{5.32}$$

where P_S is the supply pressure, Q_P is the pump flow and η its efficiency. The power dissipated is an increasing function of the underlap. Using typical values with an operating pressure of 15 bar the power required is about 150 W per damper.

5.4.2 The Feedback Chain

The friction force control system works in closed-loop mode. In a semi-active control scheme (*e.g.*, balance logic) relative displacement and velocity signals are fed back into the controller, which issues a demand signal to the valve solenoid. Position can be measured either with an inductive transducer such as an LVDT or with a resistive displacement transducer based on a potentiometric system. Velocity can be either measured or obtained numerically (by differentiating displacement). Alternatively accelerometers can be used and velocity and displacement obtained via integration.

Transducer dynamics should be included in a dynamic model. However such transducers are extremely fast relative to vehicle dynamics, and their dynamics can be either neglected or simply represented via a first-order lag in simulation studies.

5.5 Pilot Implementation of Friction Damper Control

In this section the development of a semi-active friction device for concept-proofing purposes is presented. This step is necessary before designing a more realistic vehicle FD. The scope is to have an experimental verification in a simple system that dry friction can be controlled to produce a damper force which can be exploited to reduce vibration.

The basis of this type of control is the generation of a friction force by means of a controlled normal force upon a pair of plates. As described in Chapter 4 the appeal of controlling friction is twofold: in principle any desired force can be generated, including zero force, and a control force can be produced even when the relative velocity is very low.

While any type of control logic can be used, the main interest here is in the balance logic where friction force is used to reduce or even cancel the spring force on the sprung mass. The semi-active friction device has an inherent physical limitation: it can only oppose the motion and not assist it and therefore it is not possible to apply the control force continuously, but only when the following condition is met:

$$x\dot{x} \leq 0 \tag{5.33}$$

(otherwise the control force would have the same direction as the spring force).

In order to achieve tracking, the control action must be proportional to the elastic force, hence

$$F_{\mathrm{n}}(x,\dot{x}) = \begin{cases} bk_{\mathrm{s}}|x| & \text{if } x\dot{x} < 0 \\ 0 & \text{if } x\dot{x} \geq 0. \end{cases} \tag{5.34}$$

The coefficient b is a gain defining the level of cancellation of the spring force, inversely proportional to the friction coefficient. Assuming Coulomb friction and that the friction coefficient is perfectly known, then in order to obtain perfect spring force cancellation b must be chosen to be $b = 1/\mu$. However there will always be a mismatch between the assumed friction coefficient and the actual one. Hence in reality it will be $b = 1/\mu_{\mathrm{assumed}}$ and the actual amount of spring cancellation will be dictated by the ratio $\mu/\mu_{\mathrm{assumed}}$.

Figure 5.9 depicts a simple laboratory friction damper solely for concept-proofing purposes.

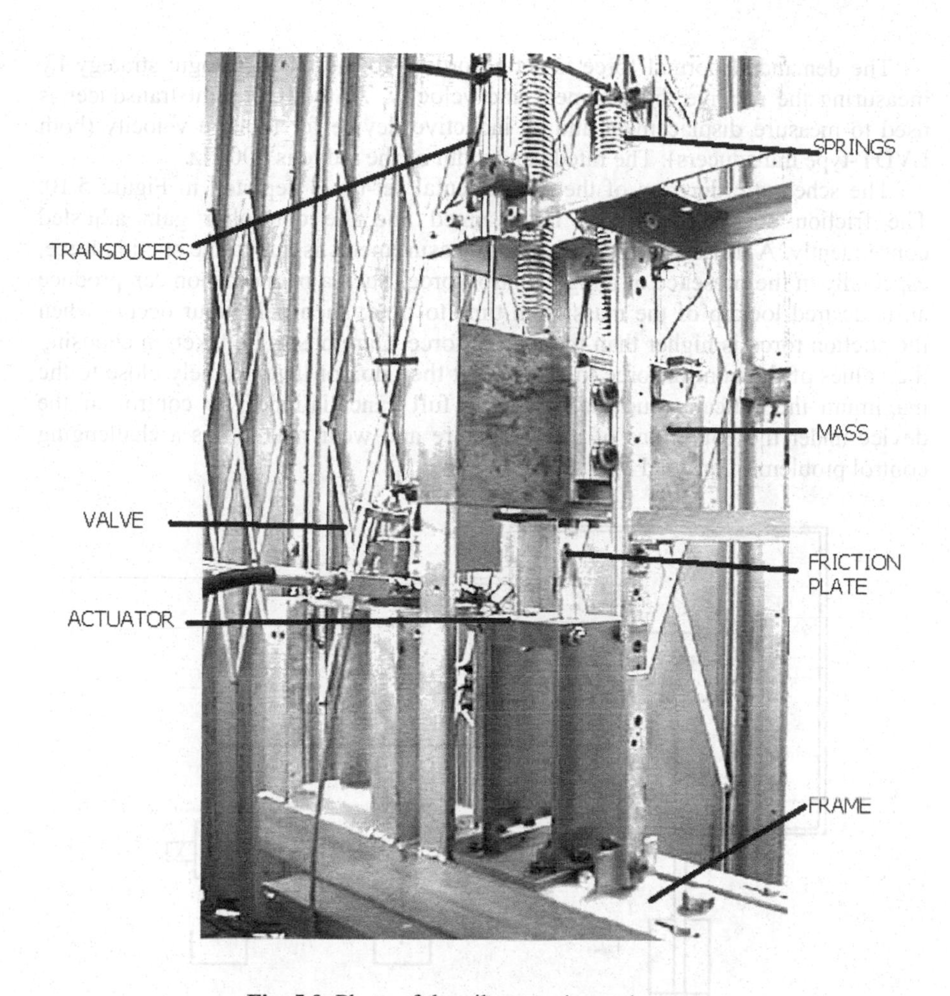

Fig. 5.9. Photo of the pilot experimental set-up

The sprung mass of a quarter car model is represented by a 33 kg mass suspended by springs, which resembles in scale the main vehicle suspension. The system has a natural frequency of 1.35 Hz, typical of vehicle suspensions. The frame from which the mass is suspended is mounted on a hydraulically activated shake table, which simulates wheel motion.

The mass runs on two vertical rails, which prevent horizontal motion, while inducing only a low level of rolling friction in the vertical direction. The controllable damping is produced by a friction pad pressed against a steel plate rigidly connected to the mass. The friction pad is mounted on a linear actuator, driven by a proportional pressure relief valve (the hydraulic drive depicted in Figure 5.1). The valve is placed in parallel with a pump, which supplies a rated constant flow. In the off position, the valve is open and the piston is in light contact with the plate. The diameter of the piston is made as small as is practical in order to increase the pressures required so that seal friction is not a significant fact.

The demanded normal force is set according to the balance logic strategy by measuring the relative displacement and velocity. An eddy-current transducer is used to measure displacement and an inductive device for relative velocity (both LVDT-type transducers). The rated bandwidth of the valve is 100 Hz.

The schematic diagram of the experimental set-up is depicted in Figure 5.10. The friction coefficient has to be assumed and the controller gain adjusted consequently. A difficulty in carrying out measurements is due to the stiction force, especially in the presence of a high control force. Such a phenomenon can produce an undesired lockup of the mass from time to time; such behaviour occurs when the stiction force is higher than the inertial force. Care has to be taken in choosing the values of the control force to ensure that they do not lie extremely close to the maximum theoretical value that produces full cancellation. The control of the device under the conditions of heat, moisture and wear makes this a challenging control problem in on-road operation.

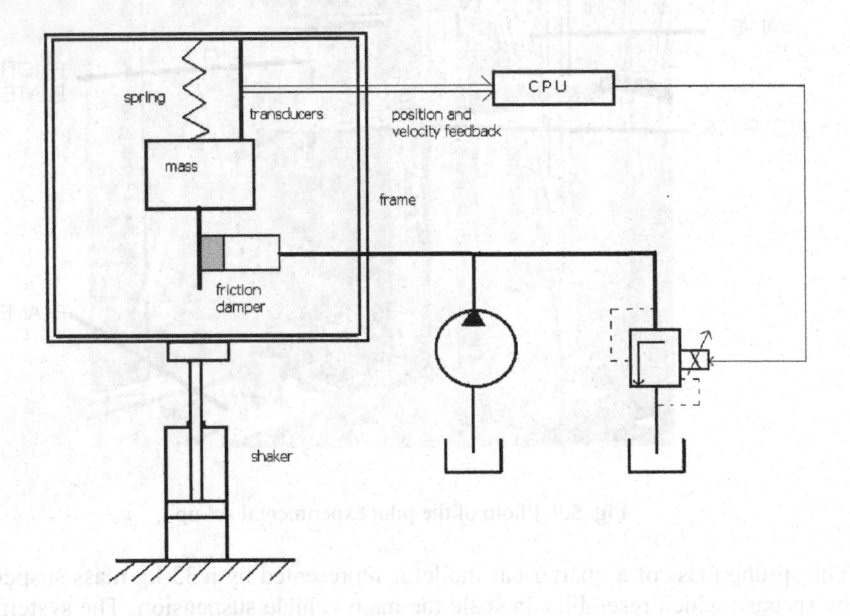

Fig. 5.10. Schematic diagram of the pilot experimental set-up [copyright IFToMM (1999), reproduced from Stammers CW, Guglielmino E, Sireteanu T, A semi-active friction system to reduce machine vibration Tenth World Congress on The Theory of Machines and Mechanisms, Oulu, Finland, used by permission]

Figure 5.11 displays position and velocity controlled transient responses following a step input. By analysing Figure 5.11 it can be noticed that firstly the control is off because position and velocity have the same sign; subsequently it is turned on. When the product of position and velocity becomes positive again, the control is turned off for the second time. Therefore the structure of the system changes three times during this transient. When the values of position and velocity become small enough to enter the chosen dead zone, the control remains switched off and the structure does not change any longer. Obviously the transition between on and off is not instantaneous, but depends upon the pump–valve–actuator dynamics.

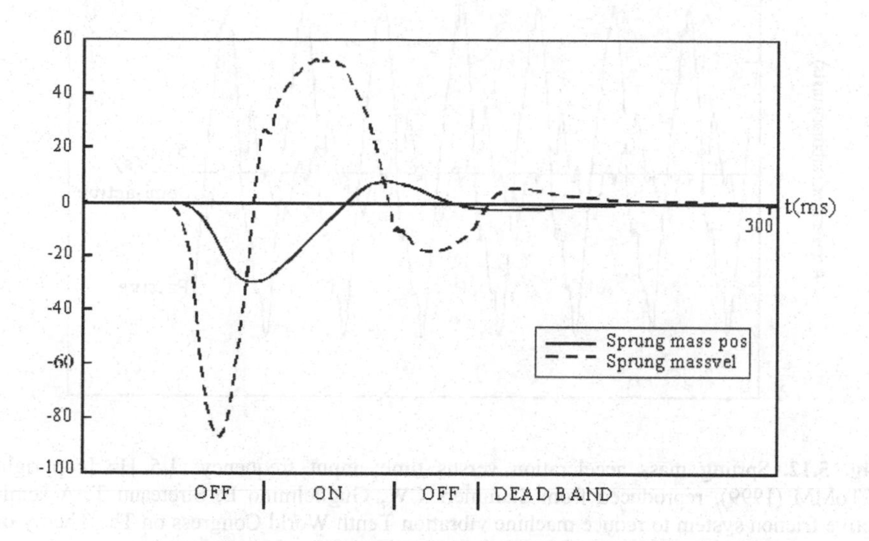

Fig. 5.11. Experimental free response [copyright IFToMM (1999), reproduced from Stammers CW, Guglielmino E, Sireteanu T, A semi-active friction system to reduce machine vibration Tenth World Congress on The Theory of Machines and Mechanisms, Oulu, Finland, used by permission]

Sinusoidal frame inputs in the range 1–5 Hz are then applied. Figure 5.12 shows, plotted on the same axis, both passive and semi-active acceleration at a frequency of 1.5 Hz. The acceleration has been obtained by numerical differentiation. The passive response is not exactly sinusoidal because of the actual non-sinusoidal motion of the shaker, of the parasitic friction effects in the rig (*e.g.*, the rolling friction in the vertical rails) and because of the noise amplification as a consequence of the numerical differentiation. The acceleration at this frequency in the semi-active case is much reduced even if it has a larger harmonic content due to the presence of the control action. It is worth remarking that qualitatively the switching control has not provoked a very harsh response. This is favourable from the point of view of comfort.

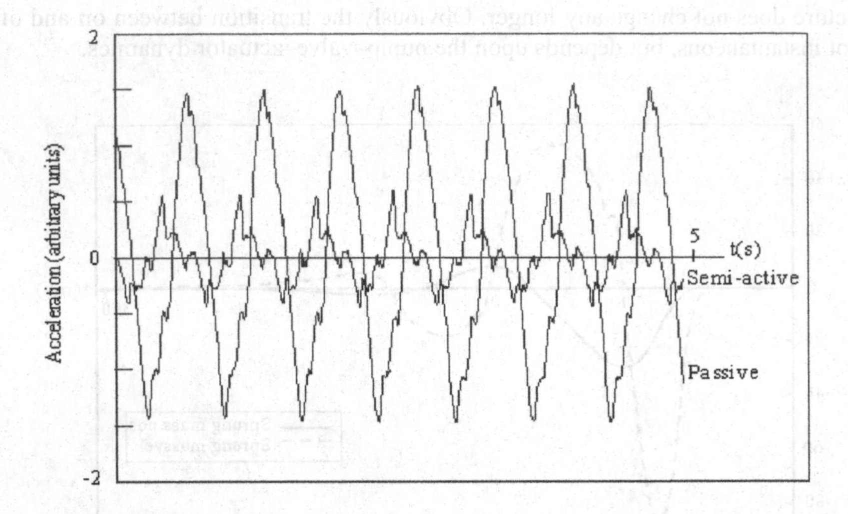

Fig. 5.12. Sprung mass acceleration versus time; input frequency: 1.5 Hz [copyright IFToMM (1999), reproduced from Stammers CW, Guglielmino E, Sireteanu T, A semi-active friction system to reduce machine vibration Tenth World Congress on The Theory of Machines and Mechanisms, Oulu, Finland, used by permission]

The amplification factor (transmissibility) of the sprung mass acceleration in the semi-active case is compared in Figure 5.13 with that in the passive case for sinusoidal inputs.

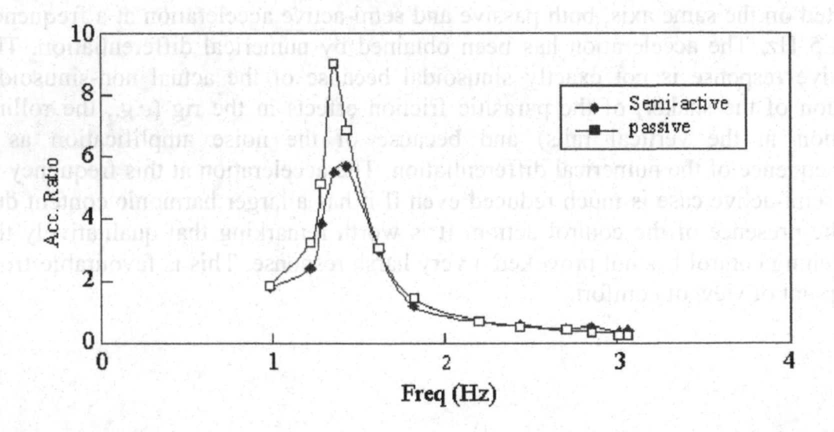

Fig. 5.13. Amplification factor for the sprung mass acceleration [copyright IFToMM (1999), reproduced from Stammers CW, Guglielmino E, Sireteanu T, A semi-active friction system to reduce machine vibration Tenth World Congress on The Theory of Machines and Mechanisms, Oulu, Finland, used by permission]

The control strategy is particularly effective in the neighbourhood of the system resonant frequency, where a reduction of the transmitted forces of about 35% with respect to the passive case is achieved. Far from the resonant frequency the behaviour is the same as in the case without control. This is a remarkable result, compared with the passive case, in which the achievement of such a big reduction at resonance (by increasing the damping) would produce the drawback of larger amplitudes at higher frequencies.

The issue of algorithm robustness can be experimentally checked for instance by adding random offsets to the feedback signals, thus emulating a situation in which feedback signals are corrupted (*e.g.*, because of a faulty transducer). The uncompensated offsets give rise to a switching condition of the form

$$(x+o_1)(\dot{x}+o_2)\leq 0\,, \tag{5.35}$$

where o_1 and o_2 are two arbitrary offsets. Therefore the condition (5.33) is not exactly respected. However the results, depicted in Figure 5.14, are very similar to those shown in Figure 5.13, except for a small increase in the frequency at which the peak transmissibility occurs. Therefore this VSC scheme has proven to be robust to corrupted feedback signals.

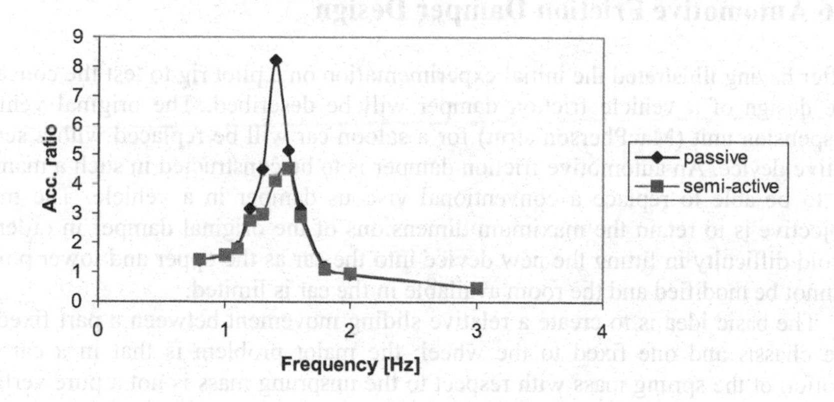

Fig. 5.14. Experimental frequency response with corrupted feedback signals

By varying the coefficient b in Equation 5.34, which defines the rate of control force applied (as mentioned before it can be thought of as related to the reciprocal of friction coefficient μ), it is possible to control the rate of cancellation and consequently vary the acceleration experienced by the mass. In order to check this, measurements have been carried out changing this parameter; the results are shown in Figure 5.15. If b is increased the acceleration ratio at that frequency decreases, tending to an asymptotic value.

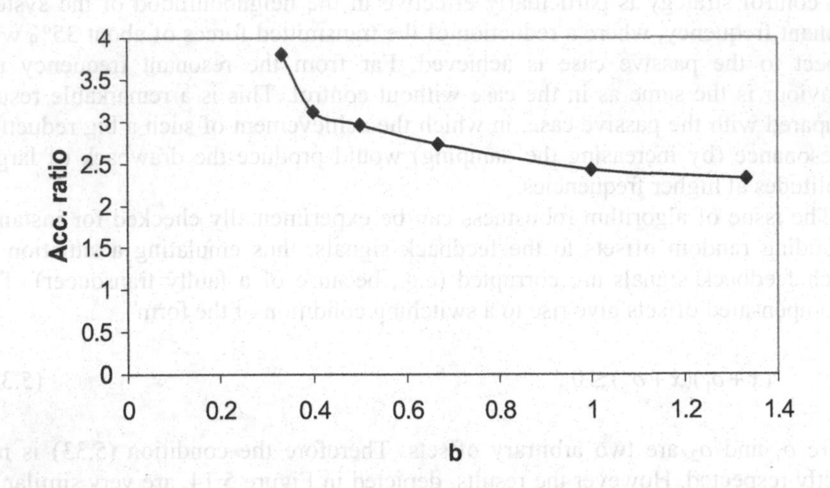

Fig. 5.15. Pilot rig RMS acceleration versus balance gain

5.6 Automotive Friction Damper Design

After having illustrated the initial experimentation on a pilot rig to test the concept, the design of a vehicle friction damper will be described. The original vehicle suspension unit (MacPherson strut) for a saloon car will be replaced with a semi-active device. An automotive friction damper is to be constructed in such a manner as to be able to replace a conventional viscous damper in a vehicle. The main objective is to retain the maximum dimensions of the original damper in order to avoid difficulty in fitting the new device into the car as the upper and lower points cannot be modified and the room available in the car is limited.

The basic idea is to create a relative sliding movement between a part fixed to the chassis and one fixed to the wheel; the major problem is that in a car the motion of the sprung mass with respect to the unsprung mass is not a pure vertical translation: the composite motion of the car prevents a good contact between the two conjugated surfaces (the pad and the plate).

A possible solution involves trying to place the friction device directly inside the existing shock absorber, installing the cylinder, the friction pad and the supply pipe inside the original cylinder of the viscous damper and leaving the external geometry unmodified. Unfortunately the damper cross-section is too small to permit this, and hence the damper diameter must be enlarged, while retaining the spring in its original position. Placing the friction damper within the spring coils eliminates the risk of contact between the car body and the modified damper, as the overall external dimensions of the original suspension unit are not modified.

The embodiment of the concept is a piston in a cylindrical housing which contains two diametrically opposed pistons to which the friction pads are bonded (Figure 5.16). The pistons are supplied with hydraulic oil through the centre of the piston rod, with the control valve mounted remotely.

The choice of the friction material is crucial. A material that is the least susceptible to wear with similar properties to those used in car brake pads is required. The material chosen has a nominal friction coefficient of 0.40.

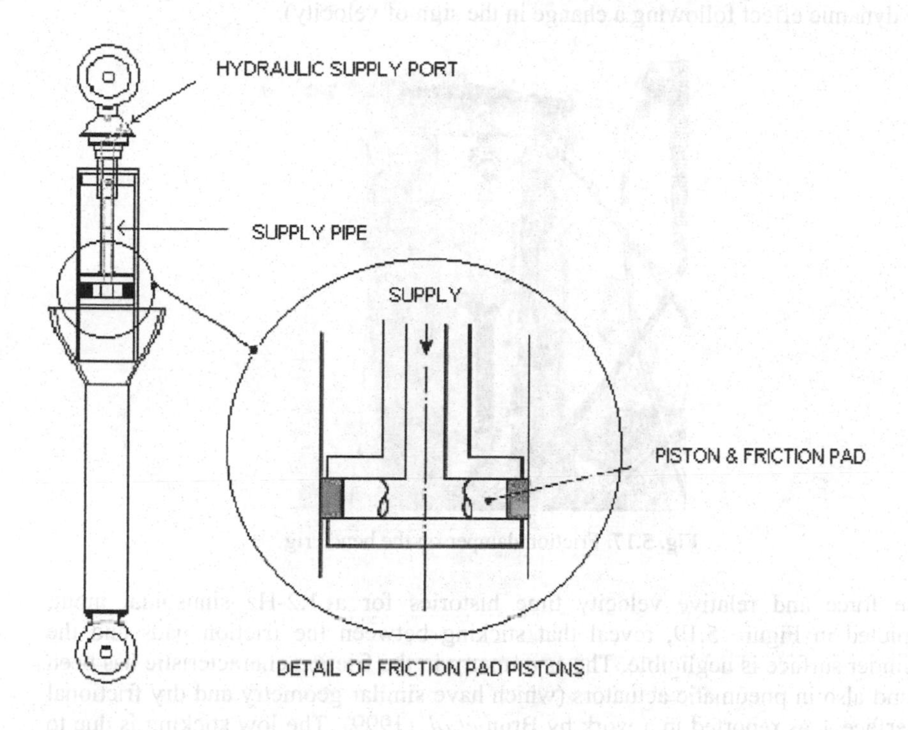

Fig. 5.16. Drawing of the friction damper prototype [copyright ASME (2004), reproduced with modifications from Ngwompo RF, Guglielmino E, Edge KA, Performance enhancement of a friction damper system using bond graphs, Proc ASME ESDA 2004, Manchester, UK, used by permission]

In order to identify a model to support controller design, a bench test assessment of the static characteristic of the device needs to be undertaken on a hydraulically powered shaker and a series of tests conducted to measure the damper response.

The device must be tested on a bench rig. The damper test rig with the damper fitted is depicted in Figure 5.17. It consists of a frame into which the damper under test is fitted. A position-controlled actuator provides the vertical excitation. The constant pressure necessary to produce the normal force is provided by an external pressure source. A load cell measures the total force and an inductive-type transducer (LVDT) measures the vertical displacement. Velocity is obtained by numerical differentiation within the data acquisition package.

The force versus velocity characteristics of the friction damper for a constant supply pressure are shown in Figure 5.18.

The overall damping characteristic is not solely due to the friction damping; some hysteresis is also present because of the elasticity of the rubber bushes (viscoelastic material) at the mounting ends and the frictional memory effect (*i.e.*, the dynamic effect following a change in the sign of velocity).

Fig. 5.17. Friction damper on the bench rig

The force and relative velocity time histories for a 1.2-Hz sinusoidal input, depicted in Figure 5.19, reveal that sticking between the friction pads and the cylinder surface is negligible. The asymmetry in the friction characteristic has been found also in pneumatic actuators (which have similar geometry and dry frictional interfaces), as reported in a work by Brun *et al.* (1999). The low sticking is due to the material employed, manufactured using low-sticking material (Wills, 1980). In response to changes in the sign of velocity, the friction force lags the demand with a delay of around 40 ms. This delay is found to be frequency independent.

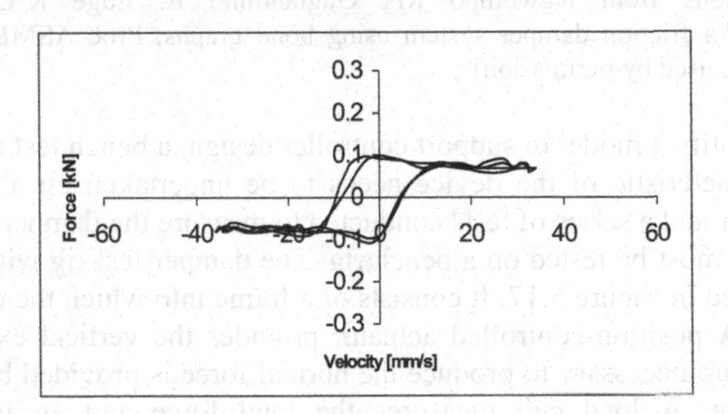

Fig. 5.18. Damping force versus velocity characteristics [copyright International Federation of Automatic Control (2005), reproduced from Guglielmino E, Stammers CW, Edge KA, Sireteanu T, Stancioiu D, Damp-by-wire: magnetorheological vs. friction dampers, IFAC 16th World Congress, Prague, Czech Republic, used by permission]

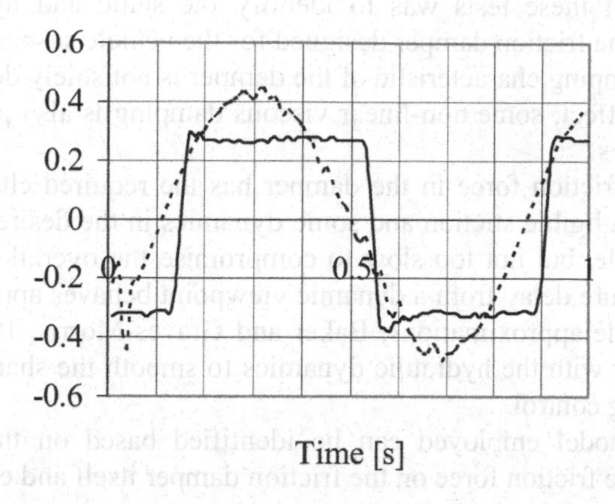

Fig. 5.19. Force (solid) and relative velocity (dashed) time histories at 1.2 Hz

It is not readily possible to test the friction force dynamic response to a dynamic change in the normal force. However as previously stated, Dupont *et al.* (1997) verified that these friction dynamics are usually fast. Therefore this dynamic behaviour is negligible for this automotive application. It is instead interesting to measure the static dependence of friction coefficient μ with pressure. In the pressure range of interest to the control (up to 20–25 bar), the variations were not very large (Figure 5.20).

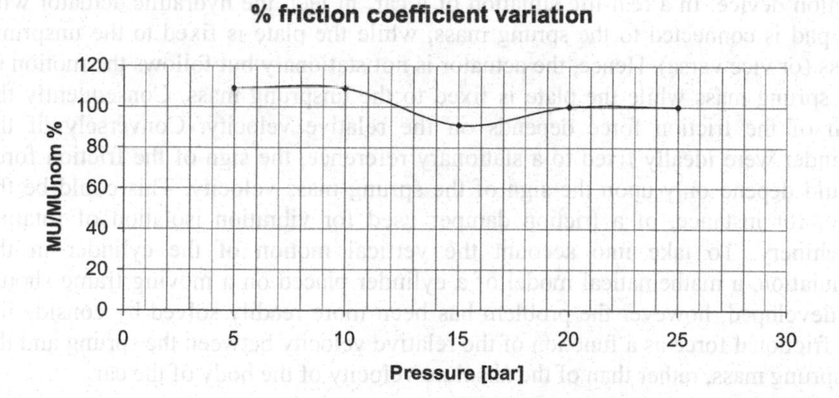

Fig. 5.20. Friction coefficient pressure dependency

An 8-hour endurance test was also carried out using a sinusoidal input and 5-bar constant pressure. The repeatability was good and wear negligible. Increase in temperature was minimal because of the good heat transfer properties of the steel wall. Besides, in a car suspension the air convection would also increase the heat dissipation when the vehicle is riding.

The purpose of these tests was to identify the static and dynamic friction characteristics of the friction damper designed for the vehicle suspension.

The overall damping characteristic of the damper is not solely dependent on the friction damping effect; some non-linear viscous damping is also present, because of the rubber bushes.

Therefore the friction force in the damper has the required characteristics for the application: negligible stiction and some dynamics in the desired range (not so fast to be negligible, but not too slow to compromise the overall performance of the system). The pure delay from a dynamic viewpoint behaves approximately as a first-order lag (Padè approximations; Baker and Graves-Morris, 1996) which can contribute together with the hydraulic dynamics to smooth the sharp transitions of the valve switching control.

The friction model employed can be identified based on the experimental measurement of the friction force on the friction damper itself and expressed by:

$$F_{\mathrm{d}}(t) = -\mu F_{\mathrm{n}}(t - T_{\mathrm{L}})\operatorname{sgn}\dot{x}, \tag{5.36}$$

where the friction coefficient μ is velocity independent and T_{L} is a delay due to the frictional memory effect. Stiction is not included in the model because it was found to be negligible (temperature effects have been not considered either in this analysis).

This model can be seen as a particular case of the more general seven-parameter model of Section 5.2.4 (a Bouc–Wen model could be used as well).

It is worth remarking that, in a vehicle, friction force is a function of relative and not of the absolute velocity. This is because of the actual mounting of the friction device. In a real-life situation of a car, in fact, the hydraulic actuator with the pad is connected to the sprung mass, while the plate is fixed to the unsprung mass (or vice versa). Hence, the actuator is not stationary but follows the motion of the sprung mass while the plate is fixed to the unsprung mass. Consequently the sign of the friction force depends on the relative velocity. Conversely, if the cylinder were ideally fixed to a stationary reference, the sign of the friction force would depend only upon the sign of the sprung mass velocity. This could be the case, for instance, of a friction damper used for vibration isolation of rotating machinery. To take into account the vertical motion of the cylinder in the simulation, a mathematical model of a cylinder placed on a moving frame should be developed; however the problem has been more readily solved by considering the frictional force as a function of the relative velocity between the sprung and the unsprung mass, rather than of the absolute velocity of the body of the car.

5.7 Switched State Feedback Control

It has previously been shown that, because of the inherent physical limitation of the semi-active friction device, which can only oppose to the motion and not assist it, it is not possible to apply the control force continuously in order to obtain spring force cancellation, but only when the condition (5.33) is fulfilled.

The second step in the controller design, after the definition of the allowed regions of the phase plane via Equation 5.33, is the definition of the analytical expression of the control logic in these two regions. Recall Equation 5.34:

$$F_{n}(x,\dot{x})=\begin{cases} bk_{s}|x| & \text{if } x\dot{x}<0 \\ 0 & \text{if } x\dot{x}\geq 0. \end{cases} \tag{5.37}$$

Equation 5.37 can be written as:

$$F_{n}=\frac{bk_{s}}{2}|x|-\frac{bk_{s}}{2}|x|\operatorname{sgn}(x\dot{x}), \tag{5.38}$$

Friction force is expressed by (for ease of notation the time delay T_L in the friction force model is omitted in the subsequent equations)

$$F_{d}=-\mu\frac{bk_{s}}{2}|x|\operatorname{sgn}\dot{x}+\mu\frac{bk_{s}}{2}|x|\operatorname{sgn}(x\dot{x})\operatorname{sgn}\dot{x}. \tag{5.39}$$

Noting that:

$$|x|\operatorname{sgn}(x\dot{x})\operatorname{sgn}\dot{x}=x \tag{5.40}$$

hence

$$F_{d}=-\mu\frac{bk_{s}}{2}x\operatorname{sgn}(x\dot{x})+\mu\frac{bk_{s}}{2}x. \tag{5.41}$$

Therefore Equation 5.41 is composed of two terms: a switching term and a state feedback controller. The product ($x\dot{x}$) can be interpreted as a particular non-linear sliding surface. With this strategy the valve mainly works in the (nearly) linear zone of its pressure gain characteristic. Additional viscous damping may be added either in the first and third quadrants of the phase plane diagram or in the second and fourth quadrants (Guglielmino and Edge, 2001). Hence the overall logic can be regarded as non-linear switched state feedback; it is actually a partial state feedback because only one state is used in the control action; the other state, the velocity, only dictates the switching condition. It is worth noting that this logic does not require pressure feedback (nor acceleration feedback).

The balance controller is designed to track the spring force. The target of the controller is to reduce chassis acceleration by performing spring force tracking, thus improving comfort.

Strictly, the tracking law ought to take into account the hydraulic system dynamic behaviour. However this would result in a very complicated control law with questionable benefit. Provided that hydraulic dynamics are fast enough, it should be possible to reduce the RMS values of the overall response, rather than perform a perfect instantaneous tracking.

Comfort is not easy to quantify and, although standards exist, it is an inherently subjective matter. Several criteria have been proposed (see Chapter 1 for a survey on the topic) based on minimising different combinations of position, velocity, acceleration and jerk. In this work chassis acceleration is used as the comfort improvement criterion.

In principle it is possible to modify the logic, taking into account also the friction force in the control loop, and possibly increasing the robustness to friction coefficient variations. Considering that spring and friction forces depend upon displacement and velocity (with opposite sign), a switching condition based on the product of spring and friction force can be defined, yielding a control force defined by

$$F_{\mathrm{n}} = \frac{bk_{\mathrm{s}}}{2}|x| - \frac{bk_{\mathrm{s}}}{2}|x|\mathrm{sgn}(F_{\mathrm{s}}F_{\mathrm{d}}), \tag{5.42}$$

where

$$F_{\mathrm{s}}F_{\mathrm{d}} = -k_{\mathrm{s}}x(M_1\ddot{x}_1 + 2\xi\omega m_1\dot{x} + k_{\mathrm{s}}x). \tag{5.43}$$

However in this case acceleration feedback is also required in order to deduce the friction force.

The controller described by Equation 5.37 does not introduce any velocity feedback (velocity needs to be measured for implementing the switching condition, but the control signal is only a function of displacement). However a small amount of velocity feedback could be advantageous to help smooth further the transition between the on and the off state. For this reason, a more general controller can be devised whose general form is:

$$F_{\mathrm{n}}(x,\dot{x}) = \begin{cases} bk_{\mathrm{s}}|x| + 2z_1\omega_1 m_1|\dot{x}| & \text{if } x\dot{x} < 0 \\ 2z_2\omega_1 m_1|\dot{x}| & \text{if } x\dot{x} \geq 0 \end{cases} \tag{5.44}$$

with b, z_1, z_2 $\geq$ 0, or alternatively:

$$F_{\mathrm{n}} = (bk_{\mathrm{s}}|x| + 2z_1\omega_1 m_1|\dot{x}|)\frac{1-\mathrm{sgn}(x\dot{x})}{2} + 2z_2\omega_1 m_1|\dot{x}|\frac{1-\mathrm{sgn}(x\dot{x})}{2}. \tag{5.45}$$

This control law combines position damping in the second and fourth quadrants of the phase plane (which results in a spring force reduction effect) blended with some pseudo-viscous damping; additional viscous damping is added in the first and third quadrants (there is no position damping in these quadrants because it would be in phase with the spring force). The latter not only introduces a viscous-like effect, but also contributes to smooth the sudden transitions between the two structures defined by (5.44). By tuning the coefficients b, z_1 and z_2 it is possible to either emphasise the position-dependent damping or the pseudo-viscous damping effect.

The tuning of the coefficients b, z_1 and z_2 is not a trivial problem. As already introduced in Subsection 4.5 by using an approach which recalls the theory of optimal control the set $\{b, z_1, z_2\}$ could be chosen to minimise a performance index such as:

$$J(b,z_1,z_2) = -\int_0^T \{[\alpha(x_2 - z_0)]^2 + [\beta(x_1 - x_2)]^2 + (\gamma\ddot{x}_1)^2\}\mathrm{d}t \,. \tag{5.46}$$

Such an index achieves a trade-off between the reduction of chassis acceleration and dynamic tyre force within the constraint of a set working space.

Formally speaking this is not an optimal control problem, because the supposedly optimal function is assumed to be defined by (5.44). Therefore it is not necessary here to employ the methods of the variational calculus but only finding the optimal value of b, z_1 and z_2 which minimises the integral (5.46).

The main problem of this optimisation method is that the optimal set depends heavily on the type of road (and also on the type of performance index).

5.8 Preliminary Simulation Results

This section is concerned with the analysis of the performance of the VSC balance control law outlined above, using a quarter car model. The performance is analysed both in the frequency and time domains. The general form of the controller (in terms of normal force) is given by Equation 5.44.

Subsequent figures show transmissibility curves for body acceleration, working space and dynamic tyre force. These curves give an evaluation of the performance of the controller in terms of RMS signals. They are plotted in the frequency range 1–5 Hz, the main range of interest in a car suspension, for the purpose of controlling chassis resonance.

Figures 5.21, 5.22 and 5.23 have been plotted for variations in the friction coefficient μ of ±20% around its nominal value (0.40). The scope of this sensitivity analysis is to assess how an uncertainty in the knowledge of the friction coefficient affects the results. Figure 5.21 shows that semi-active RMS acceleration is generally smaller than in the passive case. It is worth drawing an analogy between the behaviour of the semi-active response, when the friction coefficient varies, and the response of a linear system when the viscous damping ratio is varied: close to the resonance a highly damped system (μ = 0.5) performs better; conversely above the resonance frequency a low damped system (μ = 0.3) behaves better and, analogously to a linear system, there exists a frequency where the three trends tend to cross one another and the behaviour is almost independent of the value of the friction coefficient.

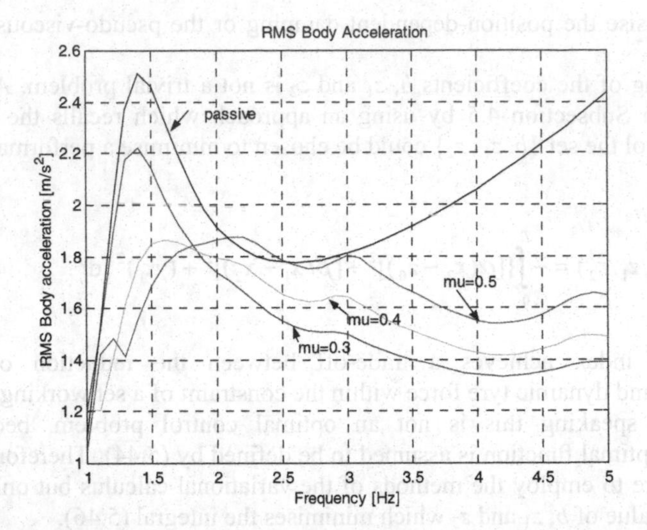

Fig. 5.21. RMS chassis acceleration transmissibility curves varying friction coefficient (0.3, 0.4, 0.5). Controller with $b = 2.5$, $z_1 = 0$, $z_2 = 0$ [copyright Kluwer (2004), reproduced from Meccanica, Vol 39, N 5, pp 395–406, Guglielmino E, Edge KA and Ghigliazza R, On the control of the friction force and used with kind permission of Springer Science and Business Media]

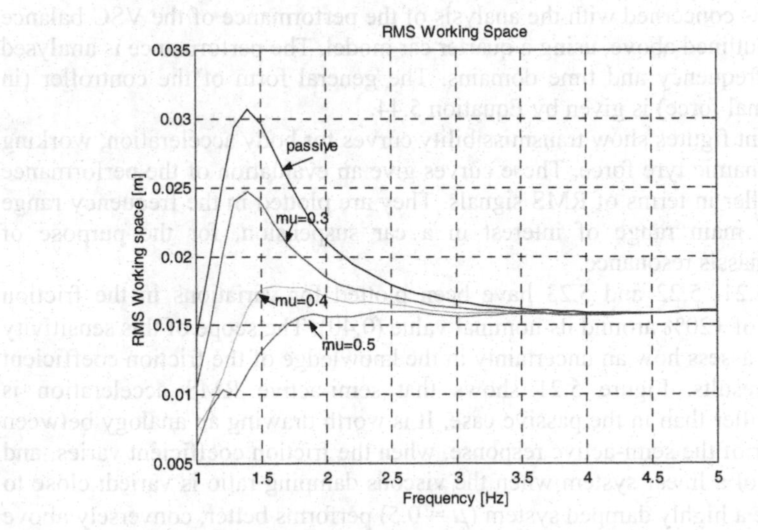

Fig. 5.22. RMS working space transmissibility curves varying friction coefficient (0.3, 0.4, 0.5). Controller with $b = 2.5$, $z_1 = 0$, $z_2 = 0$

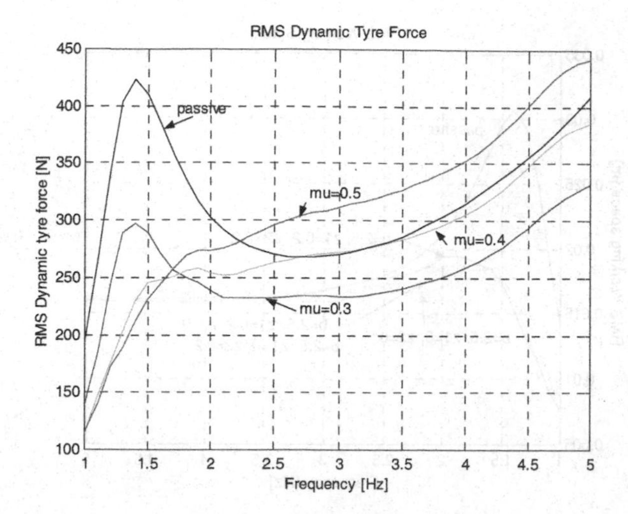

Fig. 5.23. RMS dynamic tyre force transmissibility curves varying friction coefficient (0.3, 0.4, 0.5). Controller with $b = 2.5$, $z_1 = 0$, $z_2 = 0$

The RMS semi-active working space (Figure 5.22) is smaller than the corresponding passive one in the vicinity of the resonance; after the resonance, the controlled response tends asymptotically to the passive response. The reduction in RMS working space is somewhat dependent on the value of the friction coefficient. The controlled RMS dynamic tyre force (Figure 5.23) is reduced with respect to the passive case only in the neighbourhood of the resonance, but at higher frequencies the semi-active system, depending upon the value of the friction coefficient, can be better or worse than the passive system.

In Figures 5.24, 5.25 and 5.26 the performance of various types of controllers is compared in terms of RMS frequency response.

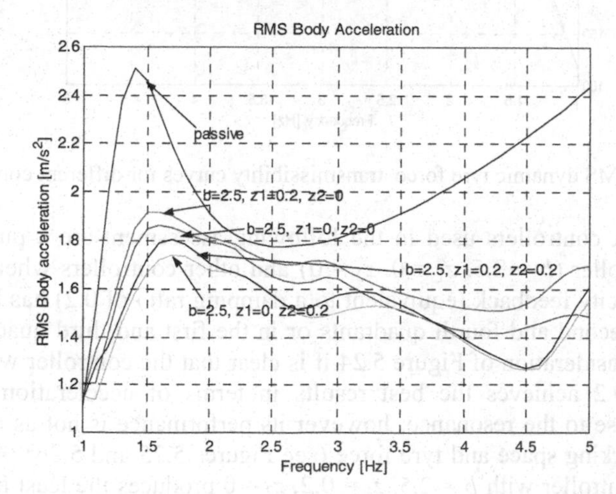

Fig. 5.24. RMS chassis acceleration transmissibility curves for different controllers

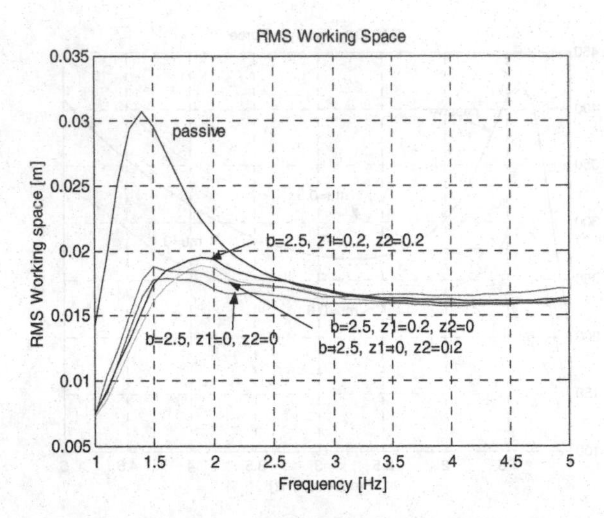

Fig. 5.25. RMS working space transmissibility curves for different controllers

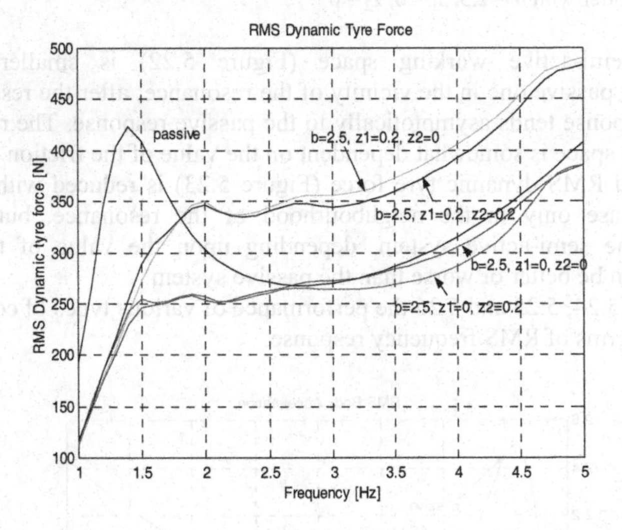

Fig. 5.26. RMS dynamic tyre force transmissibility curves for different controllers

The benchmark controllers used in the following assessment are a pure position feedback controller (b = 2.5, z_1 = 0, z_2 = 0) and other controllers where a certain amount of velocity feedback (equivalent to a damping ratio of 0.2) has been added (either in the second and fourth quadrants or in the first and third quadrants or in both). From consideration of Figure 5.24 it is clear that the controller with b = 2.5, z_1 = 0, z_2 = 0.2 achieves the best results in terms of acceleration reduction, particularly close to the resonance; however its performance is not as outstanding in terms of working space and tyre force (see Figures 5.25 and 5.26). At the other extreme the controller with b = 2.5, z_1 = 0.2, z_2 = 0 produces the least acceleration reduction among the four types of controllers investigated, but it performs better with respect to working space and tyre force. The other two controllers, the pure position feedback controller ($b = 2.5$, $z_1 = 0$, $z_2 = 0$) and that with $b = 2.5$, $z_1 = 0.2$, $z_2 = 0.2$ provide a reasonable trade-off among the requirements of minimising acceleration and reducing working space and dynamic tyre force.

Figures 5.27–5.30 depict the time trends for spring and normal forces, in order to assess qualitatively how abrupt the transition is between the two structures for the four benchmark inputs above. Sharp transitions in the normal force are an indirect and qualitative measure of the ride quality achievable with the different controllers. Abrupt transitions may result in spiky acceleration time histories.

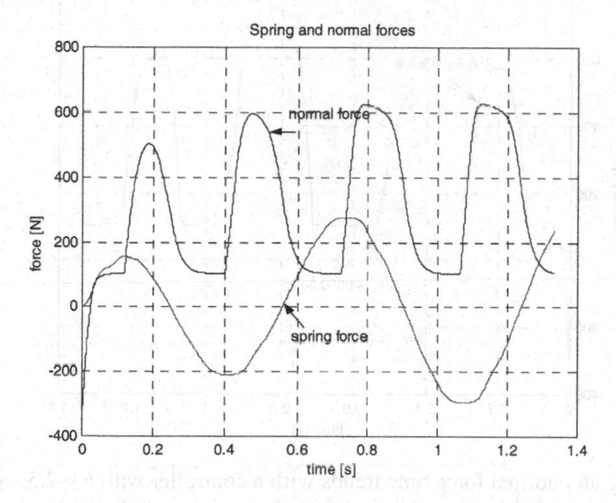

Fig. 5.27. Spring and normal force time trends with a controller with $b = 2.5$, $z_1 = 0$, $z_2 = 0$

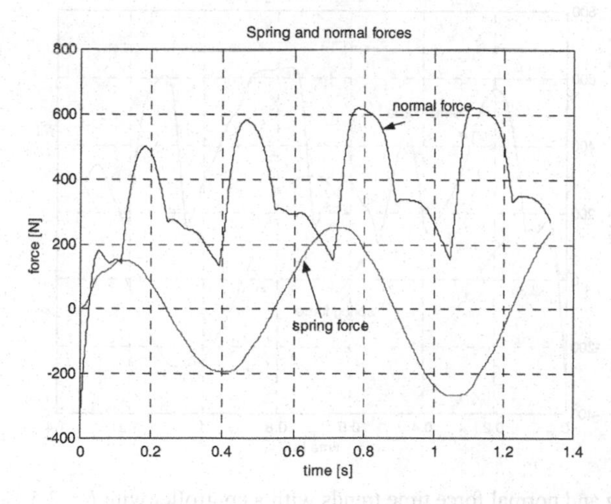

Fig. 5.28. Spring and normal force time trends with a controller with $b = 2.5$, $z_1 = 0$, $z_2 = 0.2$

The control force F_n in Figure 5.27 results from the pure spring force cancellation controller (b = 2.5, z_1 = 0, z_2 = 0); when the control is set to the off state the transition is smoothed only by the hydraulic and frictional dynamics. Some viscous damping is helpful as Figure 5.28 shows (case with b = 2.5, z_1 = 0, z_2 = 0.2). The additional viscous action in the first and third quadrants smoothes the control force trend, although it cannot compensate for the sudden rise when the controller is set to the on state.

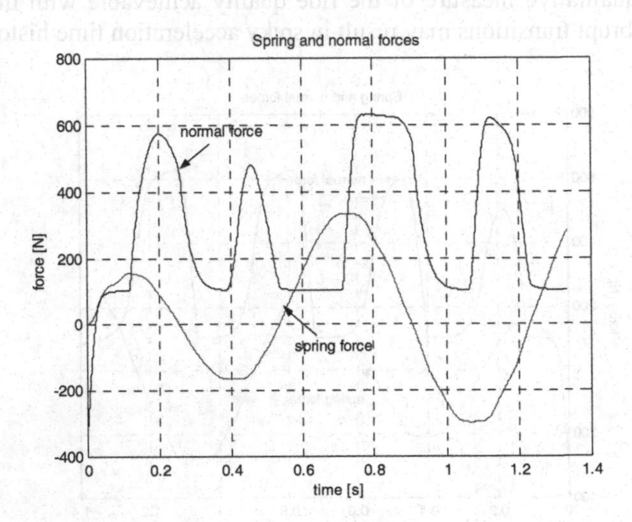

Fig. 5.29. Spring and normal force time trends with a controller with b = 2.5, z_1 = 0.2, z_2 = 0

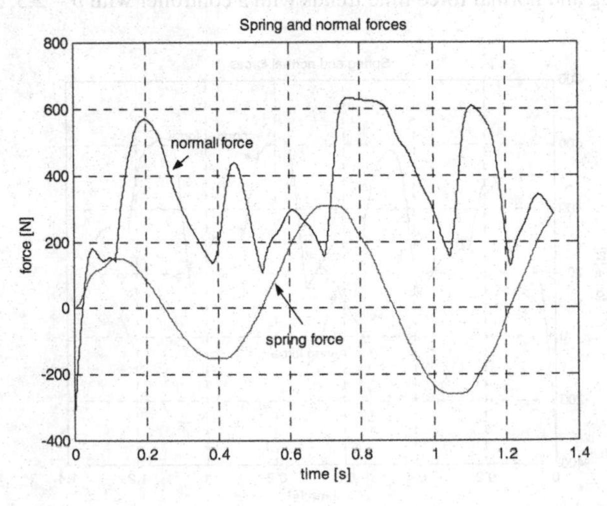

Fig. 5.30. Spring and normal force time trends with a controller with b = 2.5, z_1 = 0.2, z_2=0.2

The controller with b = 2.5, z_1 = 0.2, z_2 = 0 (Figure 5.29) is probably the worst from a ride quality point of view. The additional viscous damping in the second and fourth quadrants sharpens both the on and off transitions. The controller of Figure 5.30 ($b = 2.5$, $z_1 = 0.2$, $z_2 = 0.2$) performs fairly well, as the control force is never set to zero and the trend is relatively smooth.

A pseudo-random road input is now considered. The road model is expressed by the following relationship (Horrocks *et al.*, 1997):

$$z_0(t) = \sqrt{2}\sum_{i=1}^{45}\sqrt{\frac{\pi\,\Delta f V^{1.5}}{(\Delta f i)^{2.5}}}10^{-3}\sin[(2\pi\,\Delta f\, i)t + p(i)]\,, \tag{5.47}$$

the velocity V being in m/s.

This is a multiharmonic input where the amplitude is a decreasing function of the frequency (Figures 5.31 and 5.32); the increment Δf is 0.25 Hz and the phase $p(i)$ is a random number between $-\pi$ and π.

The spectrum defined by Equation 5.47 constitutes a discrete approximation to the continuous spectral density defined by Equation 2.34. Lower harmonics are the most important; higher harmonics, although present, are mainly filtered by the tyre.

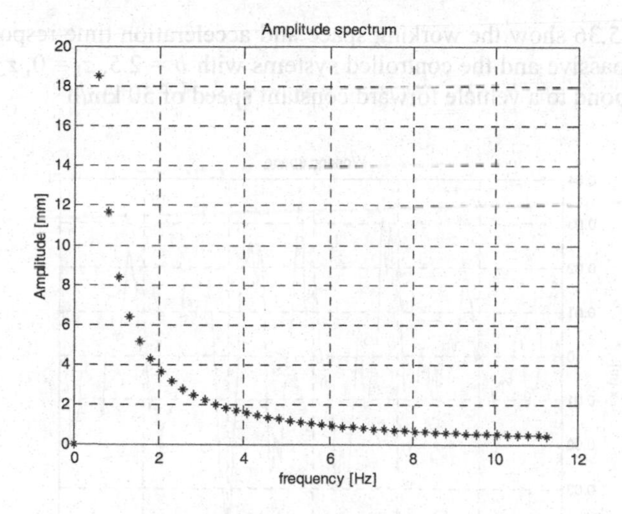

Fig. 5.31. Amplitude spectrum of the road vertical profile

It must be stressed that, with the controlled system being non-linear, the property of the response to this input cannot be inferred in any way from the properties of the response to each sinusoidal input separately, *i.e.*, the RMS and peak value percentage reductions (or increases) with respect to the passive system will be different, because the superposition principle does not hold.

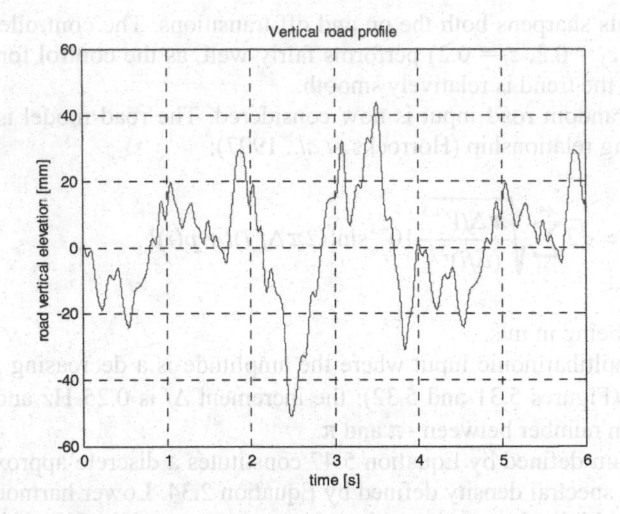

Fig. 5.32. Road vertical profile time trace; vehicle speed: 50 km/h

Figures 5.33–5.36 show the working space and acceleration time responses to this input for the passive and the controlled systems with $b = 2.5$, $z_1 = 0$, $z_2 = 0$. These results correspond to a vehicle forward constant speed of 50 km/h.

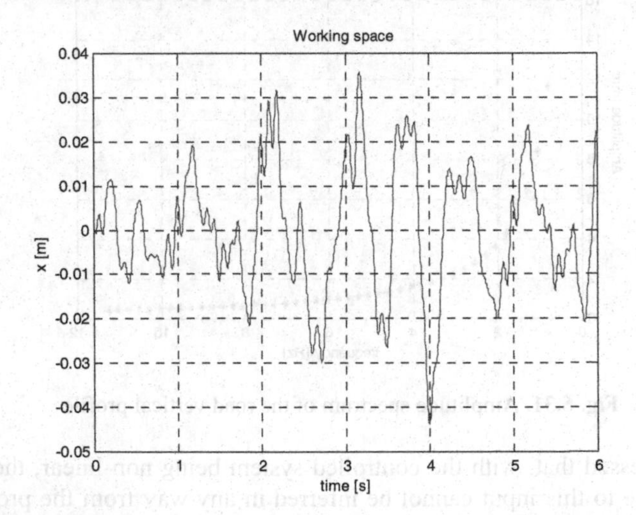

Fig. 5.33. Working space time trace for the passive system; vehicle speed: 50 km/h

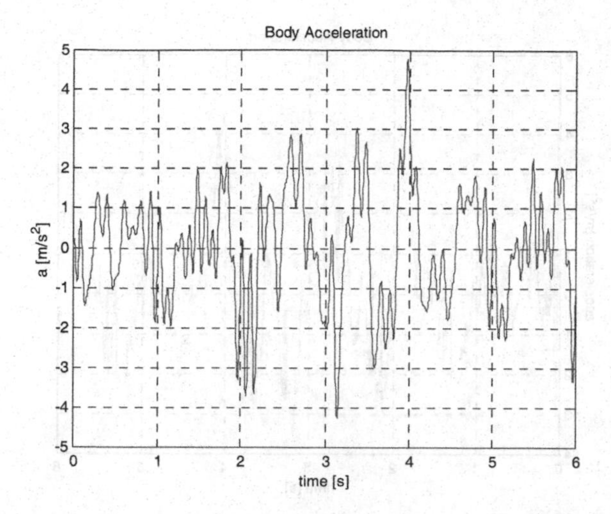

Fig. 5.34. Chassis acceleration time trace for the passive system; vehicle speed: 50 km/h

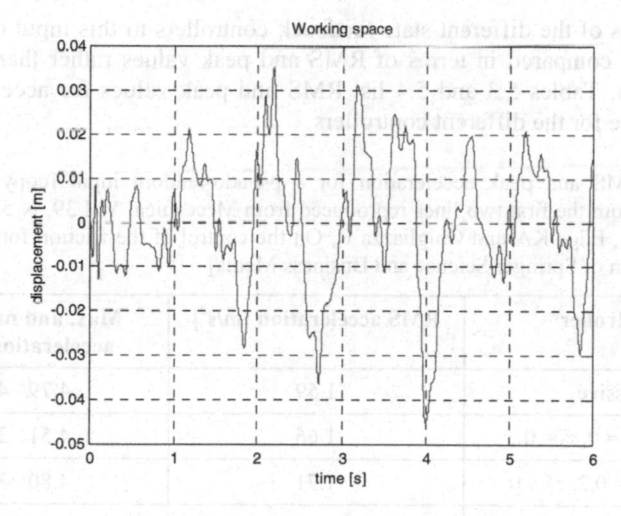

Fig. 5.35. Working space time trace for the controlled system with $b = 2.5$, $z_1 = 0$, $z_2 = 0$; vehicle speed: 50 km/h

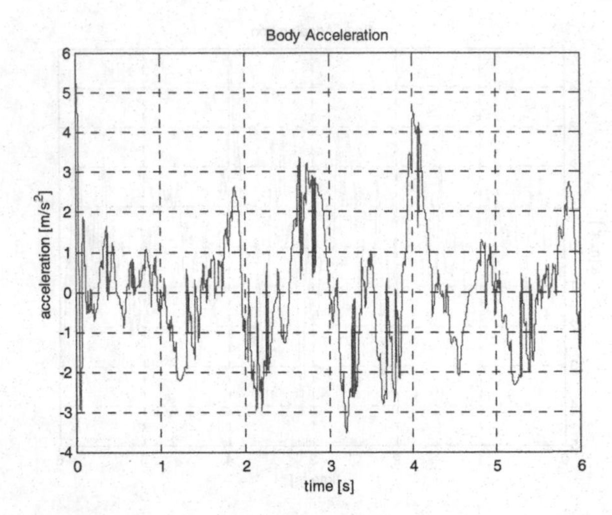

Fig. 5.36. Chassis acceleration time trace for the controlled system with $b = 2.5$, $z_1 = 0$, $z_2 = 0$; vehicle speed: 50 km/h

The responses of the different state feedback controllers to this input can be more appropriately compared in terms of RMS and peak values rather than from their time histories. Tables 5.3 and 5.4 list RMS and peak values for acceleration and working space for the different controllers.

Table 5.3. RMS and peak acceleration for a pseudo-random input [copyright Kluwer (2004), data from the first two lines reproduced from Meccanica, Vol 39, N 5, pp 395-406, Guglielmino E, Edge KA and Ghigliazza R, On the control of the friction force, used with kind permission of Springer Science and Business Media]

Controller	RMS acceleration [m/s^2]	Max. and min. peak acceleration [m/s^2]
Passive	1.59	4.79/–4.24
$b = 1, z_1 = 0, z_2 = 0$	1.66	4.51/–3.53
$B = 1, z_1 = 0.2, z_2 = 0$	1.71	4.80/–3.35
$b = 1, z_1 = 0, z_2 = 0.2$	1.41	4.46/–3.59
$b = 1, z_1 = 0.2, z_2 = 0.2$	1.45	4.65/–3.44

The pseudo-random test is a much more severe test than the sinusoidal one. The performance of the controlled system is somewhat less outstanding compared to that in response to the sinusoidal input and in some cases even worse: the RMS acceleration is only slightly better in two out of the four benchmark controllers considered. The RMS working space response is however always slightly smaller than the corresponding passive one in all but one case.

Table 5.4. RMS and peak working space for a pseudo-random input

Controller	**RMS working space [mm]**	**Max. and min. peak working space [mm]**
Passive	16.7	35.7/–43.7
$b = 1, z_1 = 0, z_2 = 0$	16.6	35.4/–45.4
$B = 1, z_1 = 0.2, z_2 = 0$	17.7	34.5/–48.3
$b = 1, z_1 = 0, z_2 = 0.2$	14.7	32.4/–40.9
$b = 1, z_1 = 0.2, z_2 = 0.2$	15.3	32.0/–42.4

Last but not least, the response to a bump must be considered. This input, which represents a discrete event in a road profile, must be analysed seperately. The relevant quantities to minimise are the peak value of the acceleration and the number of oscillations after the bump. Figures 5.37 and 5.38 show the chassis acceleration response in the two cases. The input is a 50-mm-high sinusoidal-shaped bump. The bump horizontal length is 0.5 m and the car is assumed to pass over it at a constant speed of 20 km/h. This is equivalent to saying that the car is excited with half a sinusoid having a frequency of 11.11 Hz (as the frequency is equal to the car forward velocity divided by the bump wavelength).

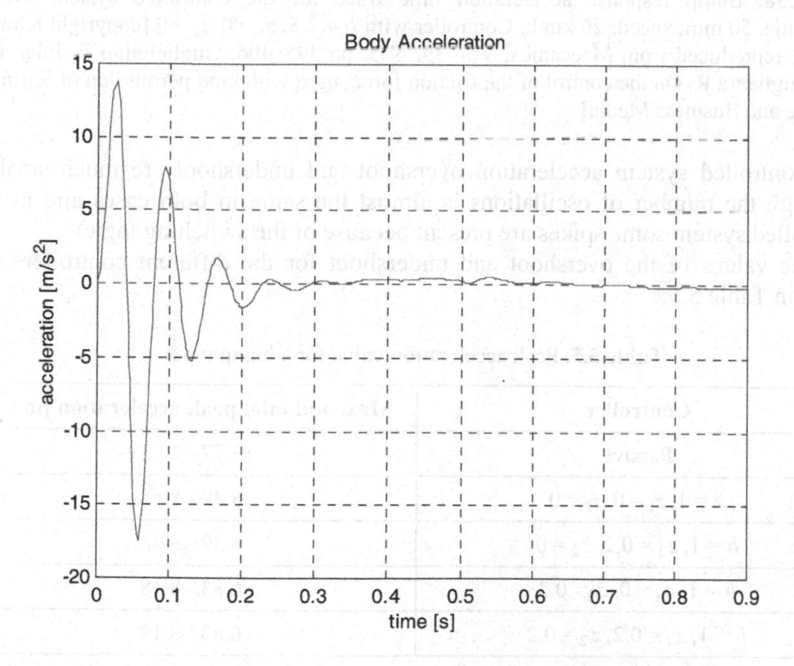

Fig. 5.37. Bump response acceleration time trace for the passive system. Bump amplitude: 50 mm; speed: 20 km/h [copyright Kluwer (2004), reproduced from Meccanica, Vol. 39, N 5, pp 395–406, Guglielmino E, Edge KA and Ghigliazza R, On the control of the friction force, used with kind permission of Springer Science and Business Media]

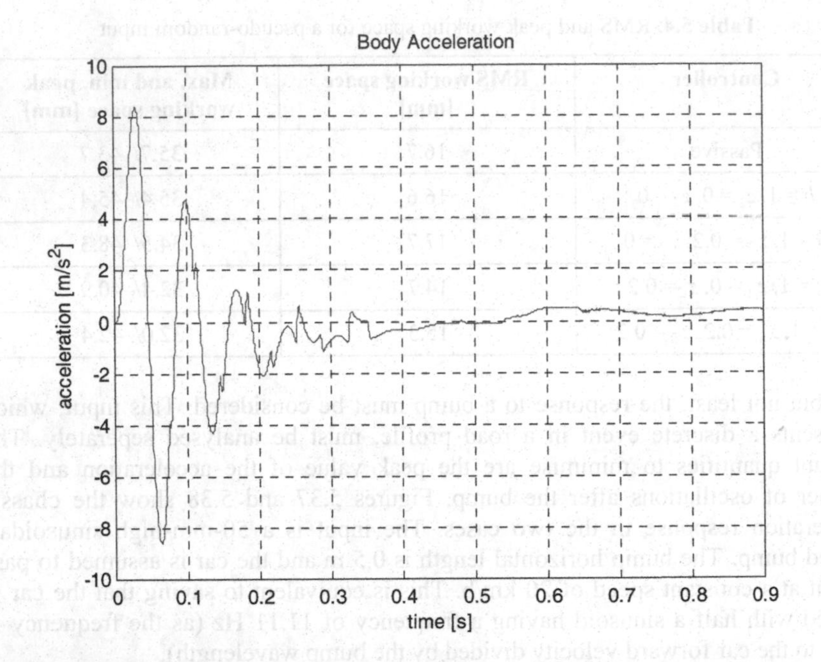

Fig. 5.38. Bump response acceleration time trace for the controlled system. Bump amplitude: 50 mm; speed: 20 km/h. Controller with $b = 2.5$, $z_1 = 0$, $z_2 = 0$ [copyright Kluwer (2004), reproduced from Meccanica, Vol. 39, N 5, pp 395–406, Guglielmino E, Edge KA and Ghigliazza R, On the control of the friction force, used with kind permission of Springer Science and Business Media]

The controlled system acceleration overshoot and undershoot are much smaller although the number of oscillations is almost the same in both cases and in the controlled system some spikes are present because of the switching logic.

The values of the overshoot and undershoot for the different controllers are listed in Table 5.5.

Table 5.5. Peak acceleration value for a bump input

Controller	Max. and min. peak acceleration [m/s^2]
Passive	13.77/–17.39
$b = 1, z_1 = 0, z_2 = 0$	8.39/–8.58
$b = 1, z_1 = 0.2, z_2 = 0$	8.39/–8.38
$b = 1, z_1 = 0, z_2 = 0.2$	6.83/–8.18
$b = 1, z_1 = 0.2, z_2 = 0.2$	6.83/–8.19

The controllers with $b = 2.5$, $z_1 = 0$, $z_2 = 0.2$ and with $b = 2.5$, $z_1 = 0.2$, $z_2 = 0.2$ provide the largest reduction of the peak values among the four types of controllers.

However a remark should be made on the validity of the results of this test. A bump can be viewed as a sort of impulsive input which produces a high acceleration peak. When a tyre undergoes high accelerations, it should be modelled taking into account its dynamic characteristics. A tyre model which includes only wheel mass and the tyre static stiffness characteristic is no longer adequate (Genta, 1993).

Furthermore during a bump the point of contact between the wheel and the road is not aligned with the vertical axis passing through the centre wheel, hence horizontal contact forces are generated that cannot be taken into account by the quarter car model (nor by the 7DOF model). Therefore the results from this test must be considered carefully and taken only as an indication of the actual behaviour of the car.

5.9 Friction Damper Electrohydraulic Drive Assessment

In automotive suspension control, there is scope to explore simple and cheap solutions, based on low-cost valves which can be appealing to the automotive market. The purpose of this section is how to assess, tune and optimise hydraulic drive performance.

It was previously concluded that the use of a proportional underlapped valve supplied by a pump in parallel with a relief valve, driving a single-chamber actuator is suitable for an effective friction damper hydraulic actuation system. Such a hydraulic circuit behaves in an analogous manner to an electric potential divider (Figure 5.5).

The valve employed in the hydraulic drive (depicted in Figure 5.3) has been designed for directional flow control; hence its performance when used in the pressure control mode is not known, (nor usually provided in manufacturers' datasheets). A potentially major limiting factor in the system is the pressure dynamic response of the valve as well as its pressure versus demand signal static characteristics (the pressure gain).

These real-life engineering issues are crucial for designing a properly working system and therefore a bench test assessment must be undertaken. This section describes a methodological approach to verify if a hydraulic system can meet the requirements of the application. Firstly the performance of a system based on a cheap 2-way valve is investigated and subsequently the same analysis is presented for a more costly proportional valve with incorporated electronics; it will also be shown that a drive using a costly valve requires an accurate optimisation process to actuate the system properly.

The solution based on a low-cost two-way valve requires an additional return orifice, the reason being that, for the control of the friction damper, a three-way valve is necessary. Hence the hydraulic circuit needs to be completed with the connection of an additional orifice to the service port of the valve, which allows the flow to tank to be damped. In this way the valve together with the orifice has a geometry equivalent to that of a three-way valve. For testing purpose the restriction can be created with an adjustable needle valve (*i.e.*, a variable orifice). The valve needs to be mounted on a flow bench test rig and both downstream pressure and spool position measured. Figure 5.39 a schematic of the measurement system shows.

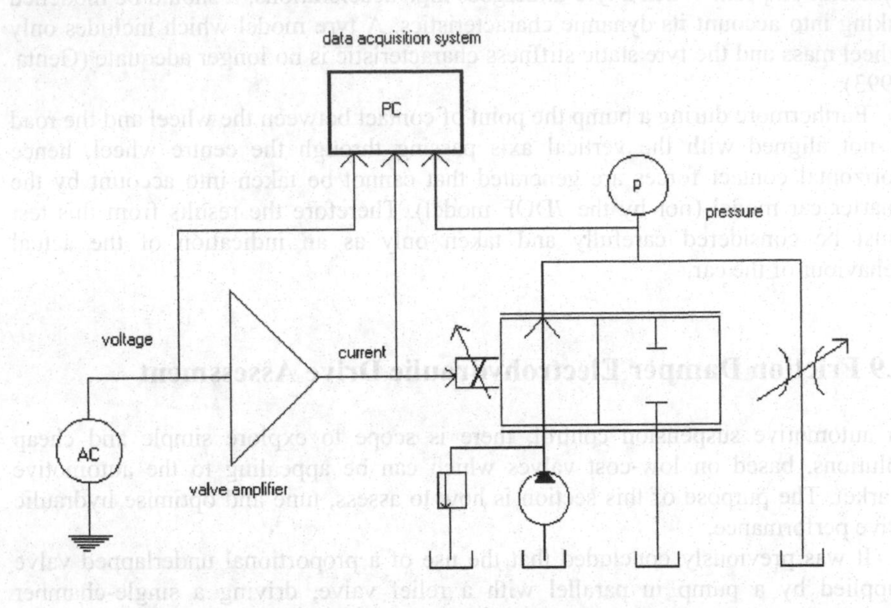

Fig. 5.39. Schematic of the measurement circuit for a two-way valve

The valve is supplied at nominally constant pressure by a volumetric pump (the pressure being around 60 bar in this initial test) and the service port connected to tank through the orifice. Downstream pressure is recorded by a variable-reluctance pressure transducer. A sinusoidal voltage source provides the input to a current amplifier. Solenoid valves are typically current driven and, if not embedded in the valve, an external current amplifier is required to provide a current driving signal. Figure 5.40 shows an example of a circuit for a current driver. It is worth noting that as the spool valve is spring-preloaded, the equilibrium position does not correspond to the null signal but to a biased constant value; therefore the amplifier needs to be supplied with two non-symmetric voltages.

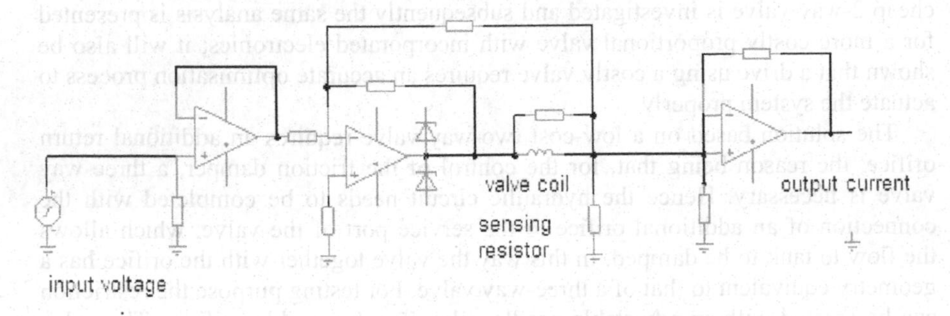

Fig. 5.40. Electrical scheme of a typical valve current amplifier

Data can be captured and stored using commercial data acquisition software, acquiring data from pressure transducers and a current probe. Good engineering practice suggests installing additional pressure gauges and oscilloscopes to check whether pressure and other signals are in the correct range.

The first characteristic to be established is the pressure versus demand curve, which defines the non-linear gain and the tracking ability of the controller. The local slope of the pressure gain around the valve null position depends upon the valve underlap *u* and the leakage coefficient k_{1s}. As this is a static characteristic, it is crucial to carry out the measurement under quasi-static conditions, otherwise the phase lag introduced at higher frequencies would result in a distortion of the characteristic. This can be achieved by driving the valve with a low-frequency sinusoidal signal. The pressure gain is expected to be saturation shaped with some hysteresis depending upon the magnetic properties of the solenoid core and on the friction within the spool. A typical measured characteristic is plotted in Figure 5.41.

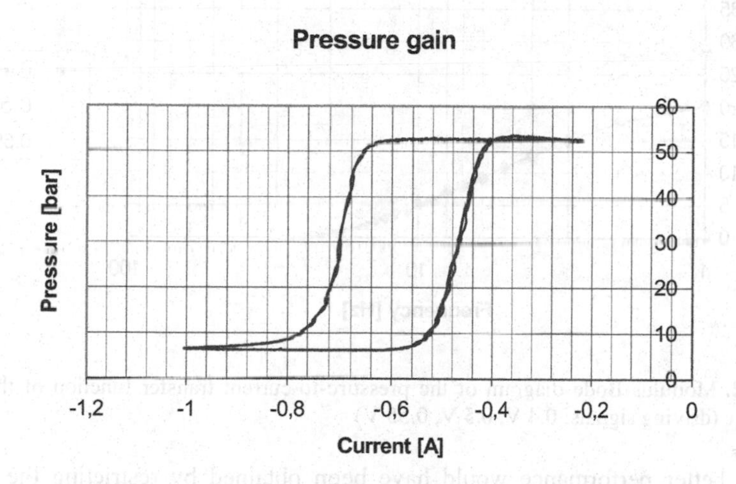

Fig. 5.41. Pressure versus current static characteristic of a two-way valve

Hysteresis is large, as expected for a low-cost valve. Such a hysteresis would compromise its usage in applications where good tracking capability is crucial. However in the FD control application good tracking is not a vital issue. The second problem is the slope of the characteristics, which is very high, meaning that the valve is almost critically lapped; this makes it more difficult to use it in a proportional fashion (it should be driven with low-amplitude signals).

Dynamic performance must be assessed as well. A very convenient way to do this is the measurement of the frequency response. The system being non-linear, its measurement requires the deduction of the first harmonic of the input and output signals and calculation of the modulus and phase of the resulting sinusoidal signals for each frequency.

Because of the strong non-linearity of the system, it is advisable to measure the frequency response for at least three different amplitudes of the driving signal, as in non-linear systems the response is not amplitude independent (as opposed to linear systems). Results are depicted in Figure 5.42. By analysing the figure it can be noticed that different input amplitudes produce frequency response traces which are not superimposable. This occurs because for each input the valve works at a different point of its non-linear pressure-flow characteristics.

The valve is very slow (the average bandwidth is less than 3 Hz). The reasons for the low bandwidth can be attributed to the combined effects of both the relatively cheap type of valve and also the likelihood of the presence of free air in the circuit.

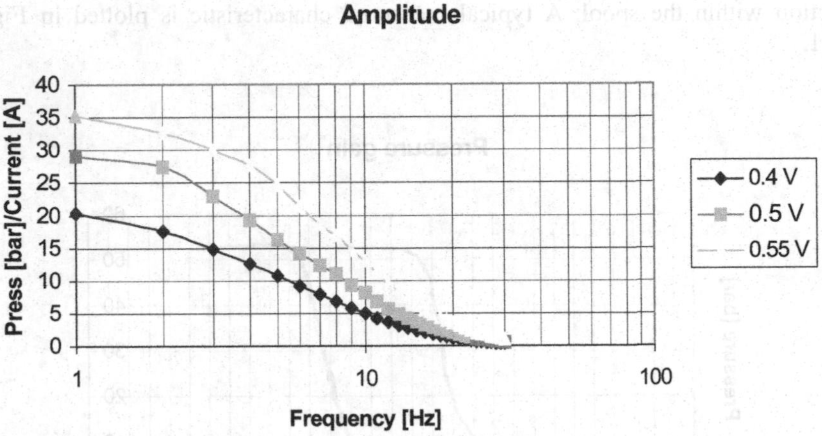

Amplitude

Press [bar]/Current [A]

40
35
30
25
20
15
10
5
0

1
10
100

Frequency [Hz]

0.4 V
0.5 V
0.55 V

Fig. 5.42. Modulus Bode diagram of the pressure-to-current transfer function of the two-way valve (driving signals: 0.4 V, 0.5 V, 0.55 V)

Slightly better performance would have been obtained by restricting the cross-sectional area of the fixed orifice (the hydraulic time constant depends on the inverse of this area), but this would have had as a drawback a larger pressure drop and hence higher energy consumption at the pump.

The tests indicate that the overall characteristics of this valve are not good enough for the application; the static characteristic is too sharp and the bandwidth is insufficient. Such a valve is not suitable for the application.

It would be reasonable to think that a more sophisticated and costly valve, namely a solenoid proportional three-way valve with incorporated conditioning electronics (the current amplifier embedded in the valve) could meet the specification. However it will be shown that this is not always the case.

It is anticipated that an unexpected behaviour at low supply pressure can arise: dynamic performance could be extremely poor with the bandwidth severely limited despite the valve being of a relatively sophisticated design.

The working supply pressure needs to be set at a value consistent with that required for the suspension application, which is about 10 bar; will be investigated the feasibility of using this system at low pressure.

Pressure gain must first be assessed. Hysteresis is expected to be negligible in this valve. Hysteresis is due to both a poor magnetic material (not the case for a fairly costly valve) and to friction in the spool and other dissipative phenomena.

However, inaccurately designed tests can result in measuring an amount of hysteresis larger than the actual one. In such cases the measurement chain must be thoroughly verified. First the inherent hysteresis of the pressure transducer must be assessed and if necessary the transducer must be replaced with a more accurate one (such as a semiconductive strain gauge transducer) with very low hysteresis. It is also useful to repeat the test by superimposing a high-frequency dither signal (square waves at 200 Hz, 300 Hz and 1 kHz) on the quasi-static signal, as dither helps reduce the amount of hysteresis. If no appreciable improvement is recorded then the input test frequency must be drastically reduced. The response plotted in Figure 5.43 depicts the static characteristic at the supply pressure of 10 bar taken at the extremely low frequency of 0.004 Hz. This test at such a low frequency was necessary because of the presence of circuits in the electronics which introduced high phase margin at very low frequency, thus resulting in an apparent hysteresis. This stresses the importance of making a test as close as possible to an ideal static test when measuring static characteristics.

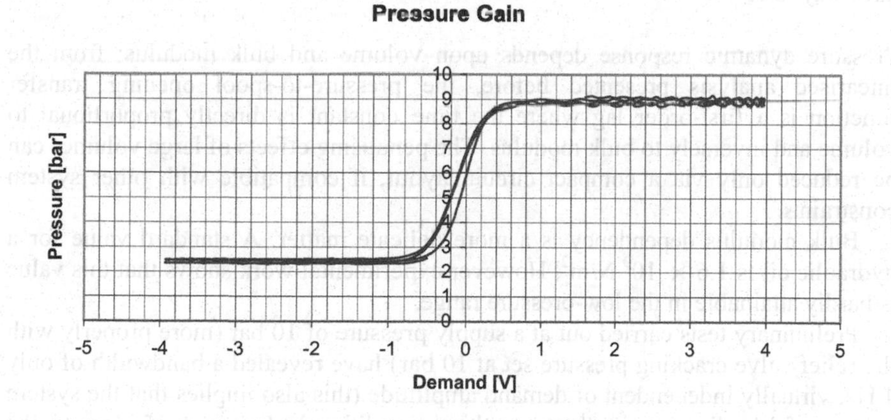

Fig. 5.43. Three-way valve pressure gain

Dynamic performance is the next step to gain an understanding of the hydraulic drive behaviour. Before evaluating pressure dynamic response it is important to measure the valve spool dynamic response (with oil, so as to account for flow forces in the spool).

Spool frequency response can be readily measured using the LVDT embedded into the valve (if an external measurement connection is present). Figure 5.44 shows a recorded spool dynamic response, which is second order, with a bandwidth of about 100 Hz, a damping ratio of about 0.60 and virtually supply pressure independent.

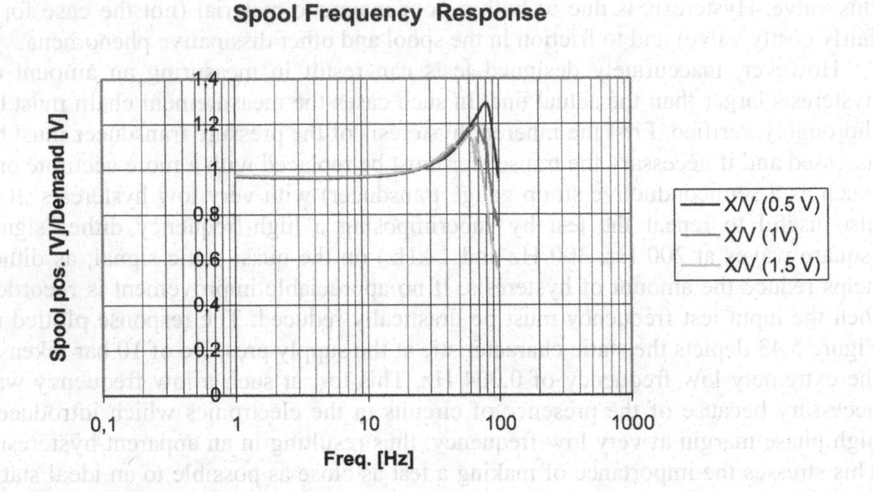

Fig. 5.44. Modulus Bode diagram of the spool position-to-voltage transfer function of the three-way valve

Pressure dynamic response depends upon volume and bulk modulus: from the linearised analysis presented before, the pressure-to-spool opening transfer function is a first-order lag where the time constant is directly proportional to volume and inversely to bulk modulus. The penalising effects of large volumes can be reduced only via a compact circuit layout, if compatible with other system constraints.

Bulk modulus dependency is a more delicate matter. A standard value for a hydraulic oil is 1.6×10^9 N/m^2. However experimental work shows that this value is hardly attainable in the low-pressure range.

Preliminary tests carried out at a supply pressure of 10 bar (more properly with the relief valve cracking pressure set at 10 bar) have revealed a bandwidth of only 4 Hz, virtually independent of demand amplitude (this also implies that the system behaves fairly linearly in those working conditions): Figure 5.45 depicts the frequency response for three different input voltages.

Such a poor bandwidth is unexpected considering that the valve is fairly sophisticated and is expected to have a bandwidth up to 100 Hz. The reasons for such a slow response need to be thoroughly investigated.

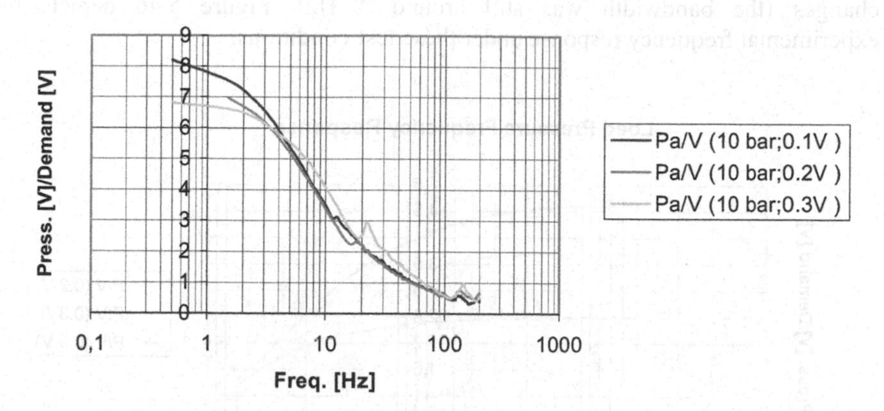

Fig. 5.45. Modulus Bode diagram of the pressure-to-voltage transfer function of the three-way valve; demand signals 0.1 V, 0.2 V, 0.3 V [copyright Elsevier (2003), reproduced with minor modifications from Guglielmino E, Edge KA, Controlled friction damper for vehicle applications, Control Engineering Practice, Vol. 12, N 4, pp 431–443, used by permission]

Experimentally testable hypotheses need to be made on the nature of this phenomenon and a set of tests designed to verify experimentally their consistency. Sensible hypotheses based on the physics of the system are:

- Presence of air bubbles trapped inside the circuit: a small free air percentage can produce a large reduction of the bulk modulus (McCloy and Martin, 1980)
- Not fully developed turbulent flow because of the small pressure drop past the valve
- Ripple and dynamic effects on the supply pressure caused by relief valve dynamics, which in turn could affect the downstream pressure

The previous test depicted in Figure 5.45 has also allowed the dependency upon the demand signal amplitude to be established (the bandwidth usually decreases if the system is driven hard). If this dependency is found to be negligible (as here) the valve has fairly linear behaviour.

Considering that the spool response is fast and independent of the measurement conditions then the problem of the limited pressure bandwidth cannot be caused by the valve itself but is a fluid mechanics problem. In this case *ad hoc* tests must be designed to test the hypotheses above and identify the possible causes.

The first hypothesis made was that of the presence of an air pocket at low pressure. This can be verified by mounting the valve upright to prevent the air stagnating at the outlet port and the mounting connection of the pressure transducer.

Different types of pressure transducers also need to be tested to ascertain whether transducer dynamics (or also a fault in the instrumentation) can be responsible for the result. An initial flushing of the valve was made before starting the measurement of the frequency response. The test resulted in no appreciable changes (the bandwidth was still around 4 Hz). Figure 5.46 depicts the experimental frequency response under these test conditions.

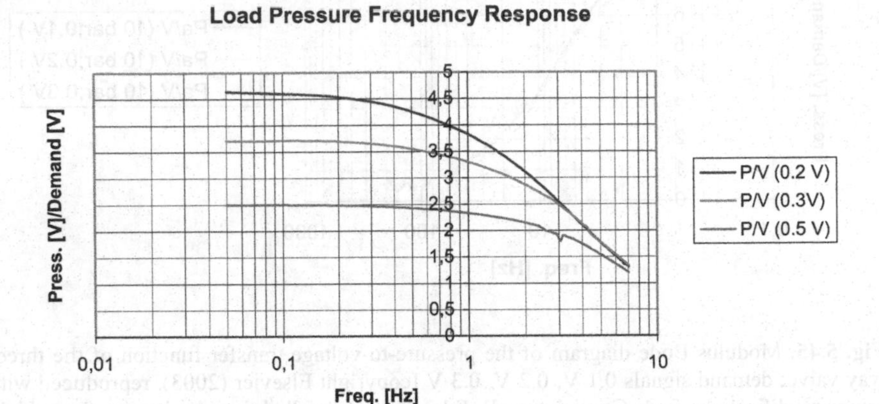

Fig. 5.46. Modulus Bode diagram of the pressure-to-voltage transfer function of the three-way valve measured in the air bubble test

In order to verify the consistency of the second hypothesis the spool must be driven with a biased signal so as to create a narrower flow path, to help promote turbulence as the small pressure drop past the orifice may not be sufficient to fully develop the turbulent flow regime. The low-Reynolds-number flow arising may be related to the small bandwidth. In order to verify this hypothesis the valve has to be driven with a biased signal so that the spool does not move symmetrically with respect to the centre of the land. This creates a narrower flow path and hence a larger pressure drop, helping to promote a turbulent flow regime. However, even in this situation the dynamic response does not show any improvement at all.

A third test to be carried out is to quantify the amount of dynamic variation in the supply pressure induced by the combined effects of the volumetric pump pressure ripple and the relief valve characteristics. If this is not negligible and is in phase opposition with the load pressure, it may slow the dynamic performance. However the measured pressure variation is negligible, as depicted in Figure 5.47. A relief valve dynamic effect (the second-order spring–mass–damper dynamics of the relief valve spool) can be excluded to be a co-cause, because relief valve dynamics are typically at much higher frequencies (around 200–300 Hz).

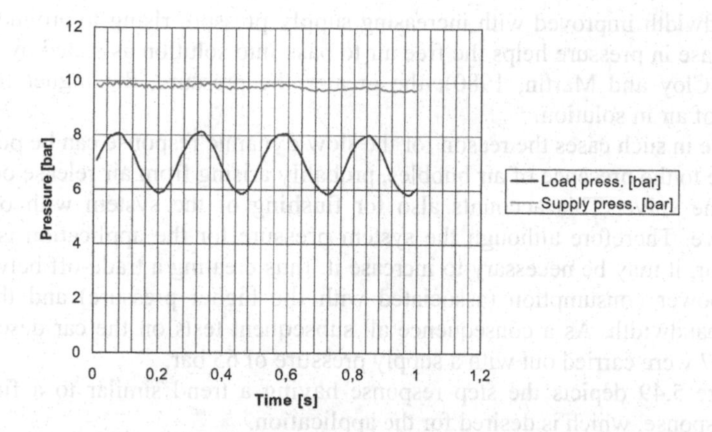

Fig. 5.47. Supply pressure dynamic variation

To double check the hypothesis a quick test consisting of replacing the relief valve with a more sophisticated one, having two stages (hence smaller pressure override) which produces a smaller ripple on the supply pressure showed that the bandwidth did not change. Hence not even the relief valve was responsible of the poor pressure dynamic performance. This test, however, also enables it to be established that a single-stage relief valve is sufficient for the application, without recourse to a more costly two-stage relief valve.

However in none of these tests an appreciable improvement has been recorded. In this case a solution is to increase the supply pressure (by appropriately setting the relief valve) up to 100 bar. This produced an increase of the bandwidth up to around 40 Hz. Under these conditions the response is finally acceptably fast: the bandwidth increases with supply pressure. Figure 5.48 shows the frequency response measured with a relief valve cracking pressure of 60, 80 and 100 bar.

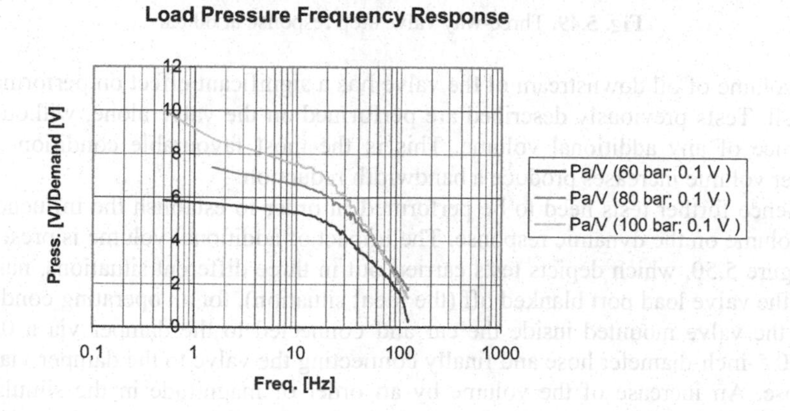

Fig. 5.48. Modulus Bode diagram of the pressure-to-voltage transfer function of the three-way valve at 60, 80, 100 bar

The bandwidth improved with increasing supply pressure rising to around 40 Hz. An increase in pressure helps the free air to pass into solution as stated by Henry's law (McCloy and Martin, 1980): the greater the pressure, the higher the mole fraction of air in solution.

Hence in such cases the reason for the slow dynamic response can be postulated to be due to the presence of air bubbles, probably arising from air release occurring within the valve. This accounts also for flushing of the system with oil being ineffective. Therefore although the system pressure for the application is around 10–20 bar, it may be necessary to increase it, thus creating a trade-off between the higher power consumption (associated with the higher pressure) and the valve system bandwidth. As a consequence all subsequent tests on the car described in Chapter 7 were carried out with a supply pressure of 65 bar.

Figure 5.49 depicts the step response having a trend similar to a first-order linear response, which is desired for the application.

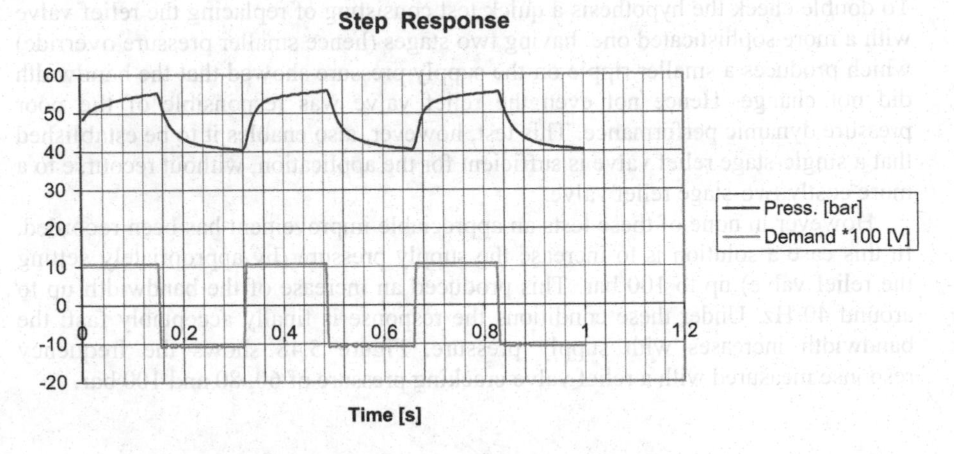

Fig. 5.49. Three-way valve step response at 60 bar

The volume of oil downstream of the valve has a significant effect on performance as well. Tests previously described are performed on the valve alone, without the presence of any additional volume. This is the most favourable condition. Any further volume increases produce a bandwidth reduction.

Hence further tests need to be performed in order to establish the influence of the volume on the dynamic response. The impact of additional volume is presented in Figure 5.50, which depicts tests carried out in three different situations, namely with the valve load port blanked off (the ideal situation), for an operating condition with the valve mounted inside the car and connected to the damper via a 0.5-m long 0.5-inch-diameter hose and finally connecting the valve to the damper via a 2-m hose. An increase of the volume by an order of magnitude in the simulation model reduces the bandwidth by about a decade. This is consistent with a linearised analysis of the system, which shows that the hydraulic time constant is directly proportional to the volume. Therefore in a final design the volume should be minimised, with the valve integrated into the assembly such that it is as close as possible to the friction pad piston(s). The minimum volume is the internal pipe inside the damper plus the volume of the hose connecting the valve to the damper.

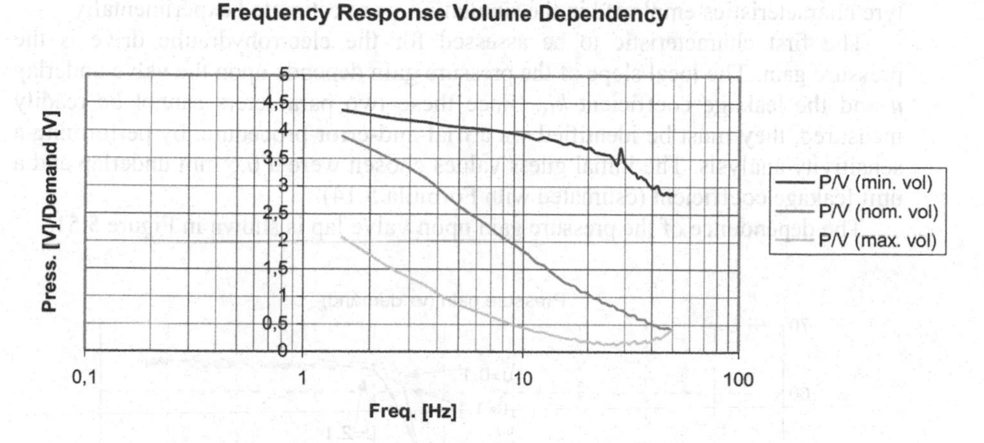

Fig. 5.50. Modulus Bode diagram of the pressure-to-voltage transfer function of the 3-way valve varying volume

Last but not least the problem of back-pressure must be mentioned. A 2-bar back-pressure is present, as is evident from the measured pressure gain trend of Figure 5.43. This creates a constant-amplitude friction force present also when the demand is set to zero. The larger the back-pressure, the bigger the residual friction force. This can produce an undesired lockup of the damper. The back-pressure is due to the loss of head in the return line and to the leakage flow. This problem can however be addressed in the design phase by mounting a pre-loaded spring between the friction pads. However from a dynamic viewpoint, this additional spring would introduce two further complex conjugated poles, which would affect the overall system response. If these poles are of the order of magnitude of the valve bandwidth, they can affect the dynamic performance of the hydraulic drive. This design enhancement will be described subsequently.

5.10 Electrohydraulic Drive Parameters Validation

In this section the validation of the experimental tests is described. A prime objective of the experimental data collected is to validate the mathematical models developed. This permits to verify that the dynamic order of the model, its complexity and the precision in the knowledge of the system parameters and inputs are sufficient to give a reasonably accurate representation of its physical behaviour over the range of the operating conditions. Backed up by this validation process, if simulation is sufficiently accurate over the working range, further optimisation work can be solely made via simulation.

From an opposite viewpoint, experimental tests on single components or parts of the suspension system also permit the identification of system parameters otherwise difficult to identify with sufficient precision (for instance, the friction characteristic). Within this context, such experimental results can be logically seen as an *a priori* stage, providing reliable input data to the simulation. The friction and tyre characteristics employed in the simulation were estimated experimentally.

The first characteristic to be assessed for the electrohydraulic drive is the pressure gain. The local slope of the pressure gain depends upon the valve underlap u and the leakage coefficient k_{1s}. Since these two parameters cannot be readily measured, they must be identified via a trial-and-error procedure, by performing a sensitivity analysis. The initial guess values chosen were a 0.6 mm underlap and a unit leakage coefficient (estimated with Formula 5.14).

The dependence of the pressure gain upon valve lap is shown in Figure 5.51.

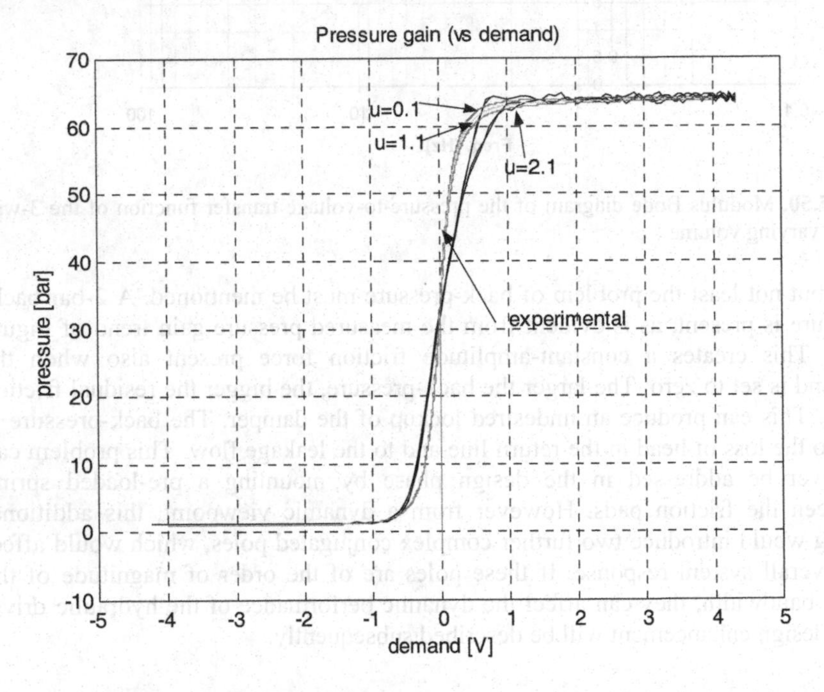

Fig. 5.51. Comparison of experimental and predicted control valve pressure gain, varying the underlap from 0.1 to 2.1 mm [copyright Elsevier (2003), reproduced from Guglielmino E, Edge KA, Controlled friction damper for vehicle applications, Control Engineering Practice, Vol. 12, N 4, pp 431–443, used by permission]

Lap values of 0.1 mm, 1.1 mm and 2.1 mm are assumed. The upper-bound region of the characteristics is affected more by a change in the underlap width than the lower-bound region, which is the one of interest. The asymmetry in the behaviour is caused by the leakage term, the relief valve pressure override, the spool dynamics and the compressibility flow. A symmetrical behaviour would have resulted if the leakage had not been considered, the supply pressure had been perfectly constant and the spool dynamics had been neglected.

In such an ideal situation, pressure gain can be expressed as a function of the demand in an analytical form and the behaviour in the upper- and lower-bound regions would be symmetrical. From the analysis of Figure 5.51, the best agreement is obtained when u is equal to 0.1 mm, although a larger value would not have affected the lower-bound region.

Figure 5.52 depicts the dependence upon the leakage coefficient, varied between 0.5 and 2.5 (taking an underlap of 0.1 mm). From consideration of the figure, the best fit is obtained when k_{1s} assumes the value of 1.5.

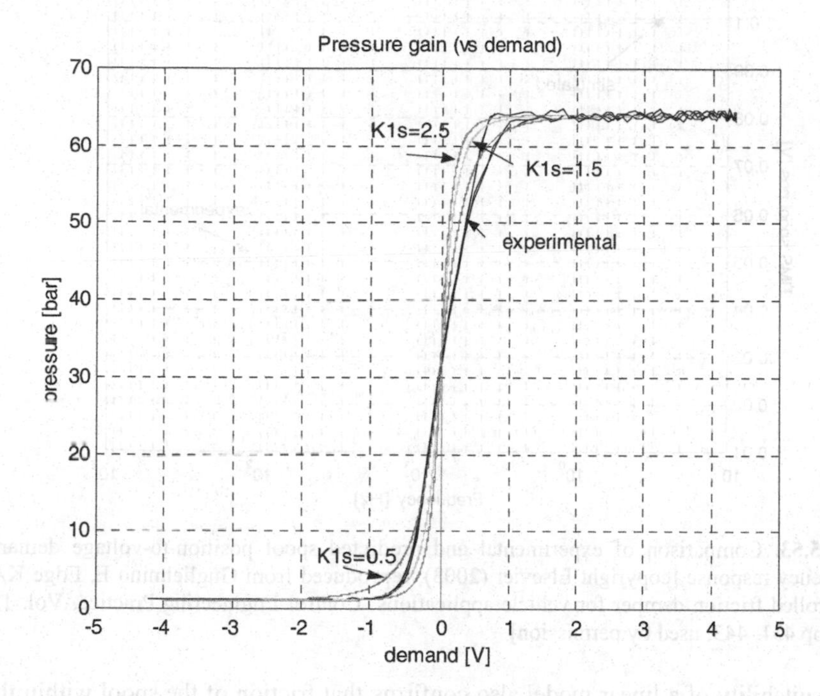

Fig. 5.52. Comparison of experimental and predicted control valve pressure gain, varying the leakage coefficient from 0.5 to 2.5 [copyright Elsevier (2003), reproduced from Guglielmino E, Edge KA, Controlled friction damper for vehicle applications, Control Engineering Practice, Vol. 12, N 4, pp 431–443, used by permission]

The dynamic response of the valve is the other critical issue. The approximations made should be recalled. While the valve experimental response is a real frequency response, obtained by extracting the first harmonic of input and output signals, the non-linear valve dynamics have been simulated and the "frequency response" obtained by computing the actual RMS, including therefore the contribution of all the harmonics. The approximation however is not so critical (and in any case comparable with that produced by a linearisation of the model) because the static characteristic of the valve is fairly linear in its central region and has negligible hysteresis, hence it is inherently low-pass and therefore the overall contribution of the higher harmonics to the RMS is small.

The trends of the frequency responses are portrayed in Figures 5.53 and 5.54. Figure 5.53 shows the spool-to-voltage demand transfer function frequency response. The measured response is close to second order: the simulation of a second-order linear model with a natural frequency of 105 Hz and a damping ratio of 0.60 matches the results well.

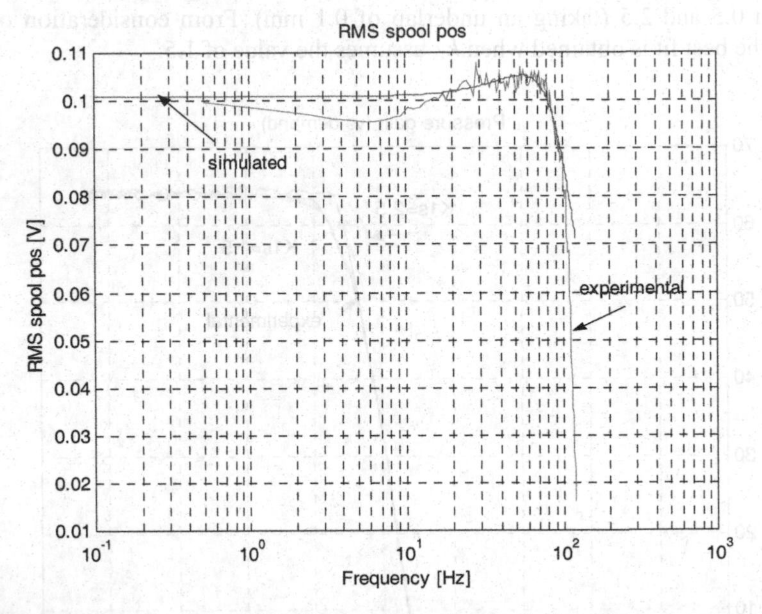

Fig. 5.53. Comparison of experimental and predicted spool position-to-voltage demand frequency response [copyright Elsevier (2003), reproduced from Guglielmino E, Edge KA, Controlled friction damper for vehicle applications, Control Engineering Practice, Vol. 12, N 4, pp 431–443, used by permission]

The suitability of a linear model also confirms that friction of the spool within the valve sleeve is negligible.

The pressure-to-valve demand frequency response presented in Figure 5.54 shows close correspondence to a first-order system with a dominant pole at around 20 Hz in cascade with the two complex conjugated poles corresponding to the spool dynamics.

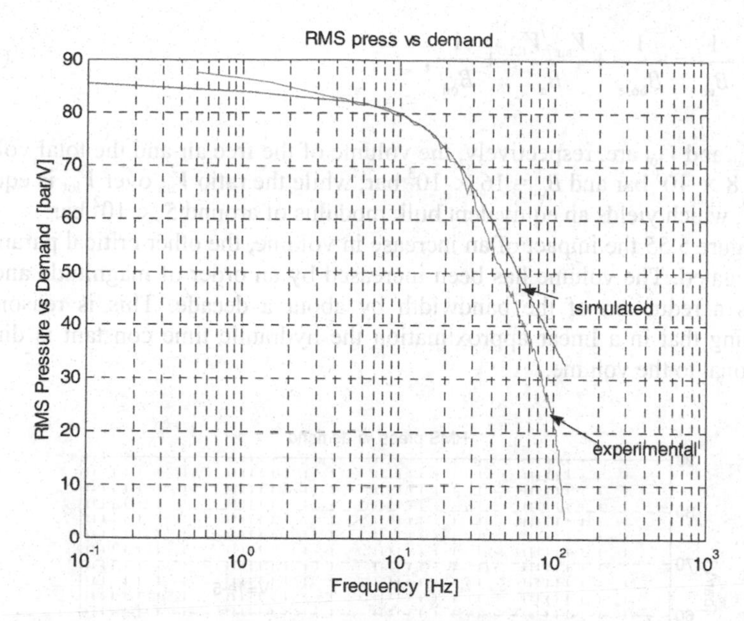

Fig. 5.54. Comparison of experimental and predicted pressure-to-voltage demand frequency response [copyright Elsevier (2003), reproduced from Guglielmino E, Edge KA, Controlled friction damper for vehicle applications, Control Engineering Practice, Vol. 12, N 4, pp 431–443, used by permission]

In the experimentation the bandwidth was substantially penalised by the presence of air bubbles formed in the oil because of the low system pressure. A simulation using standard values for the valve parameters and considering the actual volume formed by the valve with the load port blanked off results in an extremely large bandwidth, which is absolutely unrealistic when compared to the experimental results. Such a pressure response would have been even faster than the spool dynamic response. In such a situation, the pressure response would have been close to second order with its dynamics dictated by the two complex poles of the spool mechanical response, which would become dominant. In order to match the simulation with the experimental results, it is necessary to include in the dynamic model the presence of air, whose effect is dynamically manifested through a drastic reduction in the bulk modulus.

Its effect can be accounted for by a dramatic reduction in the effective bulk modulus of the fluid. A very low bulk modulus of around 5×10^7 N/m^2 (compared to 1.6×10^9 N/m^2 for the oil alone) leads to reasonable agreement between the simulated and experimental responses. The actual value is probably higher as the volume (with the port blanked off) is not precisely known (possibly underestimated). An error in the estimation of the volume would result in a bulk modulus being a few times higher. Such low values for the bulk modulus are physically possible in the presence of air (McCloy and Martin, 1980). The equivalent bulk modulus is expressed by the following equation:

$$\frac{1}{B_{\text{eq}}} = \frac{1}{B_{\text{hose}}} + \frac{V_{\text{air}}/V_{\text{tot}}}{B_{\text{air}}} + \frac{1}{B_{\text{oil}}}, \tag{5.48}$$

where V_{air} and V_{tot} are, respectively, the volume of the free air and the total volume. If $B_{\text{hose}} = 8 \times 10^3$ bar and $B_{\text{oil}} = 16 \times 10^3$ bar, while the ratio V_{air} over V_{tot} is equal to 2×10^{-2}, which yields an equivalent bulk modulus of around 5×10^2 bar.

In Figure 5.55 the impact of an increase in volume, the other critical parameter, is investigated. The volume has been increased by an order of magnitude and this produces a reduction of the bandwidth by about a decade. This is reasonable, recognising that in a linear approximation the hydraulic time constant is directly proportional to the volume.

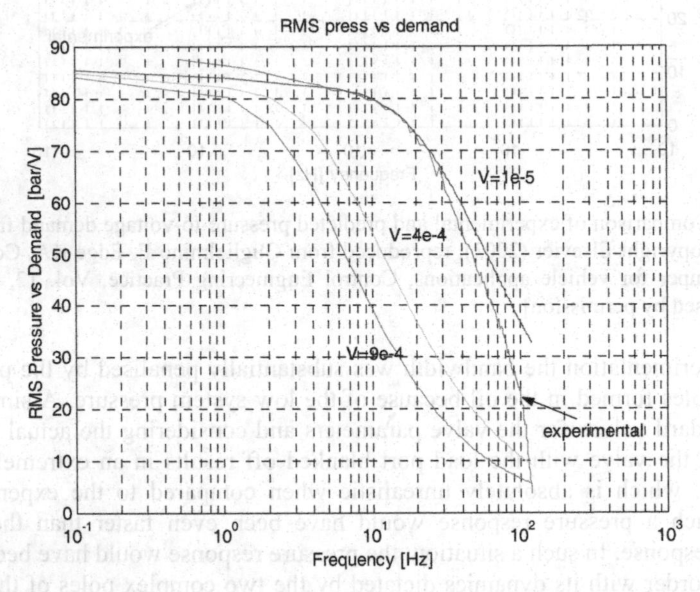

Fig. 5.55. Comparison of experimental and predicted pressure-to-voltage demand frequency response varying volume [copyright Elsevier (2003), reproduced from Guglielmino E, Edge KA, Controlled friction damper for vehicle applications, Control Engineering Practice, Vol. 12, N 4, pp 431–443, used by permission]

5.11 Performance Enhancement of the Friction Damper System

Previous experimental work showed that the theoretically predicted system performance does not entirely reflect actual system behaviour because of both the residual back-pressure, which produces an undesired constant-amplitude damping force, and reduced hydraulic system bandwidth due to presence of free air within the hydraulic oil. This can motivate the research on the improvement of the friction damper performance, outlined in the last part of this chapter.

The first issue can be addressed by spring-preloading the friction damper so as to compensate for the constant force caused by the residual back-pressure. The second issue can be tackled by closing a local pressure control loop around the control valve (at the price of an additional pressure transducer in the loop). In this section bond graphs (Karnopp *et al.*, 1990; Gawthrop and Smith, 1996) are introduced and it is shown how to employ them as a tool for modelling and designing the modifications mentioned above.

A residual back-pressure is likely to be present in the system when the controller requires flow to be returned to tank. This is due to the pressure losses in the pipework. A certain amount of back-pressure is always present and its negative impact is higher at low supply pressures. Back-pressure produces a non-controllable constant friction force superimposed on the time-varying controlled damping action. A possible solution is to insert a preloaded spring into the friction damper actuation system. In the case of the design described above, a tension spring can be conveniently placed between the friction actuators. The spring force would ideally compensate the residual back-pressure, but on the other hand another dynamic element is introduced into the system, which can affect controller performance.

The second problem is the presence of free air in the hydraulic oil, which produces a significant reduction in the value of the bulk modulus. This causes a bandwidth reduction (bandwidth is directly proportional to bulk modulus). It was previously shown that the problem occurred when using a relatively sophisticated valve.

The bulk modulus of a hydraulic oil is generally accepted to be 16,000 bar. However experimental work showed that this value is not attainable in the system under normal working conditions (with a relatively low supply pressure). The improved bandwidth attained by increasing the supply pressure is just acceptable for the application. It was shown that, in order to match simulation with experimental results, the bulk modulus has to be reduced to a value as low as 500 bar.

This experimental investigation showed that the main cause of the very low bandwidth was the presence of air bubbles at low supply pressure. Air release can occur due to the throttling action of the valve and its presence in small quantities can spoil the performance of the most sophisticated valves. If the problem cannot be solved by an appropriate re-design of the hydraulic system, then a solution can be sought at control level, by closing a pressure control loop to counteract this effect.

5.11.1 Damper Design Modification

Figure 5.56 shows a possible damper design modification to insert a spring between the two friction actuators.

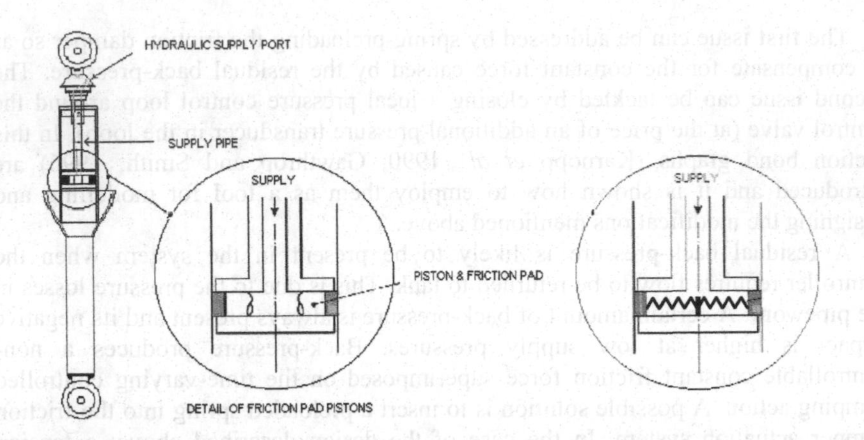

Fig. 5.56. Schematic of the original and modified friction damper [copyright ASME (2004), reproduced from Ngwompo RF, Guglielmino E, Edge KA, Performance enhancement of a friction damper system using bond graphs, Proc ASME ESDA 2004, Manchester, UK, used by permission]

In this optimisation study, a two-degree-of-freedom mass–spring–damper system modelling a quarter car is considered. If the relative displacement is defined as $x = x_1 - x_2$, the quarter car together with the friction damper is described by a system of two second-order non-linear ordinary differential equations as follows analogous to (2.7a) and (2.7b):

$$m_1\ddot{x}_1 = -\mu F_{\mathrm{n}} \operatorname{sgn} \dot{x} - 2\xi\omega_1 m_1 \dot{x} - k_{\mathrm{s}} x, \tag{5.49a}$$

$$m_2\ddot{x}_2 = \mu F_{\mathrm{n}} \operatorname{sgn} \dot{x} + 2\xi\omega_1 m_1 \dot{x} + k_{\mathrm{s}} x - k_{\mathrm{t}}(x_2 - y), \tag{5.49b}$$

where

$$F_{\mathrm{n}} = A_{\mathrm{C}} P_{\mathrm{A}}(x, \dot{x}, \ddot{x}_1) - k_{\mathrm{p}} d_0, \tag{5.50}$$

k_{p} being the preloaded spring stiffness and d_0 the spring length (*i.e.*, distance between friction pads).

The preloaded spring to compensate residual back-pressure is sized such that:

$$k_{\mathrm{p}} d_0 = A_{\mathrm{C}} P_{\mathrm{T}} \tag{5.51}$$

The bond graph model of the system (including the modified design of the friction damper as shown in Figure 5.56) is represented in Figure 5.57.

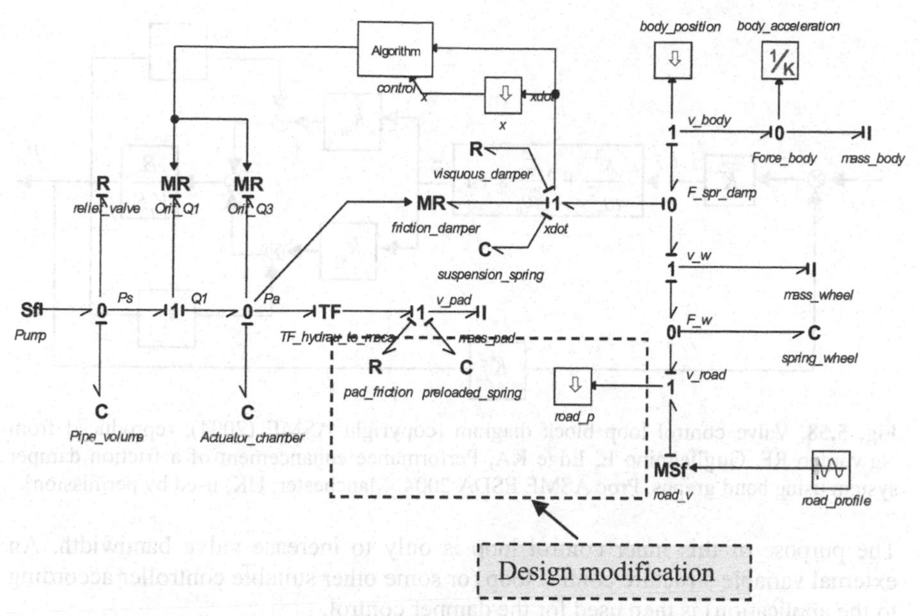

Fig. 5.57. Bond graph model of the modified friction damper [copyright ASME (2004), reproduced from Ngwompo RF, Guglielmino E, Edge KA, Performance enhancement of a friction damper system using bond graphs, Proc ASME ESDA 2004, Manchester, UK, used by permission]

The dynamics of the preloaded spring and friction pad are taken into account and represented in terms of bond graphs as a TF-element to achieve the conversion from the hydraulic domain (pressure in the actuator chamber) to the mechanical domain (force on the friction pad) connected via a 1-junction (common velocity) to I, C and R elements representing, respectively, the mass of the pad, the preloaded spring and the friction between the pad and the housing. The pump is modelled as a source of flow; each volume in the hydraulic circuit is represented by a C-element and the valve orifices as well as the friction damper are modelled as modulated resistive MR-elements, respectively controlled by a feedback signal and the actuator chamber pressure.

5.11.2 Hydraulic Drive Optimisation

The poor valve dynamics can be tackled by closing a local pressure control loop. Working on the assumption that the system can be approximated by a first-order lag, the simplest controller possible that allows to achieve an increment in bandwidth is a proportional controller. The controller as represented by the block diagram in Figure 5.58 is quite simple (a proportional controller).

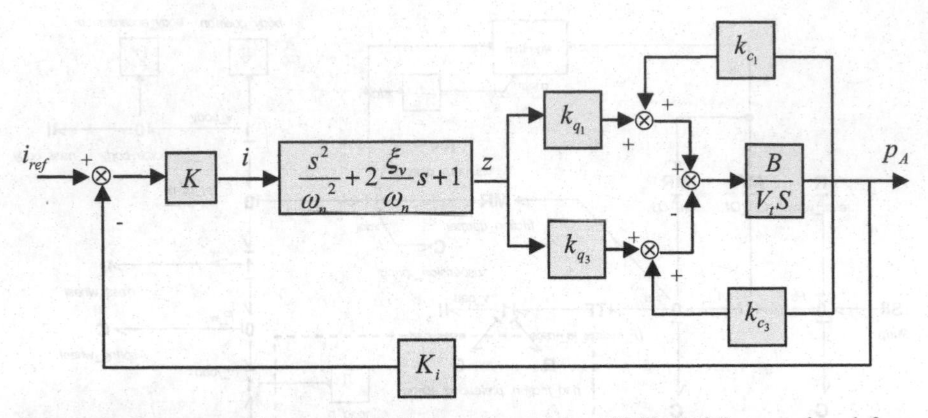

Fig. 5.58. Valve control loop block diagram [copyright ASME (2004), reproduced from Ngwompo RF, Guglielmino E, Edge KA, Performance enhancement of a friction damper system using bond graphs, Proc ASME ESDA 2004, Manchester, UK, used by permission]

The purpose of this inner control loop is only to increase valve bandwidth. An external variable structure control loop (or some other suitable controller according to the application) is then used for the damper control.

The effect of a reduction of bulk modulus on the system bandwidth is shown by the frequency response Bode diagram in Figure 5.59 (obtained with a simplified linear analysis) where a decrease of the bulk modulus from 16,000 bar to 160 bar causes a reduction of the bandwidth from 120 to 30 Hz.

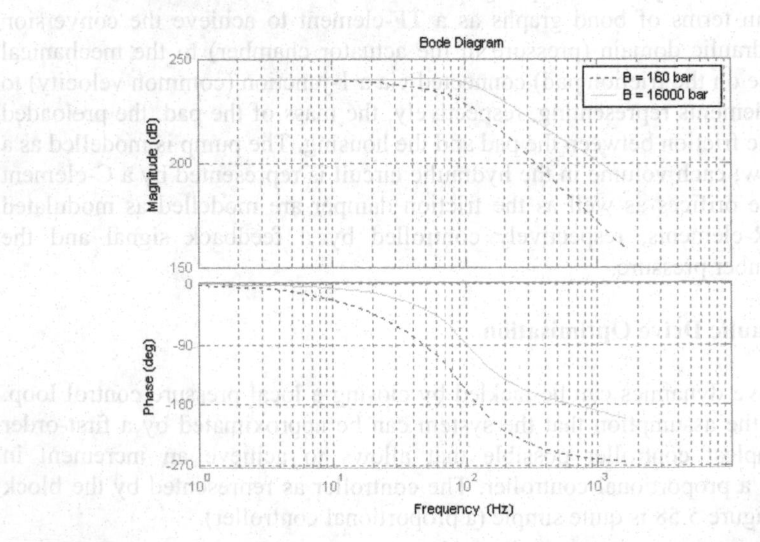

Fig. 5.59. Pressure-to-voltage frequency response of the valve actuation system for different bulk moduli [copyright ASME (2004), reproduced from Ngwompo RF, Guglielmino E, Edge KA, Performance enhancement of a friction damper system using bond graphs, Proc ASME ESDA 2004, Manchester, UK, used by permission]

Figure 5.60 shows the frequency response Bode diagrams with and without the controller for the case where the bulk modulus is reduced to 160 bar.

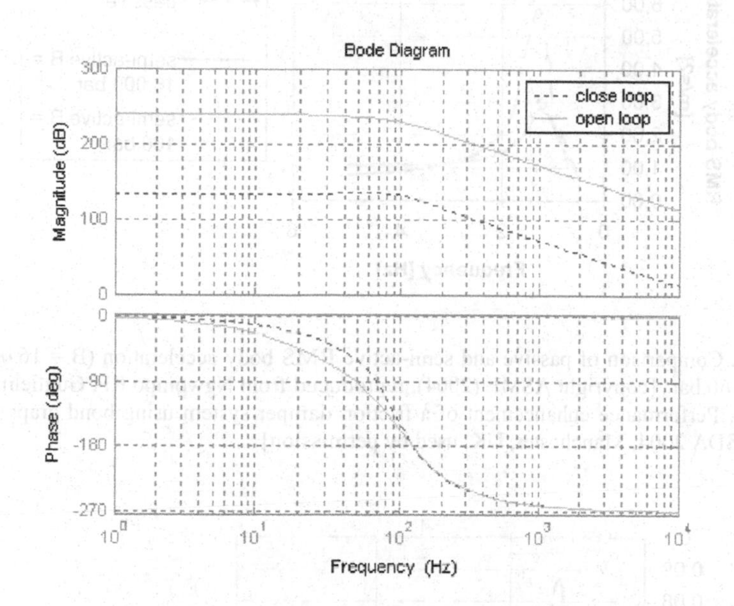

Fig. 5.60. Pressure-to-voltage frequency response of the valve actuation system in open and closed loop for B = 160 bar [copyright ASME (2004), reproduced from Ngwompo RF, Guglielmino E, Edge KA, Performance enhancement of a friction damper system using bond graphs, ASME ESDA 2004, Manchester, UK, used by permission]

As expected, the bandwidth is increased at the price of a reduction in the gain. This is a well-known trade-off of a linear proportional controller. This implies that, in order to achieve a reasonable bandwidth, the closed-loop supply pressure has to be increased with respect to the open-loop supply pressure, with a subsequent increase in the power demand. On the other hand, problems arising from the presence of air and poor bandwidth are more common in systems working at low supply pressure. Therefore, the increase in the supply pressure should not necessarily produce a large increase in the absolute value of power demand. Furthermore this particular system is a pressure control system, not a flow control system, and hence flows can be maintained at low levels.

5.11.3 Friction Damper Controller Enhancement

The following figures show the key simulation results with the enhanced friction damper using balanc logic.

Figures 5.61 and 5.62 show the RMS chassis acceleration and relative displacement versus frequency in the range 0.5–5 Hz. The response with a conventional viscous damper is compared with the semi-active response with ideal and reduced bulk modulus. The reduction in bulk modulus causes a worsening of the performance, as expected.

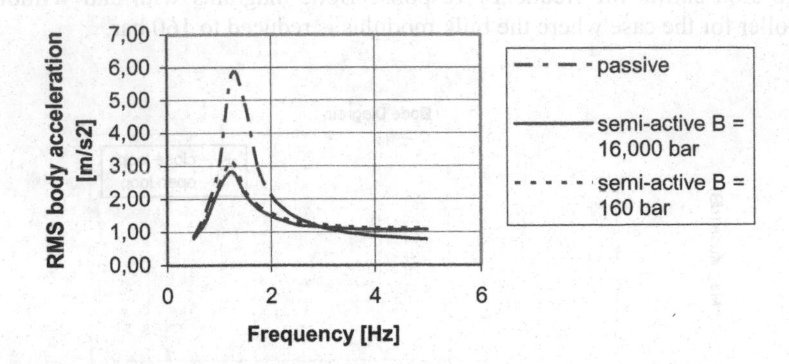

Fig. 5.61. Comparison of passive and semi-active RMS body acceleration (B = 16,000 bar and B = 160 bar) [copyright ASME (2004), reproduced from Ngwompo RF, Guglielmino E, Edge KA, Performance enhancement of a friction damper system using bond graphs, Proc ASME ESDA 2004, Manchester, UK, used by permission]

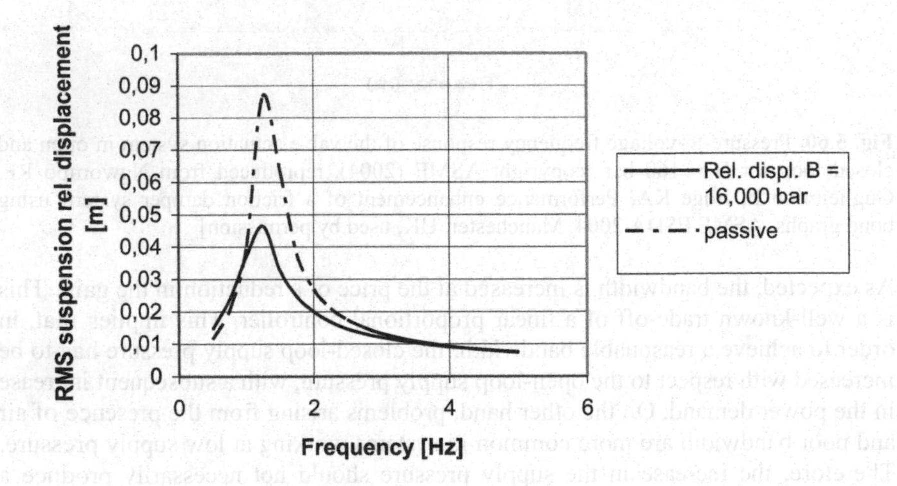

Fig. 5.62. Comparison of passive and semi-active RMS suspension relative displacement [copyright ASME (2004), reproduced from Ngwompo RF, Guglielmino E, Edge KA, Performance enhancement of a friction damper system using bond graphs, Proc ASME ESDA 2004, Manchester, UK, used by permission]

Figure 5.63 shows the impact of an imprecise residual back-pressure on the performance of the system.

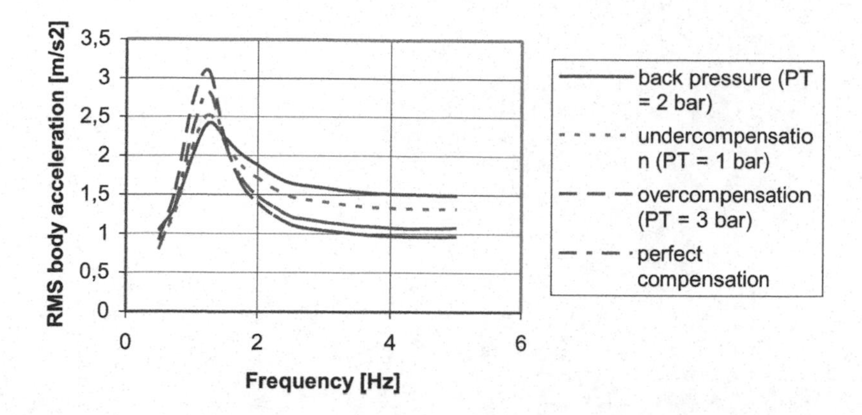

Fig. 5.63. Effect of back-pressure on the system performance [copyright ASME (2004), reproduced from Ngwompo RF, Guglielmino E, Edge KA, Performance enhancement of a friction damper system using bond graphs, Proc ASME ESDA 2004, Manchester, UK, used by permission]

Assuming a residual back-pressure $P_T = 2$ bar, various compensation scenarios are considered: (a) perfect compensation when the spring is sized according to Equation 5.51, (b) under-compensation when the stiffness of the spring is such that $k_p d_0 < A_C P_T$, and (c) the over-compensation case. In case (a) the system behaves as if there is no back-pressure, while in case (b), there is still a residual back-pressure $(k_p d_0 - A_C P_T)$ causing the pad to stick and apply a residual friction damping force. The over-compensation situation corresponds to the case where the friction pad is not in contact with the plate. This leads to an underdamped response, because no controlled frictional damping is present and the only damping present in the system is the residual viscous damping of the vehicle, which is very low.

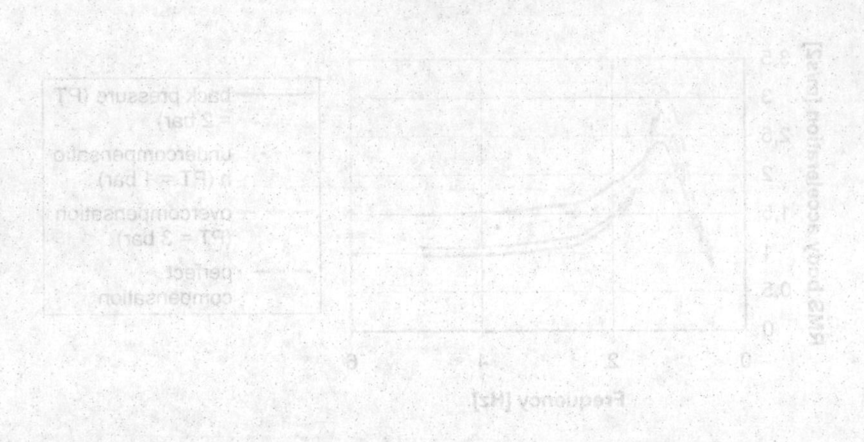

Fig. 5.65. Effect of back-pressure on the system performance [copyright ASME (2004), reproduced from Ngwompo RF, Guglielmino E, Edge KA. Performance enhancement of a friction damper system using bond graphs. Proc ASME ESDA 2004, Manchester, UK, used by permission]

Assuming a residual back-pressure $P_r = 2$ bar, various compensation scenarios are considered: (a) perfect compensation when the spring is sized according to Equation 5.51, (b) under-compensation when the stiffness of the spring is such that $k_s d_0 < A_r P_r$, and (c) the over-compensation case. In case (a) the system behaves as if there is no back-pressure, while in case (b) there is still a residual back-pressure $(k_s d_0 - A_r P_r)$ causing the pad to stick and apply a residual friction damping force. The over-compensation situation corresponds to the case where the friction pad is not in contact with the plate. This leads to an underdamped response, because no controlled frictional damping is present and the only damping present in the system is the residual viscous damping of the vehicle, which is very low.

6

Magnetorheological Dampers

6.1 Introduction

Magnetorheological dampers (MRD) are among the most promising semi-active devices used nowadays in automotive engineering. As mentioned in Chapter 1 a number of vehicles currently employs them. The key feauture of an MRD is the magnetorheological oil, a particular type of oil whose rheological properties can bc altered by applying a magnetic field; by controlling the field (*i.e.*, the current in a solenoid) variable damping can be produced.

The aim of this chapter is to provide insights into MRD modeling, characteristic identification and design. In the first section an overview of the basic properties of the MR fluids and the fluid behaviour under different flow regimes is presented. A design procedure for an MRD is then illustrated. The chapter closes with a presentation of the phenomenological models for MR dampers and their identification.

6.2 Magnetorheological Fluids

Magnetorheological fluids belong to a family of fluids whose properties depend on the strength of an electrical or magnetic field. This family includes also ferrofluids, electrorheological fluids (ER) as well as magnetorheological fluids (MR).

Ferrofluids (Melzner *et al.*, 2001) are colloidal suspensions of magnetic particles smaller than 10 nm (usually made of magnetite) floating in an appropriate carrier liquid such as water, hydrocarbons, esters *etc.* In the presence of an external magnetic field the viscosity of these fluids increases, the increase of the viscosity being proportional to the local intensity of the magnetic field.

Electrorheological and magnetorheological fluids are non-colloidal suspensions of particles having a size of the order of a few microns (5–10 μm) where Brownian motion is not present (in contrast with ferrofluids). The discovery of these fluids dates back to the 1940s (Winslow, 1947 and 1949; Rabinow, 1948).

ER and MR fluid respond to, respectively, an applied electric or magnetic fields with a dramatic change in their rheological behaviour. The main characteristic of these fluids is their ability to change reversibly from free-flowing, linear viscous liquids, to semi-solids with the yield strength swiftly and continuously controllable (milliseconds scale dynamics) when exposed to either an electric or magnetic field. In the absence of an applied field, the ER and MR fluids exhibit Newtonian-like behaviour. When an external electric or magnetic field is applied, this results in a polarisation of the suspended particles which therefore move so as to reduce the stored energy of the ensemble. A minimum-energy configuration consists of particle chains aligned in the direction of the external field. These chain-like structures modify the motion of the fluid, thereby changing its rheological properties through an increase of the yield stress in the direction perpendicular to the applied field. The mechanical energy needed to yield these chain-like structures increases as the applied field increases.

Since yield stress can be varied reversibly by controlling the field strength, both ER and MR fluids are potentially an excellent means to dynamically interface mechanical and electrical systems.

The rheological behaviour of such fluids can be separated into two distinct pre- and post-yield regimes. In the pre-yield regime, both fluids behave like elastic solids as a result of the chain stretching with some occasional ruptures, while in the post-yield there is equilibrium between chain ruptures and chain reformations and the fluids behave like a viscous Newtonian fluid.

Thus the behaviour of MR and ER fluids can be modelled as a Bingham plastic having variable yield strength (Jolly *et al.*, 1998):

$$\tau = \tau_y(\text{field})\,\text{sgn}(\dot{\gamma}) + \eta\dot{\gamma},\ \tau \geq \tau_y, \tag{6.1}$$

where *field* is the magnitude of either the electric or the magnetic field, $\dot{\gamma}$ is the fluid shear rate, η is the plastic viscosity (at zero field) and τ_y is the yield stress.

Below the yield stress (at strains of the order of 10^{-3}) the material behaves viscoelastically:

$$\tau = G\dot{\gamma},\ \tau < \tau_y, \tag{6.2}$$

where G is the complex material modulus. Note that the complex modulus is also field dependent (Weiss *et al.*, 1994).

MR fluids were initially less investigated than ER fluids but in recent years MR fluids have been extensively studied owing to a number of interesting properties, essentially linked to their robustness for real-life engineering applications. Table 6.1 (Carlson *et al.*, 1996) lists some properties of typical ER and MR fluids.

Table 6.1. Typical properties of ER and MR fluids [copyright Lord Corp. (www.lord.com), used by permission]

Property	ER fluid	MR fluid
Response time	Milliseconds	Milliseconds
Plastic viscosity, η (at 25°C)	0.2–0.3 Pa·s	0.2–0.3 Pa·s
Operable temperature range	+10 to +90°C (ionic, DC) –25 to +125°C (non-ionic, AC)	–40 to +150°C
Max. yield stress, τ_y	2–5 kPa (at 3–5 kV/mm)	50–100 kPa (at 150–250 kA/m)
Maximum field	~ 4 kV/mm	~ 250 kA/m
Power supply (typical)	2–5 kV 1–10 mA (2–50 W)	2–25 V 1–2 A (2–50 W)
η / τ_y^2	10^{-7}–10^{-8} s/Pa	10^{-10}–10^{-11} s/Pa
Density	$1 \times 10^3 - 2 \times 10^3$ kg/m^3	$3 \times 10^3 - 4 \times 10^3$ kg/m^3
Max. energy density	10^3 J/m^3	10^5 J/m^3
Stability	Cannot tolerate impurities	Unaffected by most impurities

6.3 MR Fluid Devices

6.3.1 Basic Operating Modes

Virtually all devices that use MR fluids can be classified as operating in three flow regimes:

- Pressure-driven flow mode with either fixed poles or valve mode (Figure 6.1):

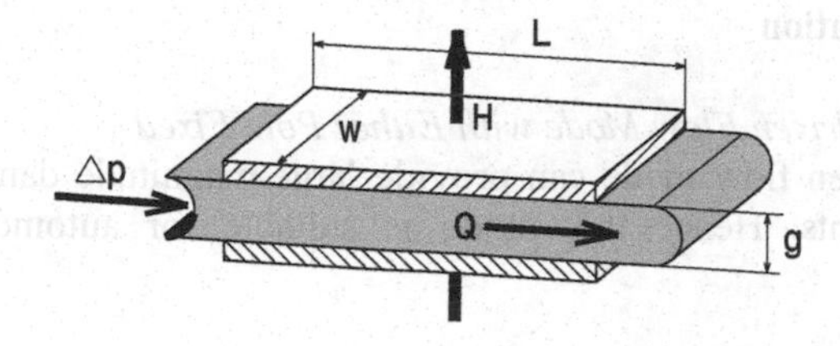

Fig. 6.1. Pressure-driven flow mode [copyright Lord Corp. (www.lord.com), used by permission]

This mode corresponds to the Hagen–Poiseuille flow. Devices that operate in this mode include dampers, shock absorbers, valves *etc*.

- Direct shear mode, with relatively movable poles, translating or rotating perpendicular to the field (Figure 6.2).
  The flow is driven by virtue of attraction forces acting on the fluid between the magnetisable particles. This mode corresponds to the Couette flow. Devices operating in this mode include clutches, brakes, locking devices, dampers for small displacements and medium-frequency applications.

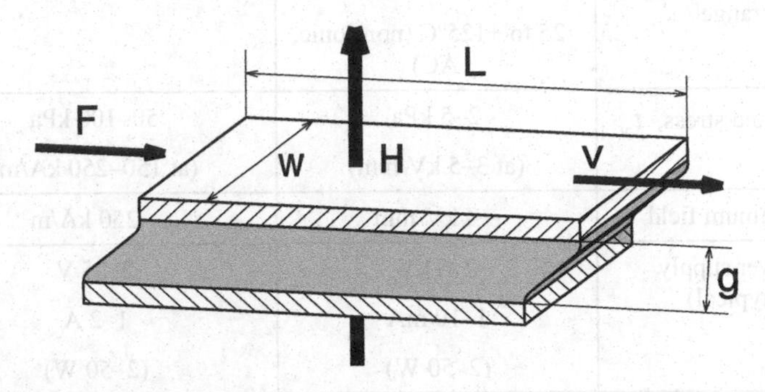

Fig. 6.2. Direct shear mode [copyright Lord Corp. (www.lord.com), used by permission]

- Squeeze-film mode with relatively movable poles top-down, in the field direction.

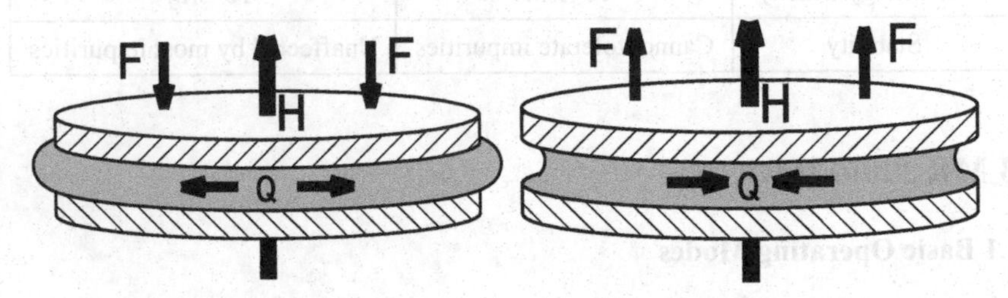

Fig. 6.3. Squeeze-film mode [copyright Lord Corp. (www.lord.com), used by permission]

Devices operating in this mode include dampers used in high-force and low-motion applications.

6.3.2 Flow Simulation

6.3.2.1 Pressure-driven Flow Mode with Either Pole Fixed

The pressure-driven flow mode can provide large-magnitude damping forces and large displacements. Hence this mode is suitable for automotive suspension applications.

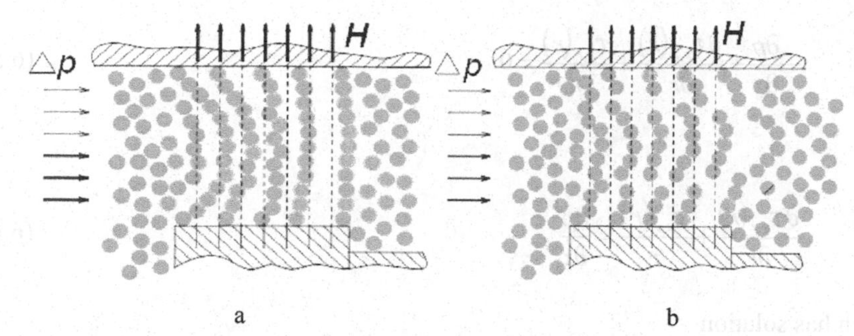

Fig. 6.4. Mode of flow through an MR fluid damper working in pressure-driven flow mode; (a) pre-yield flow; (b) post-yield flow [copyright Lord Corp. (www.lord.com), used by permission]

In Figure 6.4 the mode of flow through an MR damper working in pressure-driven flow mode is shown.

To capture the damper force–velocity behaviour for an MR damper working in pressure-driven flow mode an axisymmetric model for the flow through an annular duct can be developed (Constantinescu, 1995). This model is based on the Navier–Stokes equations for the Hagen–Poiseuille flow.

For incompressible viscous fluids the governing equations are:

$$\begin{cases} \nabla \cdot \boldsymbol{V} = 0 \quad \text{continuity equation}, \\ \rho \dfrac{\mathrm{D}\boldsymbol{V}}{\mathrm{D}t} = \rho \boldsymbol{f} - \nabla p + \nabla \boldsymbol{\tau}_{ij} \quad \text{Navier-Stokes equation}, \end{cases} \tag{6.3}$$

where $\boldsymbol{V}$ is the velocity vector, ρ MR fluid density, t is the time, $\boldsymbol{f}\boldsymbol{f}$ is the external force vector, p the pressure and $\boldsymbol{\tau}_{ij}$ is the viscous stress tensor. For axisymmetric flow $u_x = u(r)$, $u_r = 0$, $u_\omega = 0$, hence Equation 6.3 becomes:

$$\begin{cases} \dfrac{\partial u_x}{\partial x} = 0 \quad \text{continuity equation}, \\ \rho \dfrac{\partial u_x}{\partial t} = \rho f_x - \dfrac{\partial p}{\partial x} + \dfrac{\partial}{\partial x}\left(2\eta \dfrac{\partial u_x}{\partial x} + \lambda \nabla \cdot \boldsymbol{V} \right) + \dfrac{1}{r}\dfrac{\partial}{\partial r}\left(r\tau_{x,r} \right) \quad \text{Navier-Stokes eq.}, \end{cases} \tag{6.4}$$

where ρ and η are constant, $f_x = 0$, r is the radial co-ordinate, x is the longitudinal co-ordinate and $\dfrac{\partial p}{\partial x}$ the pressure gradient.

To analyse the quasi-static motion of the flow inside the damper, the fluid inertia can be neglected $\left(\dfrac{\partial u_x}{\partial t} = 0 \right)$, and hence Equation 6.4 can be reduced to

$$-\frac{\partial p}{\partial x}+\frac{\partial \tau_{x,r}(r)}{\partial r}+\frac{\tau_{x,r}(r)}{r}=0 \tag{6.5}$$

i.e.,

$$\frac{\partial \tau_{x,r}(r)}{\partial r}+\frac{\tau_{x,r}(r)}{r}=\frac{\partial p}{\partial x}, \tag{6.6}$$

which has solution

$$\tau_{x,r}=\frac{1}{2}\frac{\mathrm{d}p(x)}{\mathrm{d}x}r+\frac{C_1}{r}, \tag{6.7}$$

where C_1 is a constant which can be evaluated by the boundary conditions.

To describe the MR fluid field-dependent characteristics and shear effects the Bingham viscoplastic model is employed whose governing equation is:

$$\tau_{x,r}(r)=\tau_y(H,r)\operatorname{sgn}\left(\frac{\partial u_x(r)}{\partial r}\right)+\eta\frac{\partial u_x(r)}{\partial r}. \tag{6.8}$$

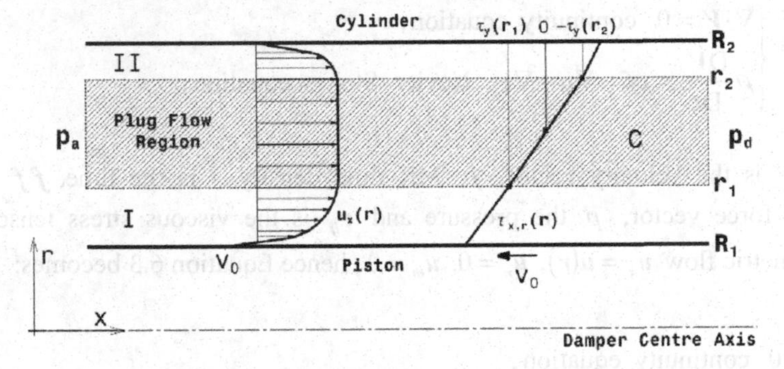

Fig. 6.5. Typical velocity profile along with a typical shear stress diagram for viscoplastic fluid flow through an annular gap for pressure-driven flow mode [copyright John Wiley & Sons Limited (1998), reproduced with minor modifications from Spencer BF, Yang G, Carlson JD and Sain MK, Smart dampers for seismic protection of structures: a full-scale study, in Proceedings of the Second World Conference on Structural Control edited by Kobori T, Inoue Y, Seto K, Iemura H and Nichitani A, reproduced with permission]

A typical velocity profile along with a typical shear stress diagram for a viscoplastic fluid (MR fluid) flow through an annular gap is depicted in Figure 6.5 (Constantinescu, 1995; Yang *et al.*, 2002).

In regions I and II, the shear stress exceeds the yield stress and is given by Equation 6.8. The velocity profile results from (6.7) and (6.8).

- region I with boundary condition at $r=R_1$, $u_x(r)=v_0$

$$u_x(r) = \frac{1}{4\eta}\frac{\mathrm{d}p}{\mathrm{d}x}\left(r^2 - R_1^2\right) + \frac{C_1}{\eta}\ln\frac{r}{R_1} - \frac{1}{\eta}\int_{R_1}^{r}\tau_y(r)\mathrm{d}r - v_0\,, \tag{6.9}$$

for $R_1 \le r \le r_1$

- region II with the boundary condition at $r = R_2$ $u_x(r) = 0$

$$u_x(r) = -\frac{1}{4\eta}\frac{\mathrm{d}p}{\mathrm{d}x}\left(R_2^2 - r^2\right) - \frac{C_1}{\eta}\ln\frac{R_2}{r} - \frac{1}{\eta}\int_{r}^{R_2}\tau_y(r)\mathrm{d}r\,, \tag{6.10}$$

for $r_2 \le r \le R_2$.

In region C (the plug flow region) the shear stress is lower than the yield stress so that no shear flow occurs. In this region the yield stress is assumed to be a monotonic function of the radius *r*. In this region $u_x(r) = \text{const.}$, $u_x(r_1) = u_x(r_2)$, $\tau_{x,r}(r_1) = \tau_y(r_1)$ and $\tau_{x,r}(r_2) = -\tau_y(r_2)$.

Since the yield stress τ_y in axisymmetric flow is related to *r*, due to the radial distribution of the magnetic field in the gap and, in general, the dimension of the gap $R_2 - R_1$ is small compared with the magnetic pole radius R_1, the variation of the yield stress with *r* in the gap can be neglected. Hence:

$$\tau_y(r_1) = \tau_y(r_2) = \tau_y(H). \tag{6.11}$$

From (6.7) and (6.11):

$$C_1 = \frac{r_1 r_2}{r_2 - r_1}\tau_y(H) \tag{6.12}$$

and, from Equation 6.6:

$$\frac{\mathrm{d}p}{\mathrm{d}x}(r_2 - r_1) = 2\tau_y(H). \tag{6.13}$$

Therefore:

$$(r_2 - r_1) = \frac{2\tau_y(H)}{\dfrac{\mathrm{d}p}{\mathrm{d}x}} \tag{6.14}$$

In (6.13) and (6.14) $\tau_y = \tau_y(H)$ is the yield stress, *H* is the value of the applied magnetic field and $(r_2 - r_1)$ is the plug thickness.

The plug thickness varies with the fluid yield stress τ_y (and with the intensity of the magnetic field). Flow can only be established when $(r_2 - r_1) < (R_2 - R_1)$ *i.e.*, the plug flow needs to be within the gap.

With the conditions (6.11) and (6.12), the equations (6.9) and (6.10) yield

$$\begin{aligned} u_x(r_1) &= \frac{1}{4\eta}\frac{\mathrm{d}p}{\mathrm{d}x}\left(r_1^2 - R_1^2\right) + \frac{C_1}{\eta}\ln\frac{r_1}{R_1} - \frac{1}{\eta}\tau_y\left(r_1 - R_1\right) - v_0\,, \\ u_x(r_2) &= -\frac{1}{4\eta}\frac{\mathrm{d}p}{\mathrm{d}x}\left(R_2^2 - r_2^2\right) - \frac{C_1}{\eta}\ln\frac{R_2}{r_2} - \frac{1}{\eta}\tau_y\left(R_2 - r_2\right). \end{aligned} \tag{6.15}$$

In the region C, from (6.15) with the condition (6.11), $u_x(r) = \text{const.}$, $u_x(r_1) = u_x(r_2)$ yields

$$\begin{aligned} &\frac{\mathrm{d}p}{\mathrm{d}x}\left[hR_{\mathrm{m}} - \frac{1}{2}\left(r_2 - r_1\right)\left(r_2 + r_1\right)\right] + 2\tau_y\frac{r_1 r_2}{r_2 - r_1}\ln\frac{r_1 R_2}{R_1 r_2} + \\ &\quad + 2\tau_y\left(2R_{\mathrm{m}} - r_1 - r_2\right) - 2\eta v_0 = 0\,, \end{aligned} \tag{6.16}$$

where

$$\begin{aligned} &R_2 - R_1 = h \;\text{(gap thickness)} \\ &R_2 + R_1 = 2R_{\mathrm{m}} \;\text{(gap average radius)} \\ &r_1 - R_1 = h_1 \;\text{(region I thickness)} \\ &r_2 - R_1 = h_2 \;\text{(region I + region C thickness)} \\ &r_2 - r_1 = h_2 - h_1 = h_{\mathrm{p}} \;\text{(plug flow thickness)} \end{aligned} \tag{6.17}$$

The volumetric flow rate Q is given by

$$Q = 2\pi\int_{R_1}^{R_2} r u_x(r)\mathrm{d}r\,. \tag{6.18}$$

Hence

$$Q = Q^{\mathrm{I}} + Q^{\mathrm{II}} + Q^{\mathrm{C}} = 2\pi\left[\int_{R_1}^{r_1} r u_x^{\mathrm{I}}(r)\mathrm{d}r + \int_{r_2}^{R_2} r u_x^{\mathrm{II}}(r)\mathrm{d}r + \int_{r_1}^{r_2} r u_x^{\mathrm{C}}(r)\mathrm{d}r\right]. \tag{6.19}$$

Substituting (6.9) and (6.10) into (6.19) and considering the boundary conditions as well as the condition (6.11) yields

$$Q = \pi R_1^2 v_0 - \pi\left[\int_{R_1}^{r_1} r^2 \frac{du_x^{\mathrm{I}}(r)}{dr} dr + \int_{r_2}^{R_2} r^2 \frac{du_x^{\mathrm{II}}}{dr} dr\right] \tag{6.20}$$

Hence

$$\begin{aligned} Q = \pi R_1^2 v_0 - \frac{\pi}{8\eta}\left\{\frac{\mathrm{d}p}{\mathrm{d}x}\left[2hR_{\mathrm{m}}\left(h^2 + 2R_1R_2\right) - \left(r_2^4 - r_1^4\right)\right] + \right. \\ \left. + \frac{4r_1r_2}{r_2 - r_1}\left[2gR_{\mathrm{m}} - \left(r_2^2 - r_1^2\right)\right] + \frac{8\tau_y}{3}\left[2R_{\mathrm{m}}\left(h^2 + R_1R_2\right) - \left(r_2^3 + r_1^3\right)\right]\right\}. \end{aligned} \tag{6.21}$$

The term $\frac{\mathrm{d}p}{\mathrm{d}x}$ and the plug thickness can be obtained numerically by using Equations (6.13) and (6.14) or (6.16) and (6.21).

If a new co-ordinate system $O_1x_1y_1$ is introduced with the origin O_1 in the surface of the piston (fixed point), where O_1x_1 is the axis oriented in direction of u_x, O_1y_1 the axis oriented in direction of the Or axis, then:

$$\frac{\partial \tau_{x,y}}{\partial y} = \frac{\partial p}{\partial x}. \tag{6.22}$$

From Equation 6.22:

$$\tau_{x,y} = \frac{\mathrm{d}p}{\mathrm{d}y} y + D_1. \tag{6.23}$$

For region C, $u_x^{\mathrm{C}} = \text{const}$. Hence

$$\left.\frac{\mathrm{d}u_x^{\mathrm{I}}}{\mathrm{d}y}\right|_{y=h_1} = \left.\frac{\mathrm{d}u_x^{\mathrm{II}}}{\mathrm{d}y}\right|_{y=h_2} = \left.\frac{\mathrm{d}u_x^{\mathrm{C}}}{\mathrm{d}y}\right|_{y=h_1}^{y=h_2} = 0. \tag{6.24}$$

Therefore

$$h_2 - h_1 = -\frac{2\tau_y}{\frac{\mathrm{d}p}{\mathrm{d}x}}. \tag{6.25}$$

Equation 6.8 becomes

$$\tau_{x,y} = \tau_y(y,H)\operatorname{sgn}\left(\frac{\mathrm{d}u_x}{\mathrm{d}y}\right) + \eta\frac{\mathrm{d}u_x}{\mathrm{d}y}. \tag{6.26}$$

Hence from (6.23) and (6.26)

$$\frac{\mathrm{d}u_x}{\mathrm{d}y} = \frac{1}{\eta}\frac{\mathrm{d}p}{\mathrm{d}x}y + \frac{1}{\eta}\left[D_1 \pm \tau_y \operatorname{sgn}\left(\frac{\mathrm{d}u_x}{\mathrm{d}y}\right)\right]. \tag{6.27}$$

The resulting velocity profiles in the three regions are:

- region I with the boundary condition at $y = 0$, $u_x(y) = v_0$ and at $y = h_1$, $u_x(y) = \text{const.} = u_x^{\mathrm{C}}$

$$u_x^{\mathrm{I}}(y) = \frac{1}{2\eta}\frac{\mathrm{d}p}{\mathrm{d}x}y(y - h_1) - \frac{v_0 - u_x^{\mathrm{C}}}{h_1}y + v_0 \tag{6.28}$$

for $0 \le y \le h_1$

- region C:

$$u_x^{\mathrm{C}} = u_x^{\mathrm{I}}(h_1) = u_x^{\mathrm{II}}(h_2) = -\frac{1}{2\eta}\frac{\mathrm{d}p}{\mathrm{d}x}(h - h_2)^2 = \text{const.} \tag{6.29}$$

- region II with the boundary condition at $y = h_2$, $u_x(y) = u_x^{\mathrm{C}}$ and at $y = h$, $u_x(h) = 0$

$$u_x^{\mathrm{II}}(y) = -\frac{1}{2\eta}\frac{\mathrm{d}p}{\mathrm{d}x}(h - y)(y - h_2) + u_x^{\mathrm{C}}\frac{h - y}{h - h_2} \tag{6.30}$$

for $h_2 \le y \le h$

In (6.28), (6.29) and (6.30) the yield stress in the gap is taken to be constant. In region C, $\frac{\mathrm{d}u_x}{\mathrm{d}y} = 0$ ($u_x^{\mathrm{C}} = \text{const.}$) and from (6.28) and (6.30)

$$\frac{\mathrm{d}p}{\mathrm{d}x} = 2\eta\frac{v_0 - u_x^{\mathrm{C}}}{h_1^2} \tag{6.31}$$

and

$$\frac{\mathrm{d}p}{\mathrm{d}x} = -2\eta \frac{u_x^{\mathrm{C}}}{(h-h_2)^2}, \tag{6.32}$$

i.e.,

$$\frac{\mathrm{d}p}{\mathrm{d}x} = -\frac{2\eta v_0}{(h-h_2)^2 - h_1^2} \tag{6.33}$$

or

$$h_1 = \frac{h}{2} + \frac{\tau_y}{\frac{\mathrm{d}p}{\mathrm{d}x}} + \frac{\eta v_0}{h\frac{\mathrm{d}p}{\mathrm{d}x} + 2\tau_y} \tag{6.34}$$

and

$$h_2 = \frac{h}{2} - \frac{\tau_y}{\frac{\mathrm{d}p}{\mathrm{d}x}} + \frac{\eta v_0}{h\frac{\mathrm{d}p}{\mathrm{d}x} + 2\tau_y}. \tag{6.35}$$

The fluid flux in the x-axis direction is:

$$q = \int_0^h u_x(y)\mathrm{d}y = \int_0^{h_1} u_x^{\mathrm{I}}(y)\mathrm{d}y + \int_{h_1}^{h_2} u_x^{\mathrm{C}}(y)\mathrm{d}y + \int_{h_2}^{h} u_x^{\mathrm{II}}(y)\mathrm{d}y, \tag{6.36}$$

i.e.,

$$q = -\frac{1}{12\eta}\frac{\mathrm{d}p}{\mathrm{d}x}\left[h_1^3 + (h-h_2)^3\right] + \frac{v_0}{2}h_1 + \frac{u_x^{\mathrm{C}}}{2}(h + h_2 - h_1). \tag{6.37}$$

The volumetric flow rate Q is

$$Q = A_{\mathrm{p}} v_{\mathrm{p}} = 2\pi R_{\mathrm{m}} q, \tag{6.38}$$

where A_{p} is the cross-sectional area of the piston head and $v_{\mathrm{p}} = v_0$ is the piston head velocity.

By considering Equation 6.38 and replacing u_x^{C}, h_1 and h_2 from (6.25), (6.29), (6.34) and (6.35) an expression can be obtained for $\frac{\mathrm{d}p}{\mathrm{d}x}$.

Introducing dimensionless variables (Yang *et al.*, 2002):

$$
\begin{aligned}
N_1 &= -\frac{\pi R_{\mathrm{m}} h v_0}{Q} = -\frac{\pi R_{\mathrm{m}} h}{A_{\mathrm{p}}}, \\
N_2 &= -\frac{\pi R_{\mathrm{m}} h^3}{6\eta Q}\frac{\mathrm{d}p}{\mathrm{d}x} = -\frac{\pi R_{\mathrm{m}} h^3}{6\eta A_{\mathrm{p}} v_{\mathrm{p}}}\frac{\mathrm{d}p}{\mathrm{d}x}, \\
N_3 &= \frac{\pi R_{\mathrm{m}} h^2 \tau_y}{6\eta Q} = \frac{\pi R_{\mathrm{m}} h^2 \tau_y}{6\eta A_{\mathrm{p}} v_{\mathrm{p}}},
\end{aligned}
\tag{6.39}
$$

The following equation can be obtained:

$$
3(N_2 - 2N_3)^2\left[N_2^3 - (1 + 3N_3 - N_1)N_2 + 4N_3\right] + N_1^2 N_2^2 N_3 = 0\,; \tag{6.40}
$$

when $|N_1| > 3(N_2 - 2N_3)^2 / N_2$, the pressure gradient is independent of the dimensionless yield stress N_3. Therefore a controllable yield stress cannot affect the resisting force of the MR damper.

Equation 6.40 cannot be solved analytically. However for $0 < N_3 < 1000$ and $-0.5 < N_1 < 0$ (*i.e.*, flow in the opposite direction to the piston velocity) an approximate solution can be found for the pressure gradient:

$$
N_2(N_1, N_3) = 1 + 2.07N_3 - N_1 + \frac{N_3}{1 + 0.4N_3}. \tag{6.41}
$$

This solution encompasses most practical design solutions with a maximum error of 4%.

The pressure drop in a device working in pressure-driven flow mode is:

$$
\Delta p = \frac{\mathrm{d}p}{\mathrm{d}x} L, \tag{6.42}
$$

where L is the length of magnetic pole.

From (6.41) the pressure drop is

$$
\Delta p = \Delta p_\eta + \Delta p_\tau = \frac{6 v_{\mathrm{p}}\left(A_{\mathrm{p}} + \pi R_{\mathrm{m}} h\right)}{\pi R_{\mathrm{m}} h^3}\eta L + \frac{c}{h}\tau_y(H)L \tag{6.43}
$$

i.e., the sum of a viscous component Δp_η and a field-dependent induced yield stress component Δp_τ.

In Equation 6.43 the parameter c is a function of the flow velocity profile. This parameter can be expressed as (Jolly *et al.*, 1998)

$$c = 2.07 + \frac{1}{1+0.4N_3} = \left(2.07 + \frac{12\eta A_{\mathrm{p}} v_{\mathrm{p}}}{12\eta A_{\mathrm{p}} v_{\mathrm{p}} + 0.4 A_{\mathrm{g}} h \tau_y} \right). \tag{6.44}$$

Its value ranges from a minimum value of 2 (for $\Delta p_\tau / \Delta p_\eta < 1$) to a maximum value of 3 (for $\Delta p_\tau / \Delta p_\eta > 100$).

Observing that $2\pi R_{\mathrm{m}} h = A_{\mathrm{g}}$ is the gap area, from (6.43) it follows that

$$\Delta p_\eta = \frac{12\eta A_{R_{\mathrm{m}}} v_{\mathrm{p}}}{A_{\mathrm{g}} h^2} L \tag{6.45}$$

where $A_{R_{\mathrm{m}}} = \pi R_{\mathrm{m}}^2 = A_{\mathrm{p}} + \frac{A_{\mathrm{g}}}{2}$.

Consequently, the force developed by a damper working in pressure-driven flow mode is:

$$F = \Delta p A_{\mathrm{p}} = \Delta p_\eta A_{\mathrm{p}} + \Delta p_\tau A_{\mathrm{p}} = F_\eta + F_\tau . \tag{6.46}$$

Hence

$$F_\eta = \frac{12\eta A_{R_{\mathrm{m}}} v_{\mathrm{p}}}{A_{\mathrm{g}} h^2} A_{\mathrm{p}} L \text{ and } F_\tau = \frac{c\tau_y(H) A_{\mathrm{p}}}{h} L \tag{6.47}$$

Defining the control ratio λ (for pressure-driven flow mode) as $\lambda = \frac{\Delta p_\tau}{\Delta p_\eta}$ and, from (6.43) and (6.45), with $W_{\mathrm{m}} = Q\Delta p_\tau$ and $k = \frac{12 A_{R_{\mathrm{m}}}}{c^2 A_{\mathrm{p}}} = \text{const.}$ the equation defining the minimum active fluid volume can be obtained as:

$$V = k\left(\frac{\eta}{\tau_y^2}\right)\lambda W_{\mathrm{m}} . \tag{6.48}$$

where $V = LA_{\mathrm{g}}$ is the active fluid volume necessary to achieve the desired control ratio λ at a required controllable mechanical power level W_{m}.

6.3.2.2 Direct Shear Mode with Relatively Movable Poles

The direct shear mode can provide small displacements and is suitable for medium-frequency applications. In Figure 6.6 the mode of flow through an MR damper working in direct shear mode is depicted.

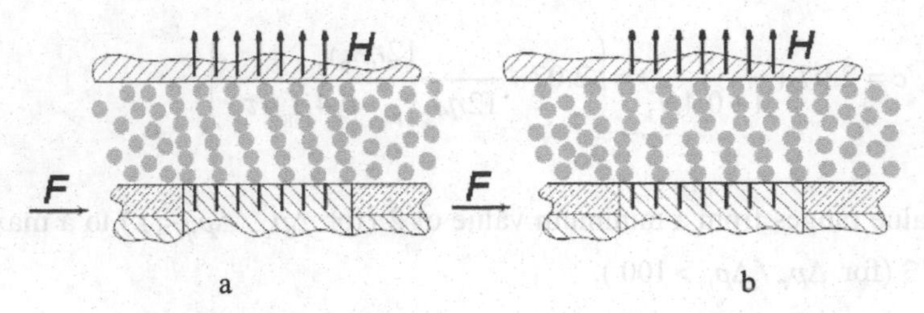

Fig. 6.6. Mode of flow through an MR damper working in direct shear mode; (a) pre-yield flow; (b) post-yield flow [copyright Lord Corp. (www.lord.com), used by permission]

A typical velocity profile along with a typical shear stress diagram for a viscoplastic fluid (MR fluid) flow through the annular gap is shown in Figure 6.7 (Constantinescu, 1995):

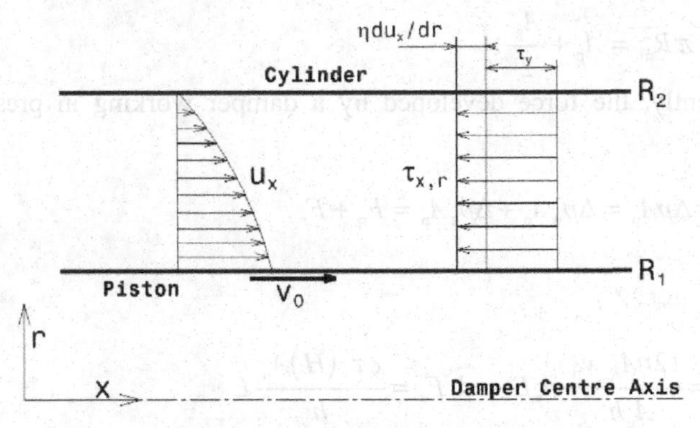

Fig. 6.7. Typical velocity profile along with a typical shear stress diagram for a viscoplastic fluid (MR fluid) flow through the annular gap in direct shear mode

In the direct shear mode $\frac{\mathrm{d}p}{\mathrm{d}x}=0$, if no external forces ($f_x = 0$) are present and in quasi-static flow motion (hence fluid inertia can be neglected, *i.e.*, $\frac{\mathrm{d}u_x}{\mathrm{d}t}=0$), Equation 6.4 becomes:

$$\begin{cases} \frac{\partial u_x}{\partial x}=0, \\ \frac{\partial}{\partial x}\left(2\eta\frac{\partial u_x}{\partial x}+\lambda\nabla\cdot V\right)+\frac{1}{r}\frac{\partial}{\partial r}\left(r\tau_{x,r}\right)=0. \end{cases} \tag{6.49}$$

Hence

$$\frac{\partial}{\partial r}\left(r\tau_{x,r}\right)=0 \text{ or } \frac{\partial\tau_{x,r}(r)}{\partial r}+\frac{\tau_{x,r}(r)}{r}=0. \tag{6.50}$$

Substituting (6.8) into (6.49) yields

$$-\frac{\partial \tau_y}{\partial r}+\frac{\partial}{\partial r}\left(\eta \frac{\mathrm{d} u_x}{\mathrm{d} r}\right)-\frac{\tau_y}{r}+\frac{\eta}{r}\frac{\mathrm{d} u_x}{\mathrm{d} r}=0 . \tag{6.51}$$

Since $\eta = \text{const.}$ and $\frac{\partial \tau_y}{\partial r}=0$

$$\frac{\mathrm{d}^2 u_x}{\mathrm{d} r^2}+\frac{1}{r}\frac{\mathrm{d} u_x}{\mathrm{d} r}-\frac{\tau_y}{\eta r}=0 \tag{6.52}$$

with boundary conditions $r=R_1$, $u_x(R_1)=v_0$ and $r=R_2$, $u_x(R_2)=0$, leading to

$$u_x=\frac{\tau_y}{\eta}\left[h\frac{\ln R_2-\ln r}{\ln R_2-\ln R_1}-R_2+r\right]+\frac{\ln R_2-\ln r}{\ln R_2-\ln R_1}v_0 \tag{6.53}$$

and shear stress

$$\tau_{x,r}=\frac{\tau_y h}{r(\ln R_2-\ln R_1)}-\frac{\eta v_0}{r(\ln R_2-\ln R_1)} \tag{6.54}$$

Considering the small gap-damper piston diameter ratio, the axisymmetric flow can be approximated as a flow through parallel plates, as shown in Figure 6.2. Hence it follows that

$$\frac{\partial \tau_{x,y}}{\partial y}=0 \text{ or } \frac{\mathrm{d}^2 u_x}{\mathrm{d} y^2}=0 \tag{6.55}$$

Equation 6.55, with boundary conditions $y=0$, $u(0)=v_0$, $y=h$ and $u_x(h)=0$, yields

$$u_x(y)=\left(1-\frac{y}{h}\right)v_0 \text{ and } \tau_{x,y}=-\tau_y-\eta\frac{v_0}{h} . \tag{6.56}$$

The force developed by a damper working in direct shear flow mode is

$$F=F_\eta+F_\tau(H)=\frac{2\pi R_m \eta v_r L}{h}+2\pi R_m L \tau_y(H), \tag{6.57}$$

where L is the length of the magnetic pole and v_r is relative velocity of magnetic pole (if the velocity $u_x(h)=0$ then $v_r = v_0$).

The control ratio λ is defined, for direct shear mode, as $\lambda = \dfrac{\Delta F_\tau}{\Delta F_\eta}$ and, from Equation 6.57, with $W_\mathrm{m} = F_\tau v_r$ and $k=1$ the equation defining the minimum active fluid volume is:

$$V = k\left(\frac{\eta}{\tau_y^2}\right)\lambda W_\mathrm{m}, \tag{6.58}$$

where $V = LA_\mathrm{g}$ is the active fluid volume necessary to achieve the desired control ratio λ at a required controllable mechanical power level W_m and the term $\dfrac{\eta}{\tau_y^2}$ (present in (6.48) and (6.58)) depends on the type of fluid.

6.3.3.3 Squeeze-film Mode

This operational mode is not analysed in this book. It should be noted that in applications where large forces are required (*e.g.*, civil engineering applications) the expected damping forces and displacement are considerably large, MR dampers operating in squeeze mode might be impractical.

6.4 MR Damper Design

After having overviewed the basics of MR fluid dynamics, the focus is now on the design of an automotive MR damper. The design of a prototype MR damper will now be described (Figure 6.8). It is essentially composed of a hydraulic cylinder through which an electromagnetic piston pumps MR fluid.

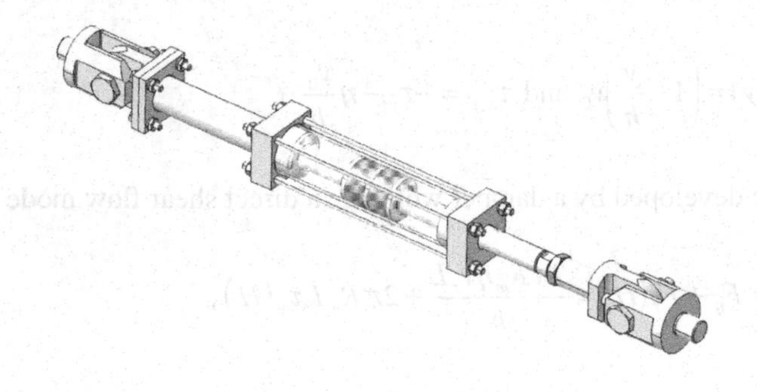

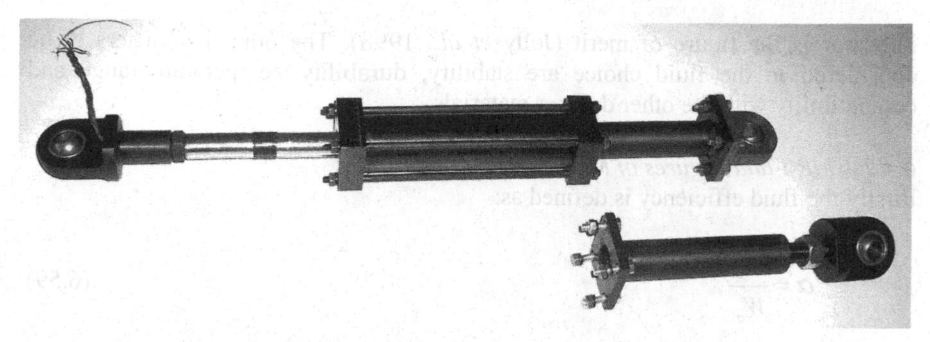

Fig. 6.8. MR damper prototype for experimental studies

The MR damper is a prototype for experimental studies. A design with two piston rods is chosen; the piston is wrapped in copper wire and forms two electromagnetic coils. When the current flows in the coils, a magnetic flux in the piston cylinder is generated and the yield stress value of the MR fluid increases.

The design procedure of the MR damper involves the following steps:

- determination of input data and choice the design solution
- choice of the working MR fluid
- determination of the optimal gap size and hydraulic design
- magnetic circuit design.

6.4.1 Input Data and Choice of the Design Solution

The input data are defined by a specification and the design solutions are chosen from operating and limit conditions. In the case described the basic design data are the following:

- controllable mechanical power level: $W_{\mathrm{m}} = 80\ \mathrm{W}$
- maximum damper controllable force: $F_{\tau} = 2700\ \mathrm{N}$
- maximum displacement of MR damper: $c^{\max} = \pm 0.05\ \mathrm{m}$
- maximum working frequency: $f^{\max} = 5\ \mathrm{Hz}$

The design conditions are

- internal diameter of damper cylinder: $d_{\mathrm{cyl}}^{\mathrm{in}} = 0.04\ \mathrm{m}$
- piston rod diameter: $d_{\mathrm{rod}} = 0.02\ \mathrm{m}$
- working temperature: $T = -20 - 150^{\circ}\mathrm{C}$
- compatibility with synthetic rubbers and elastomer

6.4.2 Selection of the Working MR Fluid

In order to choose the most appropriate MR fluid, a criterion to compare the nominal behaviour of different MR fluids must first be established. Such an indicator is the figure of merit (Jolly *et al.*, 1998). The other parameters to be considered in the fluid choice are stability, durability, temperature range and compatibility with the other damper materials.

6.4.2.1 MR Fluid Figures of Merit

Firstly the fluid efficiency is defined as:

$$\alpha = \frac{\widehat{W}_{\mathrm{m}}}{\widehat{W}_{\mathrm{e}}}, \tag{6.59}$$

where $\widehat{W} = W/V$ is the power density, $\widehat{W}_{\mathrm{m}} = \tau_y \dot{\gamma}$ the mechanical power density and $\widehat{W}_{\mathrm{e}} = \dfrac{BH}{2t_{\mathrm{c}}}$ the electrical power density, $\dot{\gamma}$ is the shear strain rate ($\dot{\gamma} \equiv v_r/h$), B the magnetic field flux density in the fluid, H is the magnetic field intensity in the fluid and t_{c} is the characteristic time for the establishment of the magnetic field within the fluid. Hence:

$$\alpha = 2t_c \dot{\gamma} \frac{\tau_y}{BH}, \tag{6.60}$$

where τ_y and B are defined at a common H. The following figures of merit can be defined:

- Figure of merit based on the active fluid volume, defined as

$$F_1 = \frac{\tau_y^2}{\eta}. \tag{6.61}$$

  This figure of merit is inversely proportional to the minimum active fluid volume V. If F_1 increases, the minimum active fluid volume V decreases. The decrease of V results in a decrease of the damper size and the electrical power consumption. For a given active fluid volume V, if F_1 increases the desired control ratio λ increases as well.

- Figure of merit based on the mass of the active fluid volume, defined as:

$$F_2 = \frac{\tau_y^2}{\eta\rho}. \tag{6.62}$$

  This figure of merit is inversely proportional to the minimum active fluid mass.

- Figure of merit based on the power efficiency, defined as:

$$F_3 = \frac{\tau_y}{BH}. \tag{6.63}$$

The maximisation of this figure of merit results in a minimisation of the electrical power consumption of the MR damper for a given delivered mechanical power.

6.4.2.2 Choice of the MR Fluid

For the purpose of evaluating the different figures of merit the Lord Corporation 132DG MR fluid (http://www.lordfulfillment.com/upload/DS7015.pdf) is chosen. The characteristics of this MR fluid are listed in Table 6.2 and in Figures 6.9, 6.10 and 6.11.

The MR fluid dampers are usually designed such that, in normal conditions, the MR fluid is magnetically saturated. It is under this condition that the fluid will generate its maximum yield stress τ_y.

Table 6.2. Rheological properties of MR fluid 132DG [copyright Lord Corp. (www.lord.com), used by permission]

Properties	Value/Limits	Properties	Value/Limits
Base Fluid	Synthetic Oil	Density	3.055 kg/m^3
System	Open or Closed	Color	Dark gray
Operating Temperature	-40÷150°C	Weight Percent Solids	80.74%
Viscosity [Pa·s] Shear Rate 10 s^{-1} Shear Rate 80 s^{-1}	 0,84 Pa·s 0,33 Pa·s	Coeficient of Thermal Expansion 0 to 50°C 50 to 100°C 100 to 150°C	Unit Volume/$^\circ$C 0,55×10^{-3} 0,61×10^{-3} 0,67×10^{-3}
Settling (dependent on device design)	The fluid is developed to settle softly, will remix with a 2-3 cycles of the device	Thermal Conductivity @ 25°C	0,25÷1,06W/m·$^\circ$C
Specific Heat @ 25°C	800 J/kg·$^\circ$C	Flash Point	>150°C

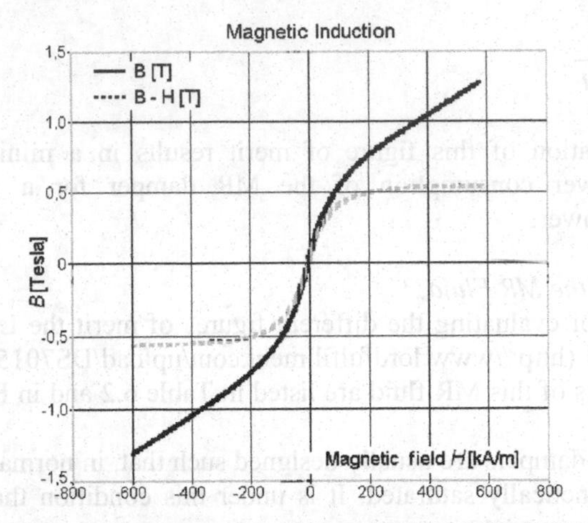

Fig. 6.9. Magnetic flux versus magnetic field for MR fluid 132DG [copyright Lord Corp. (www.lord.com), used by permission]

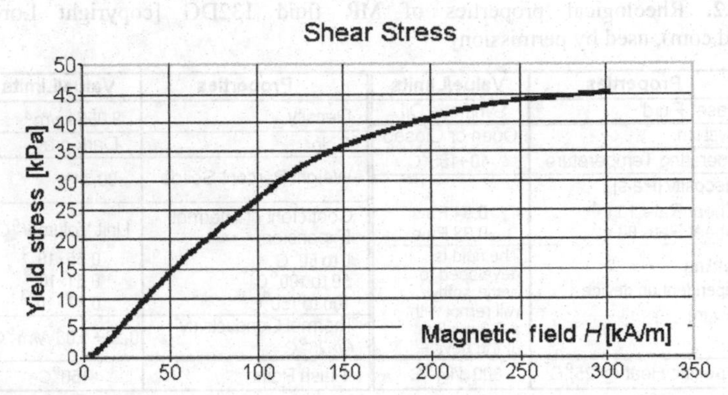

Fig. 6.10. Yield stress versus magnetic field for MR fluid 132DG [copyright Lord Corp. (www.lord.com), used by permission]

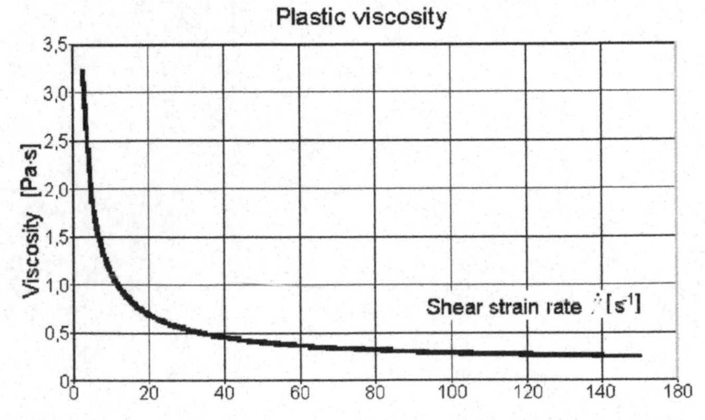

Fig. 6.11. Viscosity versus shear strain rate for MR fluid 132DG [copyright Lord Corp. (www.lord.com), used by permission]

Choosing the values $\dot{\gamma} = 140\ \mathrm{s}^{-1}$ (for the shear strain rate) and $H = 250\ \mathrm{kA/m}$ (for the magnetic field), the following parameter values can be obtained from Figures 6.9, 6.10 and 6.11:

- the magnetic flux $B \cong 0.83\ \mathrm{T}$ (from Figure 6.9)
- the yield stress $\tau_y = 44.1\ \mathrm{kPa}$ (from Figure 6.10)
- the off-state plastic viscosity $\eta = 0.25\ \mathrm{Pa{\cdot}s}$ (from Figure 6.11)

6.4.3 Determination of the Optimal Gap Size and Hydraulic Design

When selecting an MR damper the dynamic range and the maximum value of the controllable force are two key parameters.

6.4.3.1 Controllable Force and Dynamic Range

The controllable force F_τ represents the force due to the field-induced yield stress τ_y.

The dynamic range D is defined as the ratio between the total damper output force F and the uncontrollable force F_{uc} where uncontrollable forces include the fluid viscosity force F_η and the friction force F_{f}. Hence

$$D = \frac{F}{F_{\mathrm{uc}}} = \frac{F_{\mathrm{uc}} + F_\tau}{F_{\mathrm{uc}}} = 1 + \frac{F_\tau}{F_\eta + F_{\mathrm{f}}}. \tag{6.64}$$

From (6.64), considering (6.47) yields

$$D = 1 + \frac{c\tau_y L A_{\mathrm{p}}}{12\eta A_{\mathrm{R_m}} v_{\mathrm{p}} A_{\mathrm{p}} L + A_{\mathrm{g}} h^2 F_{\mathrm{f}}}, \tag{6.65}$$

where c is defined by (6.44), $A_{\mathrm{p}} = \frac{\pi}{4}\left[\left(d_{\mathrm{cyl}}^{\mathrm{in}} - 2h\right)^2 - d_{\mathrm{rod}}^2\right]$ is the piston area, F_{f} is the friction force, $A_{\mathrm{R_m}} = \frac{\pi}{4}\left[\left(d_{\mathrm{cyl}}^{\mathrm{in}} - h\right)^2 - d_{\mathrm{rod}}^2\right]$ and $A_{\mathrm{g}} = \pi\left(d_{\mathrm{cyl}}^{\mathrm{in}} - h\right)h$. It is worth noticing that the values of both the friction force F_{f} and of the magnetic pole length L do not affect the optimal value of the gap.

To maximise the effectiveness of the MR damper, the controllable range of the force should be as large as possible. A small gap size will increase the controllable force range but, when the gap size h is very small the viscous force F_η increases much faster than the controllable force F_τ and the dynamic range D decreases. If the gap size h is large, both F_η and F_τ decrease. Hence there exists an optimal gap size which maximises the dynamic range D.

In Figure 6.12 the trend of the dynamic range for the basic design data is plotted as a function of the gap size.

From Figure 6.12 $g = 0.4\times10^{-3}$ m for $D^{\max} \cong 65$. With this choice the dimensional and functional parameters of the hydraulic circuit have been determined.

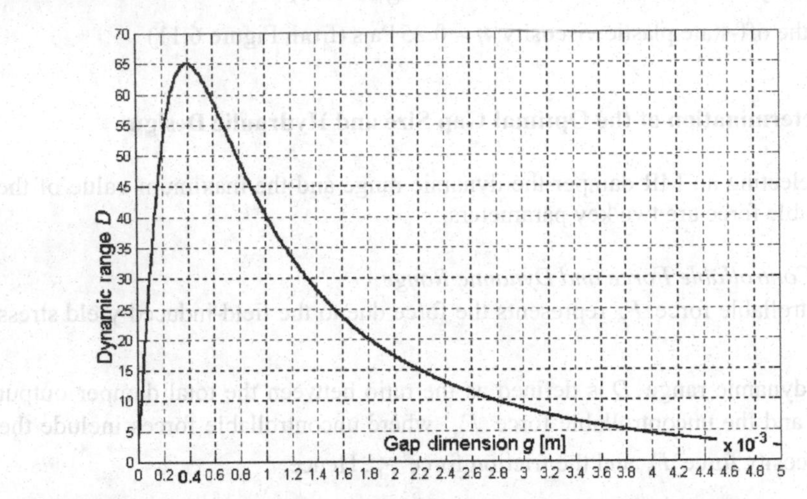

Fig. 6.12. Dynamic range D versus gap dimension g [copyright Publishing House of the Romanian Academy (2004), reproduced from Topics in Applied Mechanics, Vol. II, Ch 5, edited by Chiroiu L and Sireteanu T, used by permission]

6.4.3.2 Parameters of the Hydraulic Circuit

The damper has a fairly simple geometry in which the outer cylindrical housing is part of the magnetic circuit. The effective fluid orifice is the entire annular space between the piston outer diameter and the inside of the damper cylinder housing. The motion of the piston causes the fluid to flow through the entire annular region.

The damper is double ended (Figure 6.8). This arrangement has the advantage that a rod-volume compensator does not need to be incorporated into the damper.

The following design steps entail the choice of the remaining damper dimensions (L and h) to achieve the requirements given above.

With the aforementioned data the key design parameters of the MR damper are obtained. The parameters are given in Table 6.3. Note that seal and bearing friction and other parasitical effects may worsen the dynamic range λ.

Table 6.3. MR damper parameters

$A_p = 8.927 \times 10^{-4}$ m^2	$Q = 2.678 \times 10^{-5}$ m^3/s
$A_{R_m} = 9.175 \times 10^{-4}$ m^2	$\eta = 0.25$ Pa $\cdot$ s
$A_g = 4.976 \times 10^{-5}$ m^2	$\tau_y = 44.1 \times 10^3$ Pa
$h = 4 \times 10^{-4}$ m	$W_m = 80$ W
$v_p = 3 \times 10^{-2}$ m/s	$F_\tau = 2700$ N
$\dot{\gamma} = 140$ s^{-1}	$\Delta p_\tau = 29.87 \times 10^5$ Pa
$L = 1.2 \times 10^{-2}$ m	$\Delta p_\eta = 1.244 \times 10^5$ Pa
$\Delta p = \Delta p_\eta + \Delta p_\tau = 31.114 \times 10^5$ Pa	$\lambda = \dfrac{\Delta p_\tau}{\Delta p_\eta} = 24$
$F = F_\tau + F_\eta = 2777.55$ N	$c = 2.256$
$V = 5.82 \times 10^{-7}$ m^3	

6.4.4 Magnetic Circuit Design

The MR damper magnetic circuit is shown in Figure 6.13. With this design two magnetic flux paths can be defined. The coil design solution (two identical coils with one joint pole) allows two functional characteristics for the MR damper.

The flux in the magnetic circuit flows axially through the piston steel core (whose diameter is d_c) beneath the windings, radially through the piston poles of length L_1 and L_2, through a gap of thickness h where the MR fluid flows, and axially through the cylinder wall.

The design of a magnetic circuit requires the determination of L_c (length of coil), d_c (inner diameter of coil) and the total number of turns N in the coil.

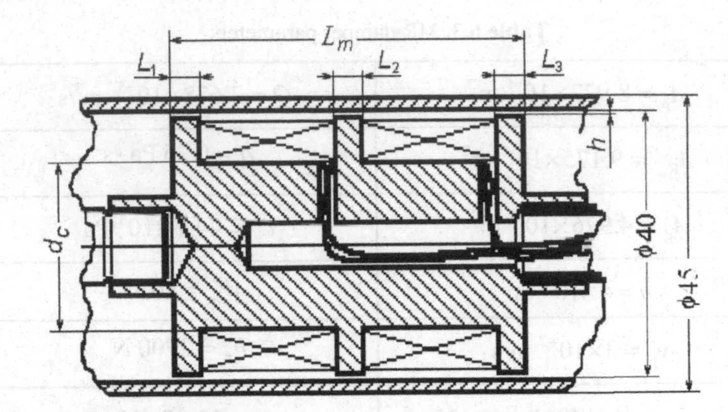

Fig. 6.13. Magnetic circuit geometry [copyright Publishing House of the Romanian Academy (2004), reproduced from Topics in Applied Mechanics, Vol. II, Ch 5, edited by Chiroiu L and Sireteanu T, used by permission]

Figure 6.14 shows the equivalent magnetic circuit.

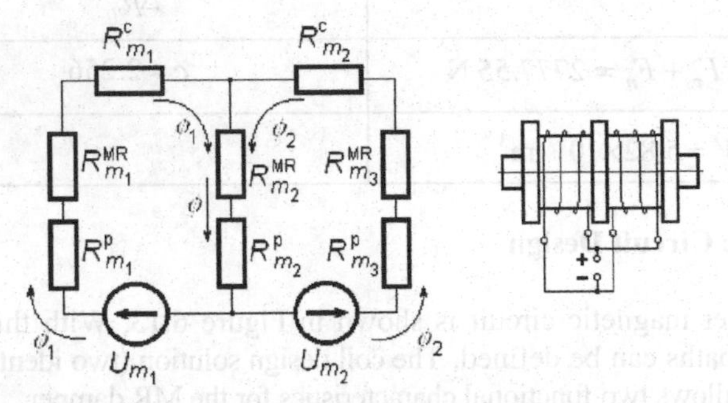

Fig. 6.14. Equivalent magnetic circuit [copyright Publishing House of the Romanian Academy (2004), reproduced from Topics in Applied Mechanics, Vol. II, Ch 5, edited by Chiroiu L and Sireteanu T, used by permission]

The design process for sizing the magnetic circuit involves the following steps:

- Selection of the operating point of the MR fluid: from consideration of Figure 6.9, in correspondence with $H = 250$ kA/m, the magnetic flux density $B = 0.83$ T is obtained.
- Choice of the steel for the piston and cylinder (for the magnetic circuit). A low-carbon steel (the carbon content of the steel should be less than 0.15%) having a high magnetic permeability is desirable.
- Choice of the magnetic pole length. For the MR damper prototype a variant with two equal poles and the middle pole having a length double that of the others was considered. The middle pole has a length $L_2 = 6\times10^{-3}$ m while both the others have a length of $L_1 = 3\times10^{-3}$ m (the total length of the pole is $L = 1.2\times10^{-2}$ m).

- Calculation of the magnetic flux density for the steel core from the continuity of magnetic flux $\phi_{\text{gap}} = \phi_{\text{steel}} = \phi$.
- The effective area of the pole is found to be $A_{\text{f}}' = 2.41\times10^{-4}\ \text{m}^2$, and $B_{\text{steel}} = \dfrac{\phi}{A_0} = \dfrac{BA_{\text{f}}'}{A_0} = 0.96\,\text{T}$.
- From the B versus H magnetic curves for the steel it can be deduced that $H_{\text{steel}} = 272.8\ \text{A/m}$.
- By using Kirchoff's law for magnetic circuits the required ampere-turns for magnetic coil can be determined: $NI = \sum H_i L_i = 124$ A-turns.
- Taking $I = 2$A, yields $N = 62$.
- Determination of the inner coil diameter for a coil with two layers: $d_{\text{c}} = 3.82\times10^{-2}$ m.
- Determination of the coil length: $L_{\text{c}} = 2.48\times10^{-2}$ m.
- Calculation of the piston length: $L_{\text{p}} = 2L_{\text{c}} + L = 6.2\times10^{-2}$ m.

It should be noticed that both the hydraulic circuit design and the magnetic circuit design involve an iterative calculation.

6.5 MRD Modelling and Characteristics Identification

In this section the modelling of MR dampers is studied. Several methods have been proposed over the years for obtaining MR damper models and identifying their parameters. Spencer and co-workers (1997) developed a phenomenological model that accurately portrays the response of an MR damper in response to cyclic excitations. This is a modified Bouc–Wen (BW) model governed by ordinary differential equations. BW-based models in semi-active seismic vibration control have proven to be easy to use and numerically amenable. Employing a different approach to Spencer's work, Kyle *et al.* (2000) modelled an MR damper in the form of a Takagi–Sugeno–Kang fuzzy inference system. The training and checking data of the model were generated using Spencer's model of MR damper.

Other authors studied the behaviour of the MR damper, emphasising the difference between the pre-yield viscoelastic region and the post-yield viscous region as a key aspect of the damper.

Starting from Spencer's studies, Butz and Stryk (1999) showed that the difference between using the modified BW model and the Bingham plastic model in a 2DOF quarter car dynamic system is very small. This justifies in many applications the interest in a simple description of the model rather than a more accurate but more complex one.

6.5.1 Experimental Data

The damper used for modelling and characteristic identification studies is the magnetorheological damper model RD 1005-3 produced by Lord Corp. (http://www.lord.com/Home/MagnetoRheologicalMRFluid/Products/MRDamper/tabid/3361/Default.aspx). Tests were carried out under the following conditions:

- the maximum current value considered was 1.75 A,
- the force variation in response to the current variations from 0.02 to 1.75 A was about 300 N,
- the measured internal resistance of the device was 5.2 Ω at 29°C.

Figure 6.15 portrays the RMS force response to sinusoidal loads versus current values for different frequencies and different excitation amplitudes. It can be readily observed that this variation is non-linear within the test range.

The force response time history of the MR damper to an imposed cyclic motion at the frequency of 1.5 Hz and 16 mm amplitude is shown in Figure 6.16 for 0.1 A, 0.2 A, 0.6 A and 1.75 A current values.

The force versus velocity loops reveal some interesting aspects (Figure 6.17). The effect of the accumulator can be noted, producing an offset in the experimental data. This effect is more evident for small current values.

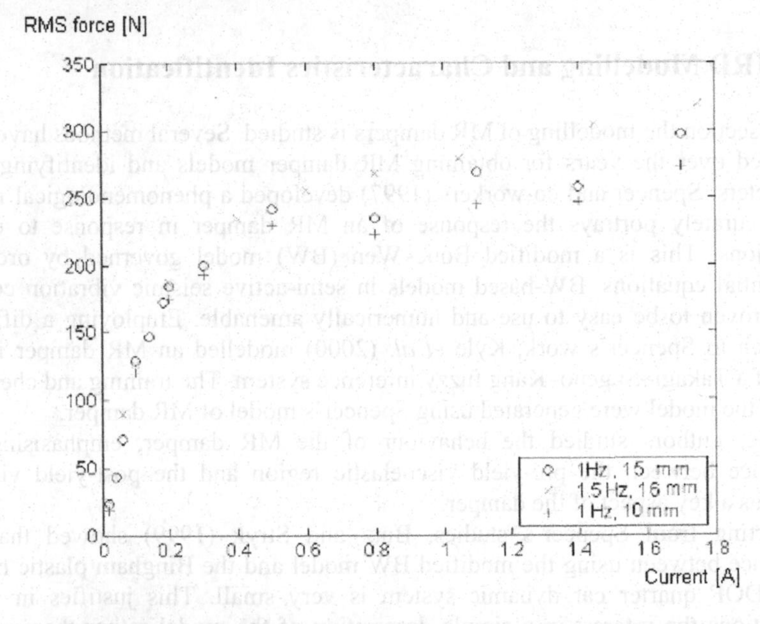

Fig. 6.15. RMS force [N] variation versus current values [A] [copyright IMechE (2004), reproduced from Giuclea M, Sireteanu T, Stancioiu D and Stammers CW, Model parameter identification for vehicle vibration control with magnetorheological dampers using computational intelligence methods, Proceedings of the Institution of Mechanical Engineers, Part I: Journal of Systems and Control Engineering, Publisher: Professional Engineering Publishing, ISSN 0959-6518, Vol. 218, N 7/2004, pp 569–581, used by permission]

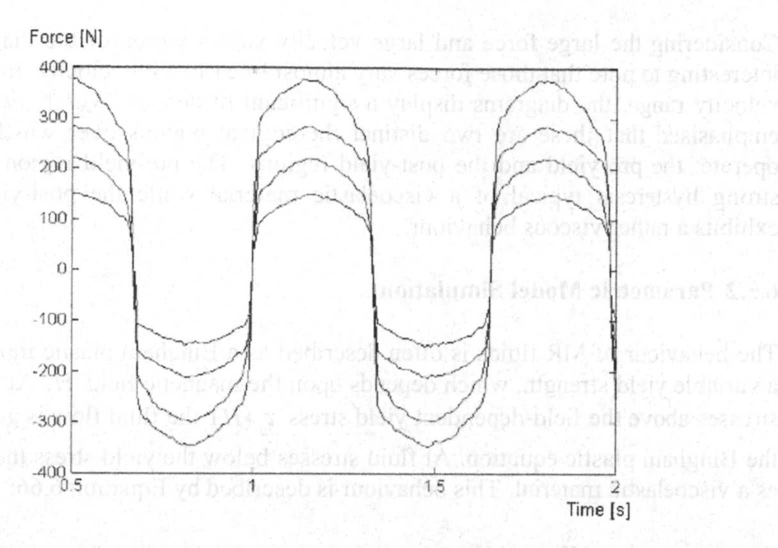

Fig. 6.16. Experimentally measured force [N] versus time [s] [copyright IMechE (2004), reproduced from Giuclea M, Sireteanu T, Stancioiu D and Stammers CW, Model parameter identification for vehicle vibration control with magnetorheological dampers using computational intelligence methods, Proceedings of the Institution of Mechanical Engineers, Part I: Journal of Systems and Control Engineering, Publisher: Professional Engineering Publishing, ISSN 0959-6518, Vol. 218, N 7/2004, pp 569–581, used by permission]

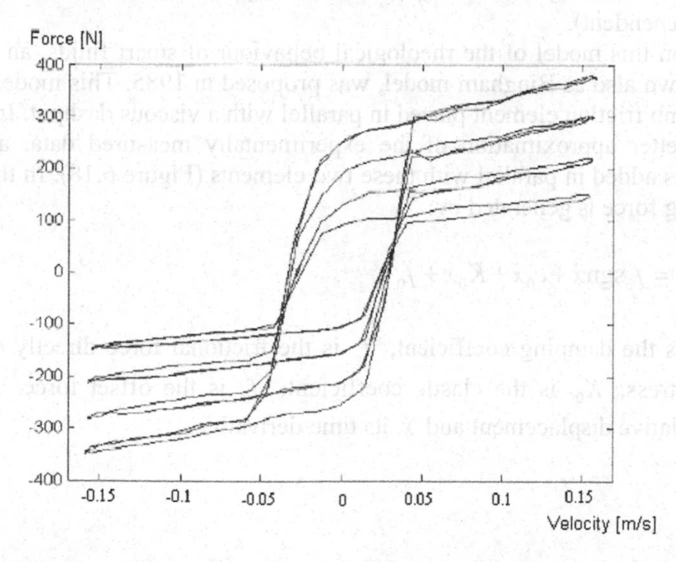

Fig. 6.17. Experimentally measured force [N] versus velocity [m/s] [copyright IMechE (2004), reproduced from Giuclea M, Sireteanu T, Stancioiu D and Stammers CW, Model parameter identification for vehicle vibration control with magnetorheological dampers using computational intelligence methods, Proceedings of the Institution of Mechanical Engineers, Part I: Journal of Systems and Control Engineering, Publisher: Professional Engineering Publishing, ISSN 0959-6518, Vol. 218, N 7/2004, pp 569–581, used by permission]

Considering the large force and large velocity values region of the diagram, it is interesting to note that those forces vary almost linearly with velocity. In the small-velocity range, the diagrams display a significant hysteresis. Werely *et al.* (1998) emphasised that these are two distinct rheological regions over which dampers operate: the pre-yield and the post-yield regions. The pre-yield region exhibits a strong hysteresis typical of a viscoelastic material while the post-yield region exhibits a rather viscous behaviour.

6.5.2 Parametric Model Simulation

The behaviour of MR fluids is often described as a Bingham plastic model having a variable yield strength, which depends upon the magnetic field H. At fluid shear stresses above the field-dependent yield stress $\tau_y(H)$ the fluid flow is governed by the Bingham plastic equation. At fluid stresses below the yield stress the fluid acts as a viscoelastic material. This behaviour is described by Equation 6.66:

$$\tau = \begin{cases} \tau_y(H) + \eta\dot{\gamma} & \tau > \tau_y \\ G\gamma & \tau < \tau_y, \end{cases} \tag{6.66}$$

where H is the magnetic field, $\dot{\gamma}$ is the fluid shear rate and η is the plastic viscosity (*i.e.*, viscosity at $H = 0$), G is the complex material modulus (which is also field dependent).

Based on this model of the rheological behaviour of smart fluids, an idealised model, known also as Bingham model, was proposed in 1985. This model consists of a Coulomb friction element placed in parallel with a viscous dashpot. In order to obtain a better approximation of the experimentally measured data, an elastic element was added in parallel with these two elements (Figure 6.18). In the model, the damping force is generated by:

$$F = f_c \operatorname{sgn}\dot{x} + c_0\dot{x} + K_0 x + f_0, \tag{6.67}$$

where c_0 is the damping coefficient, f_c is the frictional force directly related to the yield stress, K_0 is the elastic coefficient, f_0 is the offset force, x is the imposed relative displacement and $\dot{x}$ its time derivative.

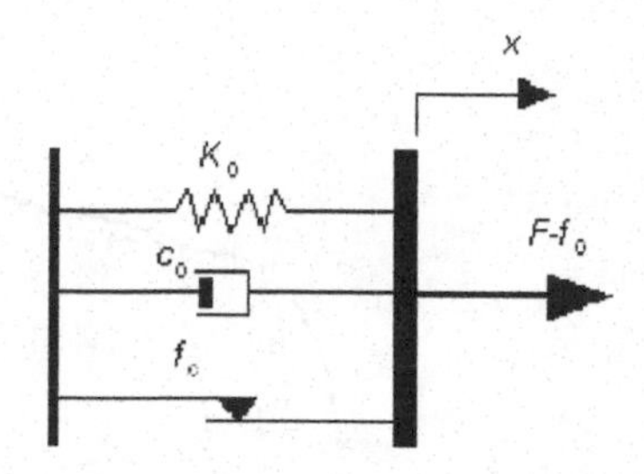

Fig. 6.18. Bingham plastic model of an MR damper [copyright IMechE (2004), reproduced with minor modifications from Giuclea M, Sireteanu T, Stancioiu D and Stammers CW, Model parameter identification for vehicle vibration control with magnetorheological dampers using computational intelligence methods, Proceedings of the Institution of Mechanical Engineers, Part I: Journal of Systems and Control Engineering, Publisher: Professional Engineering Publishing, ISSN 0959-6518, Vol. 218, N 7/2004, pp 569–581, used by permission]

Figures 6.19 and 6.20 depict the force versus time and the force versus velocity diagrams for a current value of 0.8 A (both numerically predicted and experimentally measured). The agreement between predicted and experimental data was used as a criterion for fitting the model parameters. The parameters chosen are $f_c = 200$ N, $c_0 = 0.7$ Ns/mm, $K_0 = 0.3$ N/mm, $f_0 = 0$ N.

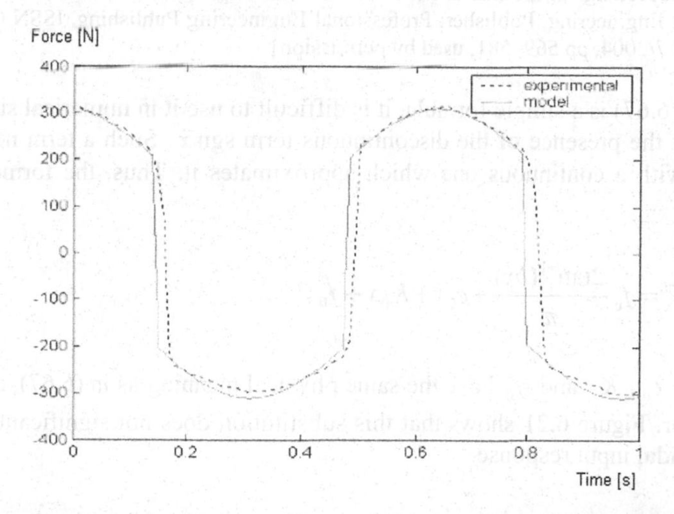

Fig. 6.19. Force [N] versus time [s], experimental data (dashed) and Bingham plastic model (solid) [copyright IMechE (2004), reproduced from Giuclea M, Sireteanu T, Stancioiu D and Stammers CW, Model parameter identification for vehicle vibration control with magnetorheological dampers using computational intelligence methods, Proceedings of the Institution of Mechanical Engineers, Part I: Journal of Systems and Control Engineering, Publisher: Professional Engineering Publishing, ISSN 0959-6518, Vol. 218, N 7/2004, pp 569–581, used by permission]

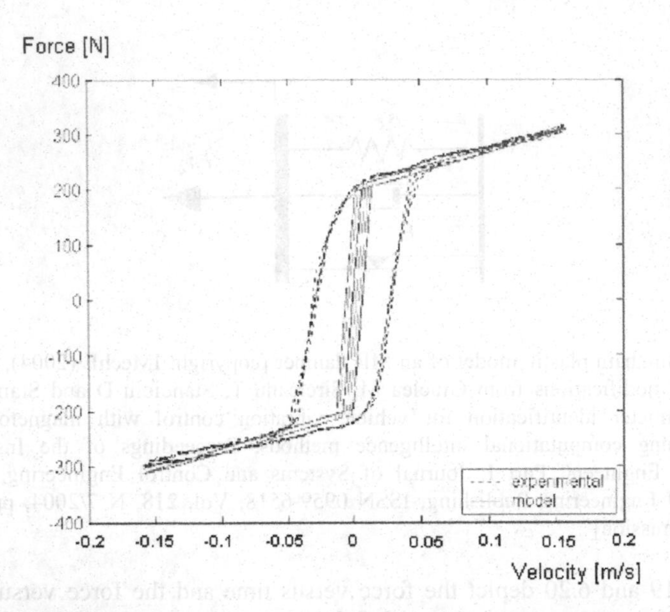

Fig. 6.20. Force [N] versus velocity [m/s], experimentally measured data (dashed) and Bingham plastic model (solid) [copyright IMechE (2004), reproduced from Giuclea M, Sireteanu T, Stancioiu D and Stammers CW, Model parameter identification for vehicle vibration control with magnetorheological dampers using computational intelligence methods, Proceedings of the Institution of Mechanical Engineers, Part I: Journal of Systems and Control Engineering, Publisher: Professional Engineering Publishing, ISSN 0959-6518, Vol. 218, N 7/2004, pp 569–581, used by permission]

Although (6.67) is a simple formula, it is difficult to use it in numerical simulations because of the presence of the discontinuous term $\operatorname{sgn} \dot{x}$. Such a term needs to be replaced with a continuous one which approximates it. Thus, the formula (6.67) becomes:

$$F = f_c \frac{2\tan^{-1}(b\dot{x})}{\pi} + c_0\dot{x} + K_0 x + f_0, \tag{6.68}$$

where f_c, c_0, K_0 and f_0 have the same physical meaning as in (6.67), and b is a form factor. Figure 6.21 shows that this substitution does not significantly change the sinusoidal input response.

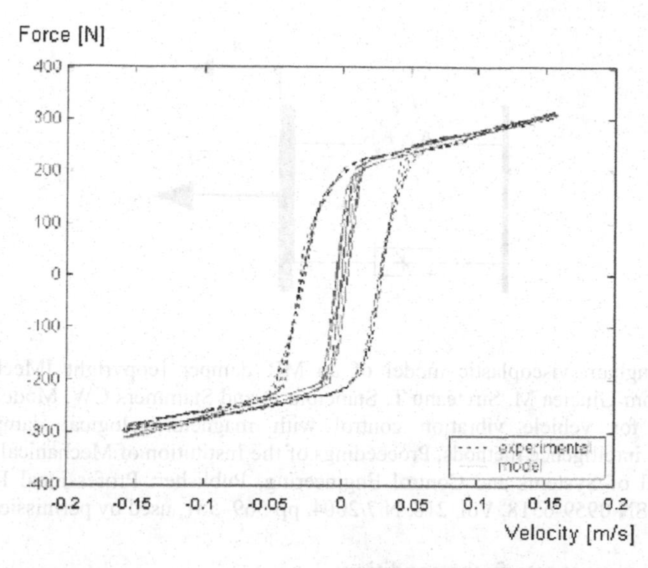

Fig. 6.21. Force [N] versus velocity [m/s], experimentally measured data (dashed) and modified Bingham plastic model (solid) [copyright IMechE (2004), reproduced from Giuclea M, Sireteanu T, Stancioiu D and Stammers CW, Model parameter identification for vehicle vibration control with magnetorheological dampers using computational intelligence methods, Proceedings of the Institution of Mechanical Engineers, Part I: Journal of Systems and Control Engineering, Publisher: Professional Engineering Publishing, ISSN 0959-6518, Vol. 218, N 7/2004, pp 569–581, used by permission]

The parameters chosen are $f_c = 210$ N, $c_0 = 0.65$Ns/mm, $K_0 = 0.3$ N/mm, $f_0 = 0$ N and $b = 10$. Clearly, the main disadvantage of this model is the poor description of the pre-yield region. However, when the velocity is large, *i.e.*, if the damper works at rather large force values, this model can be successfully employed.

Another model based on Equation 6.67 is the Bingham viscoplastic model. This model replaces the friction element by a block (Figure 6.22), which generates a force governed by the equation

$$f = \begin{cases} c_{\text{pr}}\dot{x} & |\dot{x}| < v_y \\ c_{\text{po}}\dot{x} + f_c & |\dot{x}| \geq v_y , \end{cases} \tag{6.69}$$

where c_{po} and c_{pr} are the post-yield and pre-yield damping coefficients and v_y is the yielding velocity. It should be noticed that the parameters included in this model are not independent.

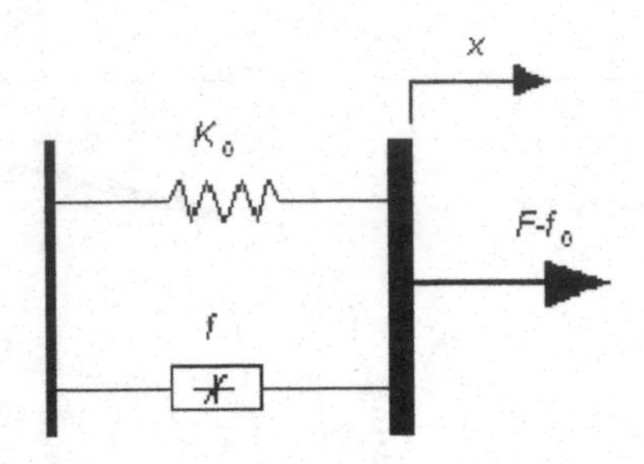

Fig. 6.22. Bingham viscoplastic model of an MR damper [copyright IMechE (2004), reproduced from Giuclea M, Sireteanu T, Stancioiu D and Stammers CW, Model parameter identification for vehicle vibration control with magnetorheological dampers using computational intelligence methods, Proceedings of the Institution of Mechanical Engineers, Part I: Journal of Systems and Control Engineering, Publisher: Professional Engineering Publishing, ISSN 0959-6518, Vol. 218, N 7/2004, pp 569–581, used by permission]

The yield force must satisfy the condition:

$$f_y(1-\frac{c_{po}}{c_{pr}})=f_c. \tag{6.70}$$

Therefore, the damping force is defined by the equation

$$F=f+K_0x+f_0, \tag{6.71}$$

where the terms K_0 and f_0 have the same meaning as in the Bingham equation (6.67). There is no significant improvement in the model description, except that a yield velocity appears. As with the Bingham plastic model, this model predicts large forces accurately.

A model that leads to results similar to the Bingham viscoplastic model is the delayed Bingham model. This model is similar to the modified Bingham model except that the velocity is delayed so that hysteresis occurs.

Figure 6.23 shows the time response of the model with a delay of 0.01 s. The force versus velocity loops (Figure 6.24) show that hysteresis is generated at low velocities. However the main disadvantage of Bingham-equation-based models is the poor accuracy in the pre-yield region.

The bi-viscous model (6.69) has also been considered for MR behaviour description. Although this model can predict well the MR damper response to sinusoidal excitations, it is a non-smooth model. This makes it difficult to use the bi-viscous model in random vibration problems. Good results can be obtained instead using the delayed model.

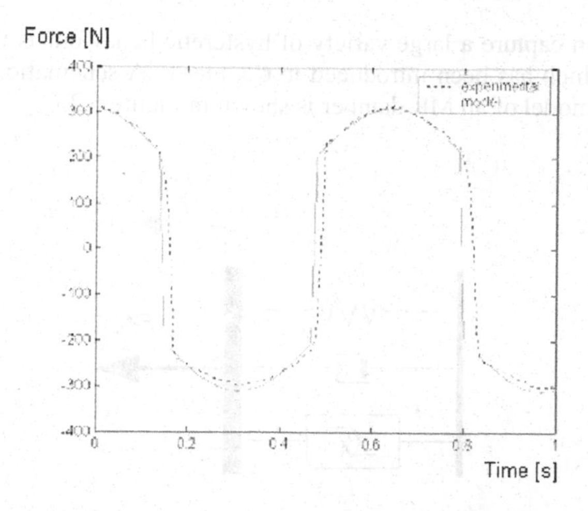

Fig. 6.23. Force [N] versus time [s], experimental data (dashed) and modified Bingham plastic model (solid) [copyright IMechE (2004), reproduced from Giuclea M, Sireteanu T, Stancioiu D and Stammers CW, Model parameter identification for vehicle vibration control with magnetorheological dampers using computational intelligence methods, Proceedings of the Institution of Mechanical Engineers, Part I: Journal of Systems and Control Engineering, Publisher: Professional Engineering Publishing, ISSN 0959-6518, Vol. 218, N 7/2004, pp 569–581, used by permission]

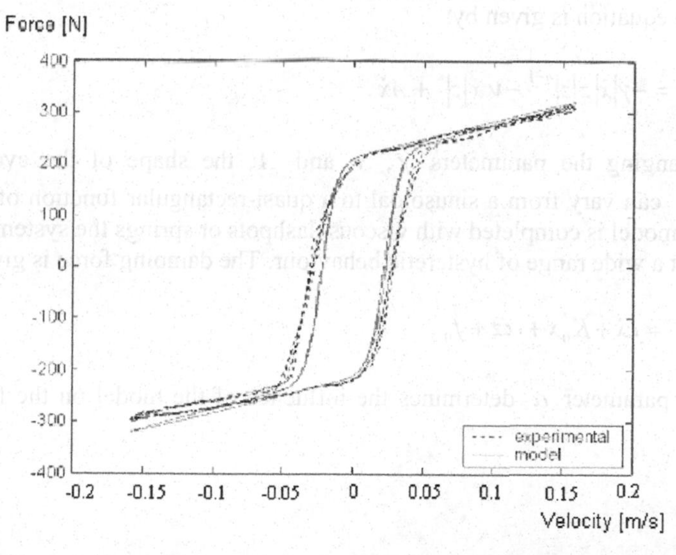

Fig. 6.24. Force [N] versus velocity [m/s], experimental data (dashed) and delayed Bingham plastic model (solid) [copyright IMechE (2004), reproduced from Giuclea M, Sireteanu T, Stancioiu D and Stammers CW, Model parameter identification for vehicle vibration control with magnetorheological dampers using computational intelligence methods, Proceedings of the Institution of Mechanical Engineers, Part I: Journal of Systems and Control Engineering, Publisher: Professional Engineering Publishing, ISSN 0959-6518, Vol. 218, N 7/2004, pp 569–581, used by permission]

A model that can capture a large variety of hysteretic behaviour is the Bouc–Wen (BW) model, which has been introduced in Chapter 2. A schematic representation of a BW-based model of an MR damper is shown in Figure 6.25.

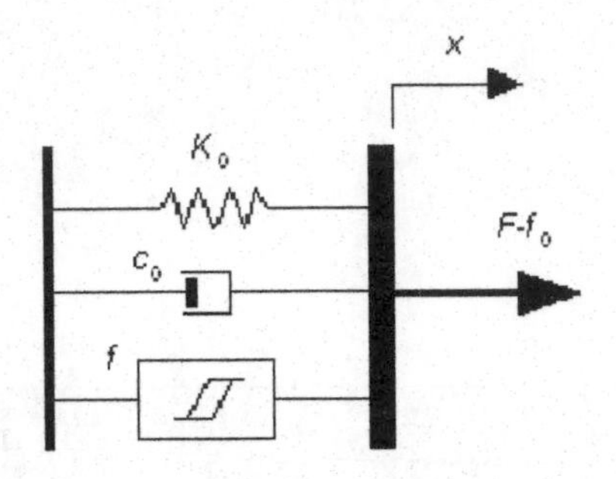

Fig. 6.25. Schematic of Bouc–Wen model of an MR damper [copyright IMechE (2004), reproduced from Giuclea M, Sireteanu T, Stancioiu D and Stammers CW, Model parameter identification for vehicle vibration control with magnetorheological dampers using computational intelligence methods, Proceedings of the Institution of Mechanical Engineers, Part I: Journal of Systems and Control Engineering, Publisher: Professional Engineering Publishing, ISSN 0959-6518, Vol. 218, N 7/2004, pp 569–581, used by permission]

The model equation is given by:

$$\dot{z} = -\gamma|\dot{x}|\,z\,|z|^{n-1} - \nu\,\dot{x}\,|z|^{n} + A\dot{x}. \tag{6.72}$$

By changing the parameters γ, ν and A, the shape of the evolutionary variable z can vary from a sinusoidal to a quasi-rectangular function of the time. When the model is completed with viscous dashpots or springs the system response can predict a wide range of hysteretic behaviour. The damping force is given by

$$F = c\dot{x} + K_0 x + \alpha z + f_0, \tag{6.73}$$

where the parameter α determines the influence of the model on the final force value.

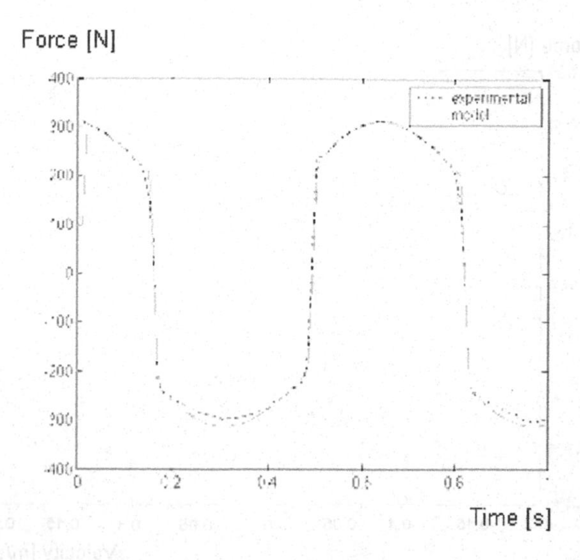

Fig. 6.26. Force [N] vs. time [s] experimental data (dashed) and Bouc–Wen (solid) model [copyright IMechE (2004), reproduced from Giuclea M, Sireteanu T, Stancioiu D and Stammers CW, Model parameter identification for vehicle vibration control with magnetorheological dampers using computational intelligence methods, Proceedings of the Institution of Mechanical Engineers, Part I: Journal of Systems and Control Engineering, Publisher: Professional Engineering Publishing, ISSN 0959-6518, Vol. 218, N 7/2004, pp 569–581, used by permission]

To determine the parameters that fit the MR damper model response to the experimentally measured response at a frequency of 1.5 Hz and with a 16 mm amplitude sinusoidal input, a least-mean-squares nonlinear algorithm was used. The set of parameters chosen to fit the 0.8 A current value response was $n = 2$, $\gamma = 1.2\ \text{mm}^{-2}$, $\nu = 1\ \text{mm}^{-2}$, $A = 15$, $\alpha = 80$ N/mm, $c_0 = 0.65$ Ns/mm, $K_0 = 0.3$ N/mm, $f_0 = 0$ N.

A comparison between the experimentally determined data and the BW-based model response is shown in Figure 6.26. The predicted response fits the measured data well, even in the pre-yield region. The force versus velocity loops are displayed in Figure 6.27. It is worth noting that BW model provides also a smooth transition between the viscoelastic state and the post-yield state.

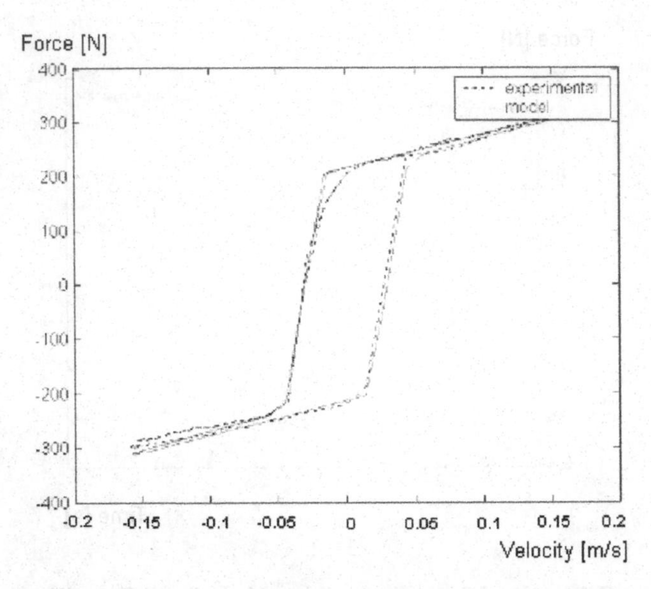

Fig. 6.27. Force [N] versus velocity [m/s] experimentally measured data (dashed) and Bouc–Wen (solid) model [copyright IMechE (2004), reproduced from Giuclea M, Sireteanu T, Stancioiu D and Stammers CW, Model parameter identification for vehicle vibration control with magnetorheological dampers using computational intelligence methods, Proceedings of the Institution of Mechanical Engineers, Part I: Journal of Systems and Control Engineering, Publisher: Professional Engineering Publishing, ISSN 0959-6518, Vol. 218, N 7/2004, pp 569–581, used by permission]

6.5.3 Fuzzy-logic-based Model

The use of a fuzzy-logic-based modelling scheme offers a number of benefits, the first and foremost being that is suitable to incorporate the qualitative aspects of human experience within its mapping laws. A neuro-adaptive learning technique can be used to train a family of membership functions (MF) to simulate a non-linear mapping function. The neuro-adaptive learning technique works similarly to neural networks. This technique provides a method for the fuzzy modelling algorithm to learn information on a data set, in order to fit the membership function parameters to the fuzzy inference system to track the input/output data. The parameters associated with the membership functions will change through the learning process. Data for training and checking are obtained from experimentally measured data. The fuzzy-logic model is based on a Takagi–Sugeno–Kang architecture. Displacement and velocities were used as input functions. The number of MFs was set at two for displacements and five for velocities. The after-learning input MFs that fit the experimentally measured data are plotted in Figure 6.28.

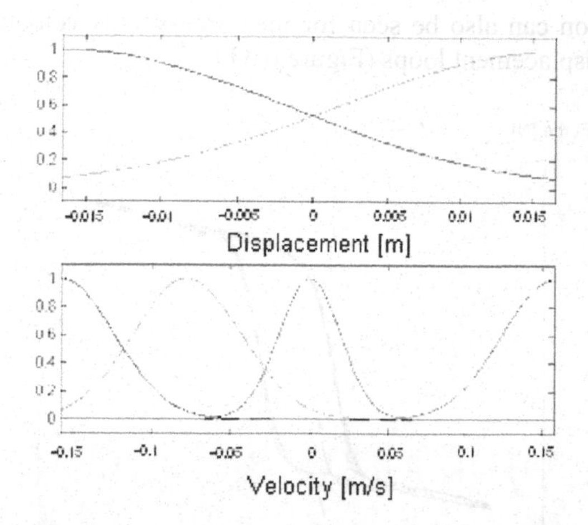

Fig. 6.28. Membership functions after training [copyright IMechE (2004), reproduced from Giuclea M, Sireteanu T, Stancioiu D and Stammers CW, Model parameter identification for vehicle vibration control with magnetorheological dampers using computational intelligence methods, Proceedings of the Institution of Mechanical Engineers, Part I: Journal of Systems and Control Engineering, Publisher: Professional Engineering Publishing, ISSN 0959-6518, Vol. 218, N 7/2004, pp 569–581, used by permission]

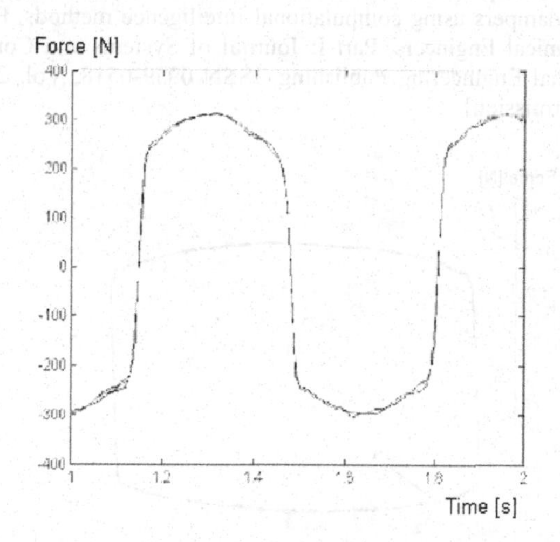

Fig. 6.29. Force [N] versus time [s] experimental (dashed) and fuzzy-logic-based model (solid) [copyright IMechE (2004), reproduced from Giuclea M, Sireteanu T, Stancioiu D and Stammers CW, Model parameter identification for vehicle vibration control with magnetorheological dampers using computational intelligence methods, Proceedings of the Institution of Mechanical Engineers, Part I: Journal of Systems and Control Engineering, Publisher: Professional Engineering Publishing, ISSN 0959-6518, Vol. 218, N 7/2004, pp 569–581, used by permission]

The result of simulation in terms of the force-time dependence is shown in Figure 6.29. It is possible to see that an almost perfect fit is obtained. The good accuracy of the representation can also be seen for the force versus velocity (Figure 6.30) and force versus displacement loops (Figure 6.31).

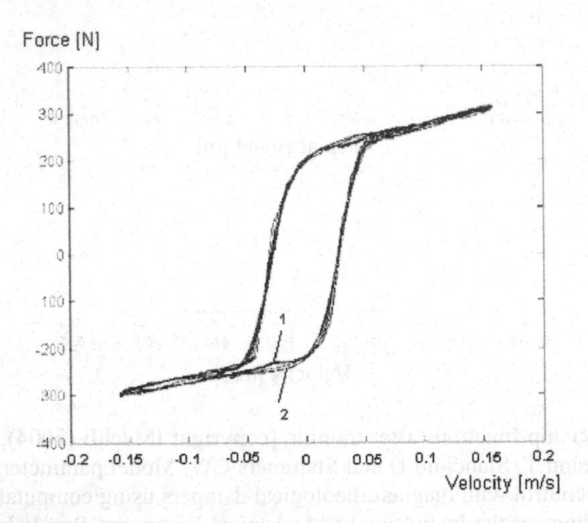

Fig. 6.30. Force [N] versus velocity [m/s] experimental (1) and fuzzy-logic-based model (2) [copyright IMechE (2004), reproduced from Giuclea M, Sireteanu T, Stancioiu D and Stammers CW, Model parameter identification for vehicle vibration control with magnetorheological dampers using computational intelligence methods, Proceedings of the Institution of Mechanical Engineers, Part I: Journal of Systems and Control Engineering, Publisher: Professional Engineering Publishing, ISSN 0959-6518, Vol. 218, N 7/2004, pp 569–581, used by permission]

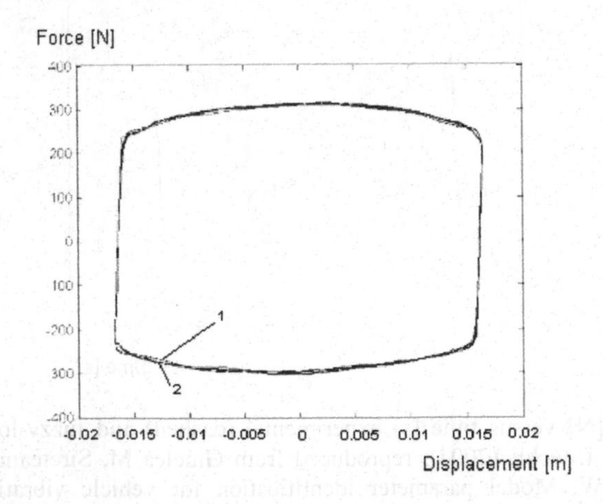

Fig. 6.31. Force [N] versus displacement [m] experimental (1) and fuzzy-logic-based model (2) [copyright IMechE (2004), reproduced from Giuclea M, Sireteanu T, Stancioiu D and Stammers CW, Model parameter identification for vehicle vibration control with magnetorheological dampers using computational intelligence methods, Proceedings of the Institution of Mechanical Engineers, Part I: Journal of Systems and Control Engineering, Publisher: Professional Engineering Publishing, ISSN 0959-6518, Vol. 218, N 7/2004, pp 569–581, used by permission]

Therefore, it can be stated that the fuzzy-logic-based model predicts well the behaviour of the damper in all the regions, including the pre-yield one.

6.5.4 Modelling the Variable Field Strength

To take full advantage of the MR damper models in active or semi-active control applications the effect of a continuously variable magnetic field must be considered. Spencer *et al.* (1997) generalised the model for such cases by imposing for the four parameters a linear dependence on the applied voltage. Seven parameters were considered to fit the measured data. None of the BW equation parameters was assumed to be voltage dependent. With this approach the model of the hysteresis loops and of the transition between the pre-yield and post-yield states are *de facto* unchanged when voltage varies and this is also in agreement with measured data.

Kyle and Roschke (2000) used a fuzzy-logic-based model with three inputs: displacement, voltage and velocity. The results obtained are very accurate, but the modelling is based on data generated by simulation. It is also worth noting that an increase in neuro-fuzzy modelling precision takes into consideration even the measured noise.

In order to determine the parametric models for variable field, it must be assumed that some of the parameters chosen are continuously variable functions of the current. From the experimentally measured data, it is possible to see that the hysteresis loop is not field dependent (Figure 6.17). Hence, the BW model parameters γ, ν and A are considered to be independent of the current.

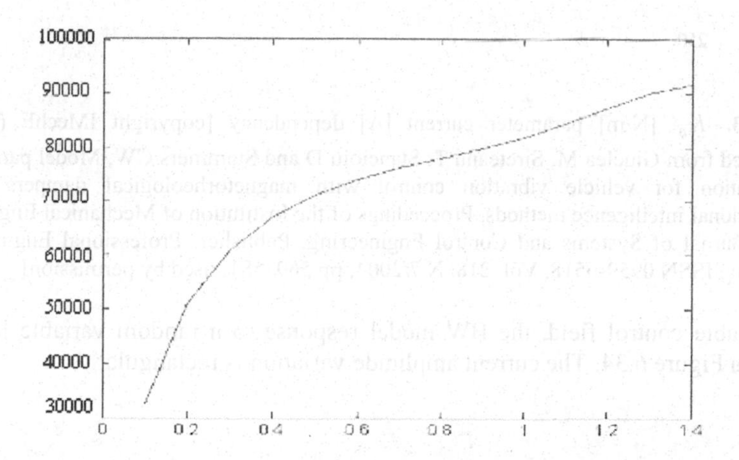

Fig. 6.32. α [N/m] parameter current [A] dependency [copyright IMechE (2004), reproduced from Giuclea M, Sireteanu T, Stancioiu D and Stammers CW, Model parameter identification for vehicle vibration control with magnetorheological dampers using computational intelligence methods, Proceedings of the Institution of Mechanical Engineers, Part I: Journal of Systems and Control Engineering, Publisher: Professional Engineering Publishing, ISSN 0959-6518, Vol. 218, N 7/2004, pp 569–581, used by permission]

Assuming that these model parameters are constant, the α parameter, which determines the yield force, is field dependent. Its current dependency can be supposed to be polynomial, having almost a linear variation for current values above 0.6 A (Figure 6.32).

The parameter c_0 that describes the damping dashpot has also a nonlinear variation. The current dependency of this parameter is similar to that of the α parameter shown in Figure 6.32.

The spring elastic constant, denoted K_0 in the BW-based model (Figure 6.33), exhibits an almost linear variation with the supplied current. Therefore the field-dependent parameters for the BW-based model were set to be α, c_0, K_0 and f_0. Polynomial variation functions having different degrees were used to fit the model.

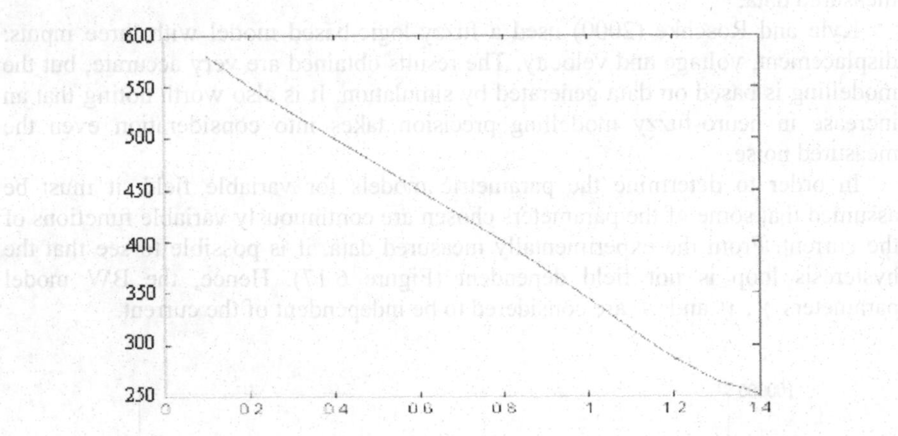

Fig. 6.33. K_0 [N/m] parameter current [A] dependency [copyright IMechE (2004), reproduced from Giuclea M, Sireteanu T, Stancioiu D and Stammers CW, Model parameter identification for vehicle vibration control with magnetorheological dampers using computational intelligence methods, Proceedings of the Institution of Mechanical Engineers, Part I: Journal of Systems and Control Engineering, Publisher: Professional Engineering Publishing, ISSN 0959-6518, Vol. 218, N 7/2004, pp 569–581, used by permission]

For variable control field, the BW model response to a random variable load is shown in Figure 6.34. The current amplitude variation is rectangular.

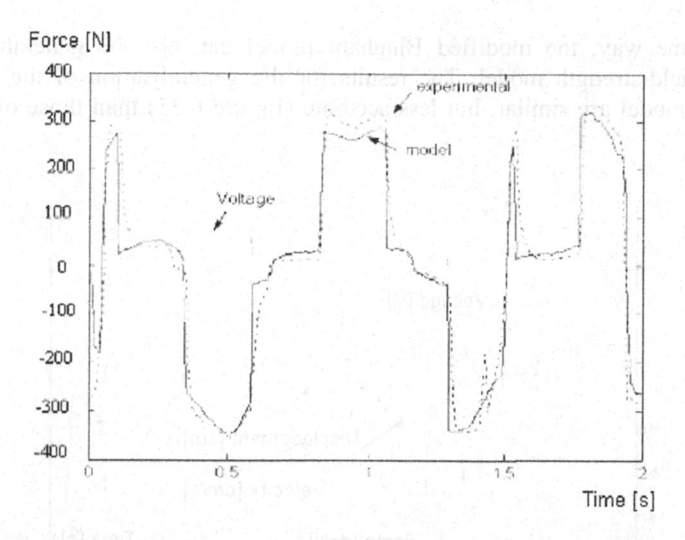

Fig. 6.34. Bouc–Wen model for variable control field. Force [N] vs. time [s] [copyright IMechE (2004), reproduced from Giuclea M, Sireteanu T, Stancioiu D and Stammers CW, Model parameter identification for vehicle vibration control with magnetorheological dampers using computational intelligence methods, Proceedings of the Institution of Mechanical Engineers, Part I: Journal of Systems and Control Engineering, Publisher: Professional Engineering Publishing, ISSN 0959-6518, Vol. 218, N 7/2004, pp 569–581, used by permission]

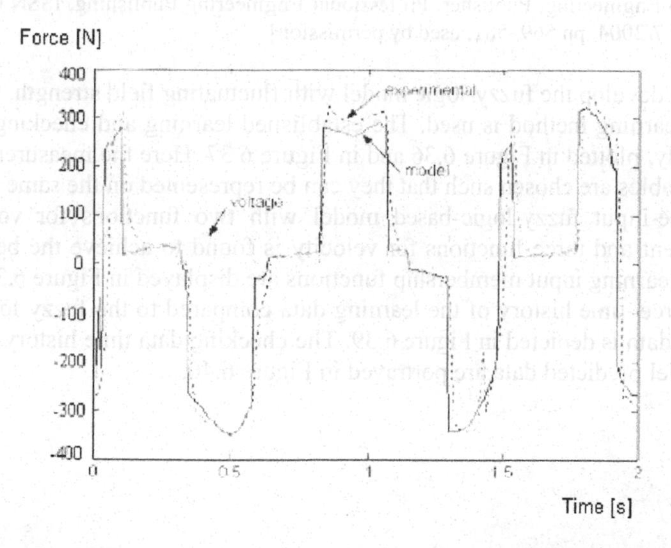

Fig. 6.35. Modified Bingham model for variable control field. Force [N] versus time [s]; [copyright IMechE (2004), reproduced from Giuclea M, Sireteanu T, Stancioiu D and Stammers CW, Model parameter identification for vehicle vibration control with magnetorheological dampers using computational intelligence methods, Proceedings of the Institution of Mechanical Engineers, Part I: Journal of Systems and Control Engineering, Publisher: Professional Engineering Publishing, ISSN 0959-6518, Vol. 218, N 7/2004, pp 569–581, used by permission]

In the same way, the modified Bingham model can also be generalised to a variable-field-strength model. The results for the generalisation of the modified Bingham model are similar, but less accurate (Figure 6.35) than those of the BW model.

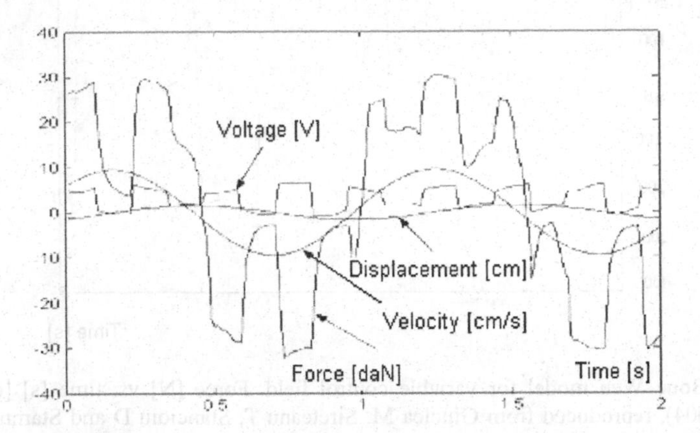

Fig. 6.36. The learning data [copyright IMechE (2004), reproduced from Giuclea M, Sireteanu T, Stancioiu D and Stammers CW, Model parameter identification for vehicle vibration control with magnetorheological dampers using computational intelligence methods, Proceedings of the Institution of Mechanical Engineers, Part I: Journal of Systems and Control Engineering, Publisher: Professional Engineering Publishing, ISSN 0959-6518, Vol. 218, N 7/2004, pp 569–581, used by permission]

In order to develop the fuzzy-logic model with fluctuating field strength, the neuro-adaptive learning method is used. The established learning and checking data are, respectively, plotted in Figure 6.36 and in Figure 6.37. Here the measurement units of the variables are chosen such that they can be represented on the same graph.

A three-input fuzzy-logic-based model with two functions for voltage and displacement and three functions for velocity is found to achieve the best results. The after-learning input membership functions are displayed in Figure 6.38.

The force–time history of the learning data compared to the fuzzy-logic model generated data is depicted in Figure 6.39. The checking data time history compared to the model predicted data are portrayed in Figure 6.40.

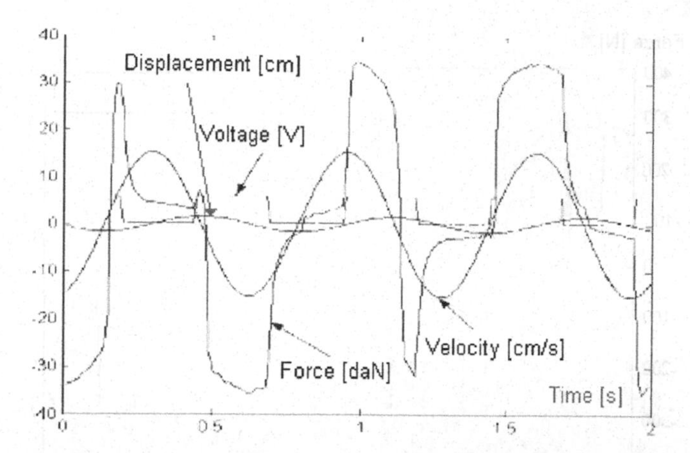

Fig. 6.37. The checking data [copyright IMechE (2004), reproduced from Giuclea M, Sireteanu T, Stancioiu D and Stammers CW, Model parameter identification for vehicle vibration control with magnetorheological dampers using computational intelligence methods, Proceedings of the Institution of Mechanical Engineers, Part I: Journal of Systems and Control Engineering, Publisher: Professional Engineering Publishing, ISSN 0959-6518, Vol. 218, N 7/2004, pp 569–581, used by permission]

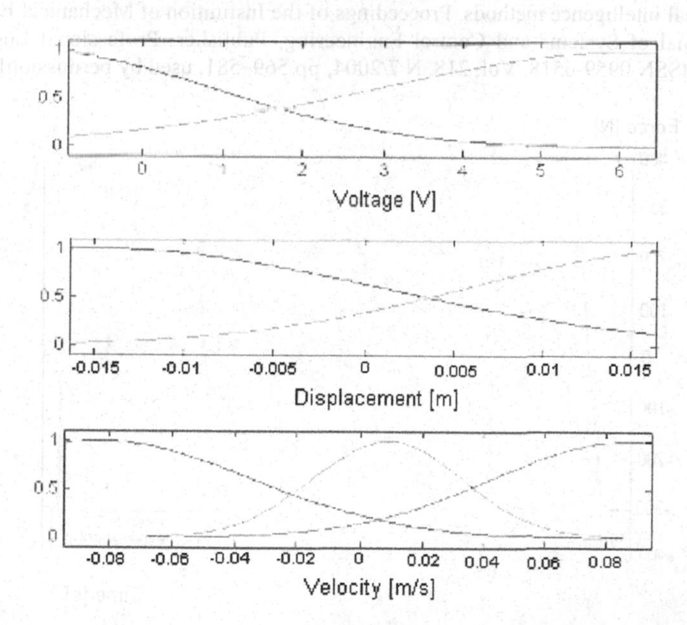

Fig. 6.38. The after-learning membership functions (voltage–displacement–velocity); [copyright IMechE (2004), reproduced from Giuclea M, Sireteanu T, Stancioiu D and Stammers CW, Model parameter identification for vehicle vibration control with magnetorheological dampers using computational intelligence methods, Proceedings of the Institution of Mechanical Engineers, Part I: Journal of Systems and Control Engineering, Publisher: Professional Engineering Publishing, ISSN 0959-6518, Vol. 218, N 7/2004, pp 569–581, used by permission]

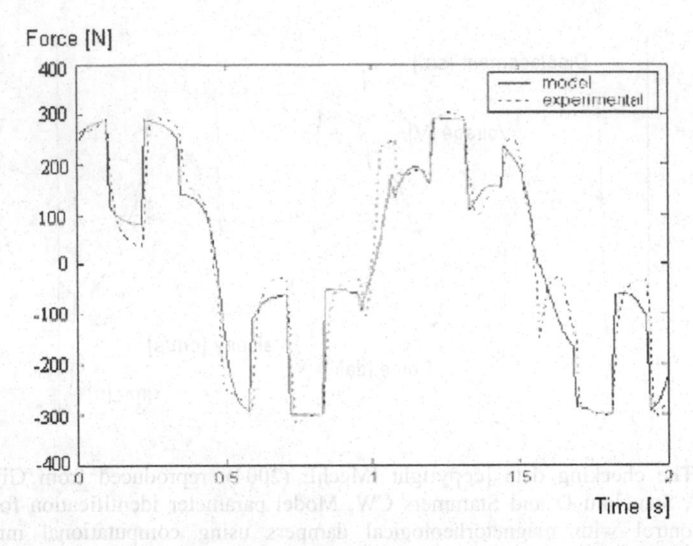

Fig. 6.39. The learning data validation. Force [N] versus time [s] [copyright IMechE (2004), reproduced from Giuclea M, Sireteanu T, Stancioiu D and Stammers CW, Model parameter identification for vehicle vibration control with magnetorheological dampers using computational intelligence methods, Proceedings of the Institution of Mechanical Engineers, Part I: Journal of Systems and Control Engineering, Publisher: Professional Engineering Publishing, ISSN 0959-6518, Vol. 218, N 7/2004, pp 569–581, used by permission]

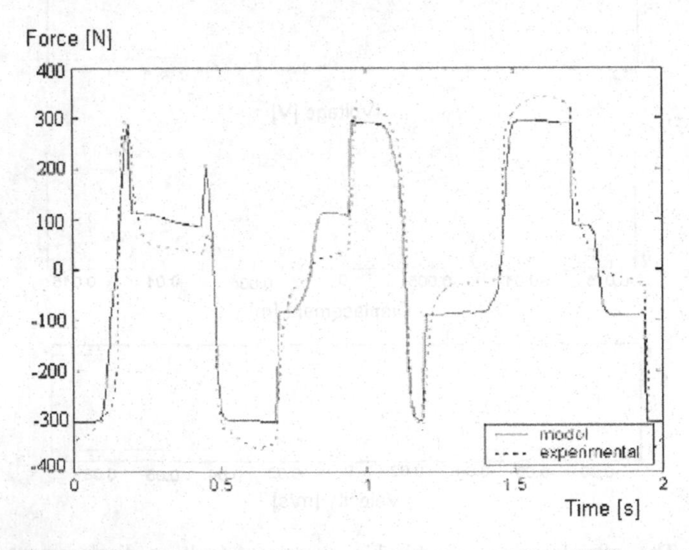

Fig. 6.40. The checking data validation. Force [N] versus time [s] [copyright IMechE (2004), reproduced from Giuclea M, Sireteanu T, Stancioiu D and Stammers CW, Model parameter identification for vehicle vibration control with magnetorheological dampers using computational intelligence methods, Proceedings of the Institution of Mechanical Engineers, Part I: Journal of Systems and Control Engineering, Publisher: Professional Engineering Publishing, ISSN 0959-6518, Vol. 218, N 7/2004, pp 569–581, used by permission]

6.5.5 GA-based Method for MR Damper Model Parameters Identification

As was shown by Spencer *et al.* (1997), the best results portraying the hysteretic behaviour of MR dampers are obtained with the BW-modified model, which is depicted in Figure 6.41.

In this subsection, a method for finding the BW-modified model parameters is proposed by using a genetic algorithm (GA) optimisation procedure. Initially, the values of the parameters are found for a set of constant values of the applied current and an imposed cyclic motion, such that the predicted response optimally fit the experimental data. The second step consists in obtaining the variation law for each parameter as a function of the current, considering the corresponding values from the first step. The resulting model is validated by comparison of the predicted and experimentally obtained responses for some cases with variable control current.

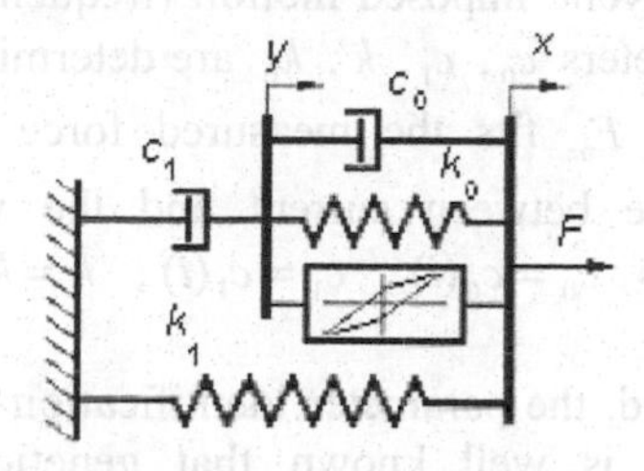

Fig. 6.41. BW-modified mechanical model [copyright IMechE (2004), reproduced from Giuclea M, Sireteanu T, Stancioiu D and Stammers CW, Model parameter identification for vehicle vibration control with magnetorheological dampers using computational intelligence methods, Proceedings of the Institution of Mechanical Engineers, Part I: Journal of Systems and Control Engineering, Publisher: Professional Engineering Publishing, ISSN 0959-6518, Vol. 218, N 7/2004, pp 569–581, used by permission]

In this model the damping force generated by the device is given by

$$F = c_1\dot{y} + k_1(x - x_0)\,, \tag{6.74}$$

where x is the total relative displacement and x_0 the initial deflection of the accumulator gas spring with stiffness k_1. The partial relative displacement y and the evolutionary variable z are governed by the coupled differential equations:

$$\dot{y} = \frac{1}{c_0 + c_1}[c_0\dot{x} + k_0(x - y) + kz]\,, \tag{6.75}$$

$$\dot{z} = -\gamma|\dot{x} - \dot{y}|\,z|z|^{n-1} - \nu(\dot{x} - \dot{y})|z|^n + A(\dot{x} - \dot{y})\,, \tag{6.76}$$

where k is a stiffness coefficient associated with the displacement z. The parameters n, γ, ν, A and k_1 are considered fixed, and the parameters c_0, c_1,

k and k_0 are assumed to be functions of the delayed current $i = i(t)$ applied to the MR damper: $c_0 = c_0(i)$, $c_1 = c_1(i)$, $k = k(i)$, $k_0 = k_0(i)$. If $i_0(t)$ is the supplied current, then the dynamics of the MR fluid to reach its rheological equilibrium are modelled by a first-order lag (T is the time constant):

$$\frac{\mathrm{d}i}{\mathrm{d}t} = -\frac{1}{T}(i - i_0). \tag{6.77}$$

The BW-modified model includes the above-mentioned parameters (c_0, c_1, k, k_0) that depend on the applied current. In order to determine the corresponding functional dependencies, a GA-based inverse method is applied. For different constant current values (i = 0.02 A, 0.06 A, 0.1 A, 0.2 A, 0.4 A, 0.6 A, 0.8 A, 1.05 A, 1.45 A, 1.75 A) and cyclic imposed motion (frequency 1.5 Hz, amplitude 16 mm), the values of parameters c_0, c_1, k, k_0 are determined by using a GA, such that the predicted force F_p fits the measured force F_exp. Subsequently, by analysing the dependence between current and the values obtained for the parameters, the functions $c_0 = c_0(i)$, $c_1 = c_1(i)$, $k = k(i)$, $k_0 = k_0(i)$ can be determined.

In the method proposed, the parameter identification is treated as a black-box optimisation problem. It is well known that genetic algorithms are robust probabilistic search techniques with very good results in black-box optimisation problems (Goldberg, 1989). They are based on the mechanism of natural genetics and natural selection, starting with an initial population of (encoded) problem solutions and evolving towards better solutions. For the considered optimisation procedure, a real-coded GA is employed with four real genes corresponding to the four coefficients c_0, c_1, k, k_0. By using appropriate scaling factors, it can be assumed that the parameters take on values within the interval [0, 1]. The other characteristics of the GA applied are an averaged crossover with probability 0.8, uniform mutation, Monte Carlo selection and an objective function of the RMS error between the predicted and experimental response.

For numerical simulations the following values of the fixed coefficients were chosen: n = 2, d = 500 N/m, g = 613000 m^{-2}, α = 30.56, k_1 = 540 N/m. By applying the proposed GA, after about 500 generations, the values given in Table 6.4 were produced. The accuracy of GA optimisation can be evinced from Figures 6.42, 6.43, 6.44 and 6.45, showing the predicted and experimentally obtained responses for 0.06 A and 1.75 A. Since the parameter k_0 has a fairly irregular variation with respect to the current the average value k_0 =1050 N/m was considered.

Table 6.4. Parameter values obtained by GA [copyright IMechE (2004), reproduced from Giuclea M, Sireteanu T, Stancioiu D and Stammers CW, Model parameter identification for vehicle vibration control with magnetorheological dampers using computational intelligence methods, Proceedings of the Institution of Mechanical Engineers, Part I: Journal of Systems and Control Engineering, Publisher: Professional Engineering Publishing, ISSN 0959-6518, Vol. 218, N 7/2004, pp 569–581, used by permission]

$i[\mathrm{A}]$	$c_0\left[\frac{\mathrm{Ns}}{\mathrm{m}}\right]$	$c_1\left[\frac{\mathrm{Ns}}{\mathrm{m}}\right]$	$k\left[\frac{\mathrm{N}}{\mathrm{m}}\right]$	$k_0\left[\frac{\mathrm{N}}{\mathrm{m}}\right]$
0.02	121	10300	2950	527
0.06	340	8350	15300	306
0.10	465	13900	23900	22.3
0.20	966	33600	34000	468
0.40	1690	83900	58500	988
0.06	2880	93300	89800	1990
0.80	3220	101800	104900	1240
1.05	3500	107800	114500	1330
1.45	4730	111600	114400	1630
1.75	4050	122500	133900	2010

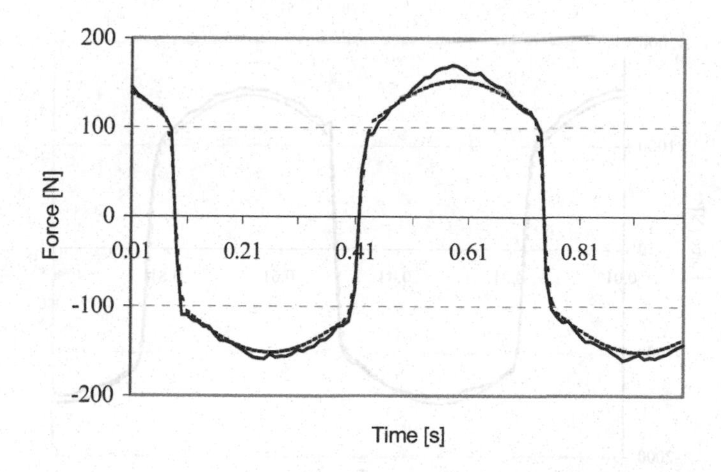

Fig. 6.42. Force vs. time for 0.06 A ____ experimental; …… predicted [copyright IMechE (2004), reproduced from Giuclea M, Sireteanu T, Stancioiu D and Stammers CW, Model parameter identification for vehicle vibration control with magnetorheological dampers using computational intelligence methods, Proceedings of the Institution of Mechanical Engineers, Part I: Journal of Systems and Control Engineering, Publisher: Professional Engineering Publishing, ISSN 0959-6518, Vol. 218, N 7/2004, pp 569–581, used by permission]

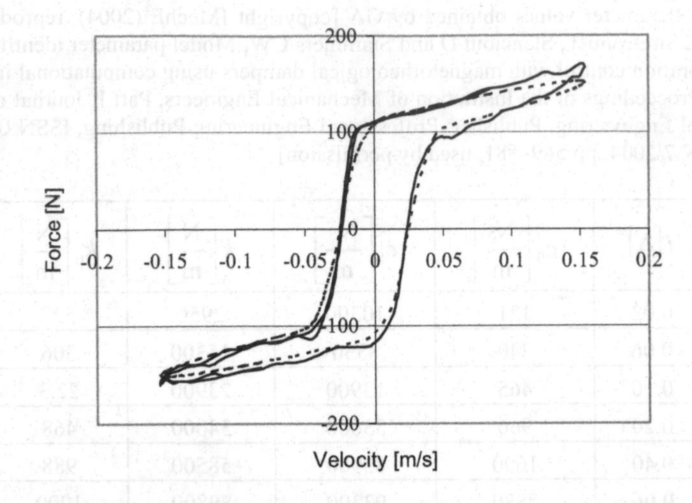

Fig. 6.43. Force [N] versus velocity [m/s] for 0.06 A ____ experimental; …… predicted [copyright IMechE (2004), reproduced from Giuclea M, Sireteanu T, Stancioiu D and Stammers CW, Model parameter identification for vehicle vibration control with magnetorheological dampers using computational intelligence methods, Proceedings of the Institution of Mechanical Engineers, Part I: Journal of Systems and Control Engineering, Publisher: Professional Engineering Publishing, ISSN 0959-6518, Vol. 218, N 7/2004, pp 569–581, used by permission]

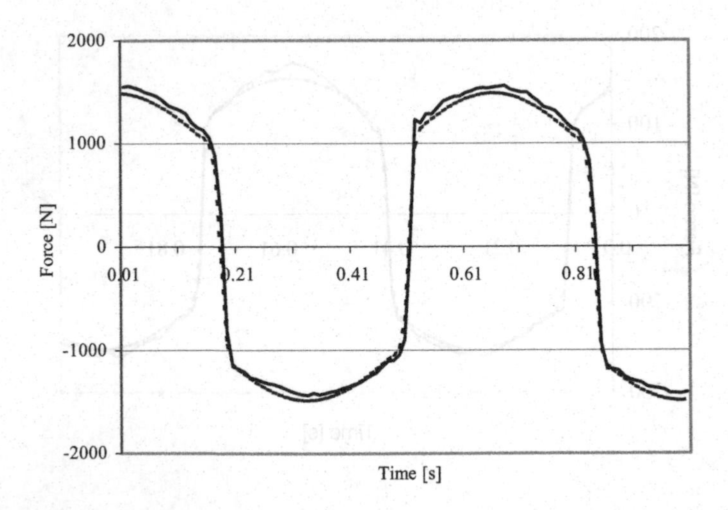

Fig. 6.44. Force [N] versus time [s] for 1.75 A ____ experimental; …… predicted [copyright IMechE (2004), reproduced from Giuclea M, Sireteanu T, Stancioiu D and Stammers CW, Model parameter identification for vehicle vibration control with magnetorheological dampers using computational intelligence methods, Proceedings of the Institution of Mechanical Engineers, Part I: Journal of Systems and Control Engineering, Publisher: Professional Engineering Publishing, ISSN 0959-6518, Vol. 218, N 7/2004, pp 569–581, used by permission]

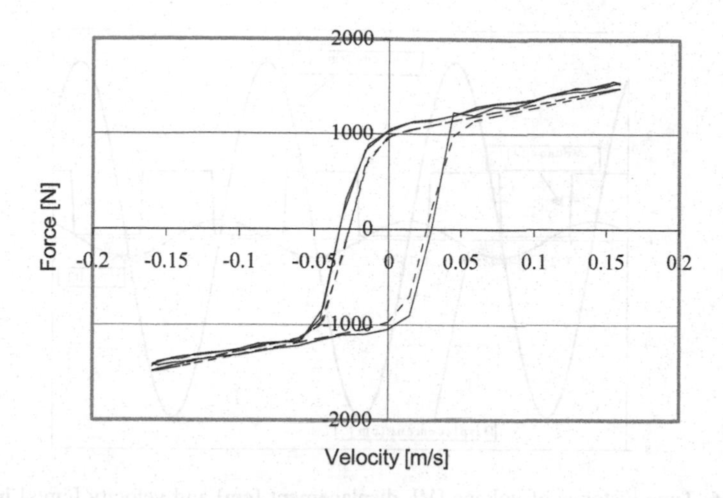

Fig. 6.45. Force [N] versus velocity [m/s] for 1.75 A ______ experimental; ……. predicted [copyright IMechE (2004), reproduced from Giuclea M, Sireteanu T, Stancioiu D and Stammers CW, Model parameter identification for vehicle vibration control with magnetorheological dampers using computational intelligence methods, Proceedings of the Institution of Mechanical Engineers, Part I: Journal of Systems and Control Engineering, Publisher: Professional Engineering Publishing, ISSN 0959-6518, Vol. 218, N 7/2004, pp 569–581, used by permission]

The functional approximations obtained for the remaining parameters c_0, c_1, k, are

$$c_0(i) = 26.134\, i + 5.164\,, \tag{6.78}$$

$$c_1(i) = 1150 \tanh(1.95\, i)\,, \tag{6.79}$$

$$k(i) = 1297.2 \tanh(1.3\, i)\,. \tag{6.80}$$

Therefore, the BW-modified model is completely defined and can be used for different loading conditions and variation patterns of the applied current.

In the remainder of this section, a comparison is made between the measured and predicted forces computed for the identified model. Simulations were performed using the experimentally determined displacement x and the calculated velocity $\dot{x}$ of the piston rod in determining the force generated by the damper model. The effectiveness of the model in describing the dynamic behaviour of the MR damper has been investigated in a variety of deterministic tests. When compared with experimental data, the model is shown to accurately predict the response of the MR damper over a wide range of operating conditions.

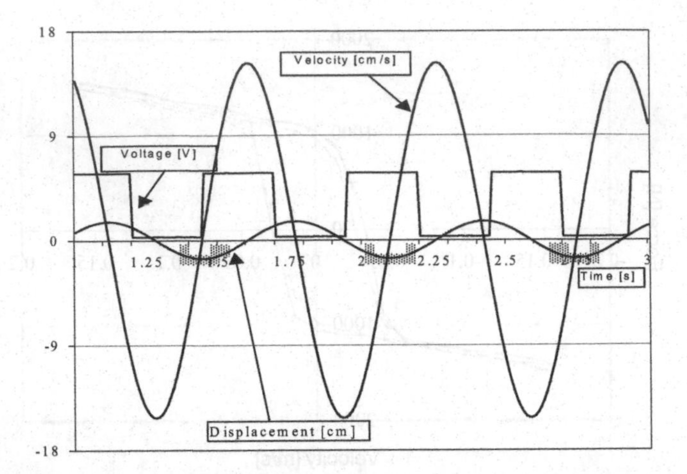

Fig. 6.46. Time histories of voltage [V], displacement [cm] and velocity [cm/s] in the first test case [copyright IMechE (2004), reproduced from Giuclea M, Sireteanu T, Stancioiu D and Stammers CW, Model parameter identification for vehicle vibration control with magnetorheological dampers using computational intelligence methods, Proceedings of the Institution of Mechanical Engineers, Part I: Journal of Systems and Control Engineering, Publisher: Professional Engineering Publishing, ISSN 0959-6518, Vol. 218, N 7/2004, pp 569–581, used by permission]

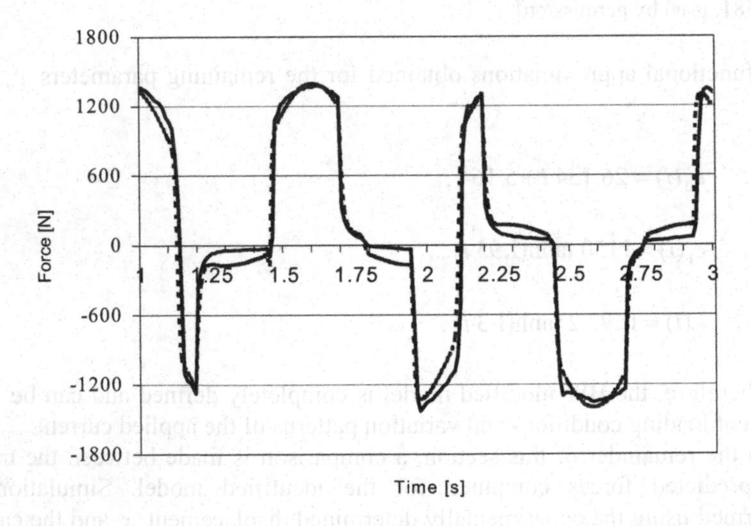

Fig. 6.47. Force [N] versus time [s] in the first test case ______ experimental; …… predicted [copyright IMechE (2004), reproduced from Giuclea M, Sireteanu T, Stancioiu D and Stammers CW, Model parameter identification for vehicle vibration control with magnetorheological dampers using computational intelligence methods, Proceedings of the Institution of Mechanical Engineers, Part I: Journal of Systems and Control Engineering, Publisher: Professional Engineering Publishing, ISSN 0959-6518, Vol. 218, N 7/2004, pp 569–581, used by permission]

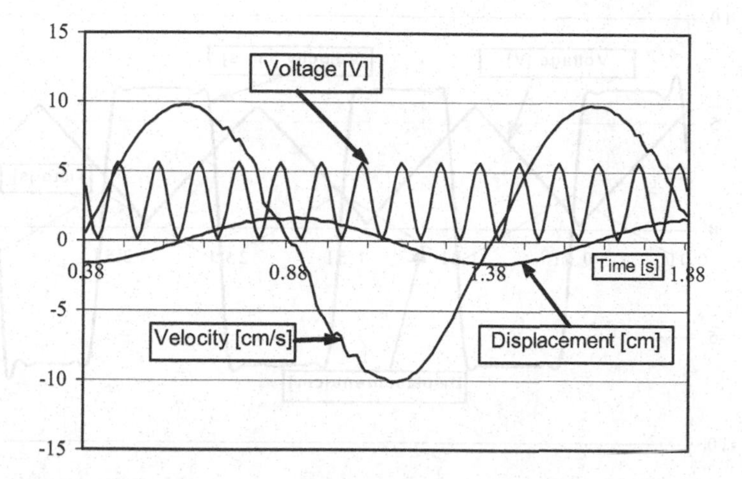

Fig. 6.48. Time histories of voltage [V], displacement [cm] and velocity [cm/s] in the second test case [copyright IMechE (2004), reproduced from Giuclea M, Sireteanu T, Stancioiu D and Stammers CW, Model parameter identification for vehicle vibration control with magnetorheological dampers using computational intelligence methods, Proceedings of the Institution of Mechanical Engineers, Part I: Journal of Systems and Control Engineering, Publisher: Professional Engineering Publishing, ISSN 0959-6518, Vol. 218, N 7/2004, pp 569–581, used by permission]

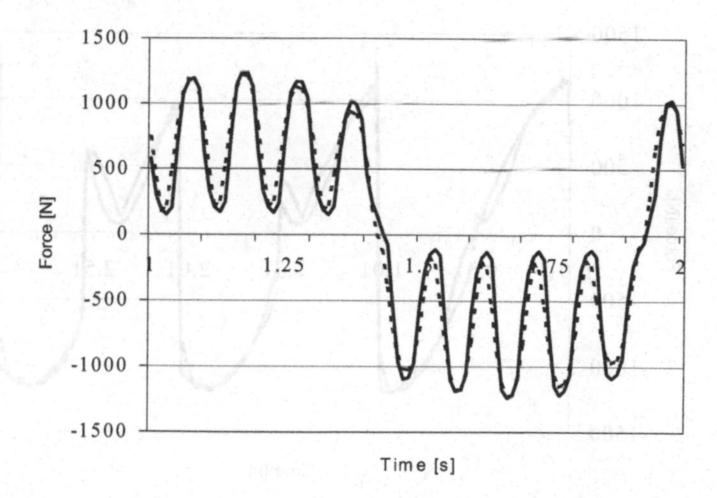

Fig. 6.49. Force [N] versus time [s] in the second test case _____ experimental;...... predicted [copyright IMechE (2004), reproduced from Giuclea M, Sireteanu T, Stancioiu D and Stammers CW, Model parameter identification for vehicle vibration control with magnetorheological dampers using computational intelligence methods, Proceedings of the Institution of Mechanical Engineers, Part I: Journal of Systems and Control Engineering, Publisher: Professional Engineering Publishing, ISSN 0959-6518, Vol. 218, N 7/2004, pp 569–581, used by permission]

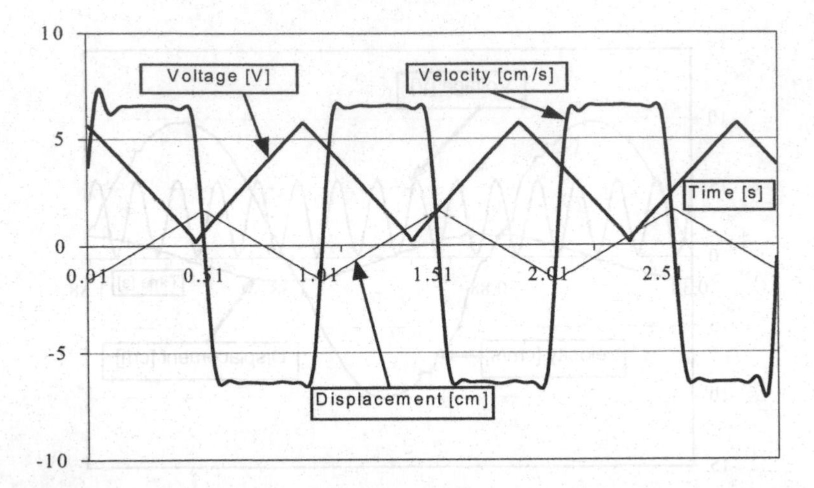

Fig. 6.50. Voltage [V], displacement [cm] and velocity [cm/s] for the third test case [copyright IMechE (2004), reproduced from Giuclea M, Sireteanu T, Stancioiu D and Stammers CW, Model parameter identification for vehicle vibration control with magnetorheological dampers using computational intelligence methods, Proceedings of the Institution of Mechanical Engineers, Part I: Journal of Systems and Control Engineering, Publisher: Professional Engineering Publishing, ISSN 0959-6518, Vol. 218, N 7/2004, pp 569–581, used by permission]

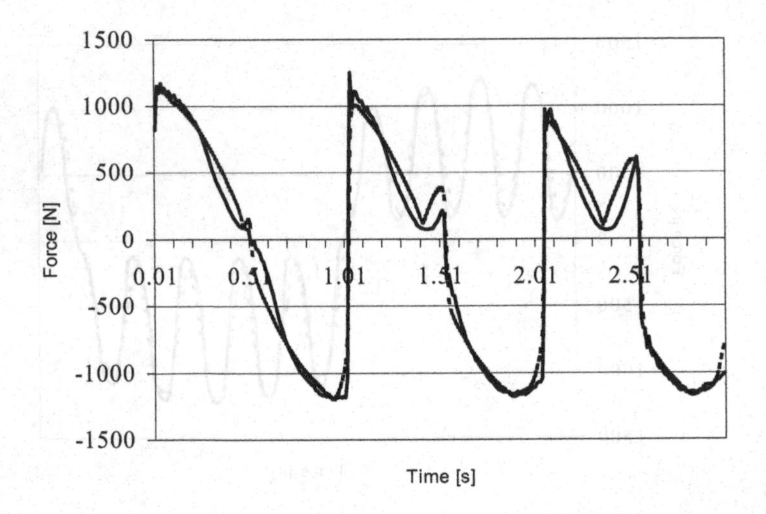

Fig. 6.51. Force [N] versus time [s] in the third test case_____ experimental; …… predicted [copyright IMechE (2004), reproduced from Giuclea M, Sireteanu T, Stancioiu D and Stammers CW, Model parameter identification for vehicle vibration control with magnetorheological dampers using computational intelligence methods, Proceedings of the Institution of Mechanical Engineers, Part I: Journal of Systems and Control Engineering, Publisher: Professional Engineering Publishing, ISSN 0959-6518, Vol. 218, N 7/2004, pp 569–581, used by permission]

To illustrate this statement, the results obtained in three test cases are presented. The time histories of the displacement, velocity and of the voltage applied to the MR damper controller for each test case are plotted in Figures 6.46–6.51. The current i_0 supplied by the device controller is proportional to the applied voltage, the output–input ratio being 0.4 A/V. The units of measurement of Figures 6.36, 6.37, 6.46, 6.48 and 6.50 (daN, cm, cm/s) have been chosen to allow all the relevant quantities to be plotted on the same graph.

7

Case Studies

7.1 Introduction

A number of case studies are now presented to show applications of the concepts illustrated in the previous chapters. After a succinct introduction of some aspects of data acquisition and digital control, three case studies are presented, namely a friction damper-based suspension unit for a saloon car, a magnetorheological damper-based seat suspension for vehicles not equipped with primary suspensions and a semi-active suspension system with magnetorheological dampers for heavy vehicles where the emphasis is not only on ride comfort but also on road damage reduction.

7.1.1 Some Aspects of Data Acquisition and Control

The reader is assumed to be familiar with the fundamental theoretical concepts of computer-based data acquisition and digital control (Shannon theorem, Nyquist theorem and digital-to-analogue and analogue-to-digital conversion). For furthering the topic the reader can refer to the numerous textbooks on the subject (*e.g.*, Leigh, 1992). Data acquisition and real-time control can be implemented using an embedded microprocessor architecture, a PC or a programmable logic controller (PLC) equipped with one or more I/O cards. Commercial data acquisition and control cards are made to acquire a variety of input signals and issue a variety of output signals including not only digital and analogue (*e.g.*, 0–10 V, ± 5 V, 4–20 mA) signals but also more specialised signals (*e.g.*, those produced by thermocouples, thermoresistances, solenoid valves, servovalves, vibration sensors, frequency modulated signals *etc.*). Signals are properly conditioned (filtered, amplified, modulated, demodulated *etc.*) by appropriate hardware within the card or via software.

In the case studies described an off-the-shelf card was used. The card was capable of receiving up to 16 single-ended channels with 12-bit resolution. Depending upon the transducers output and actuators input range, the card I/O range could be configured. For instance in the applications described, analogue inputs were configured to have a unipolar range of 0–10 V and analogue outputs a bipolar range of ±5 V. This implies that the mean value of the quantisation error is 0.025 V, which is negligible with respect to the range of the converter[2] and in the suspension application does not produce any appreciable deterioration of the controller performance.

The card can be made to operate in clocked D/A mode; in this mode its clock is set up to produce the required sampling frequency. In the light of the Shannon theorem, considering the frequency content of the vehicle dynamic response, the sampling frequency can be set to 100 Hz. The total delay introduced in the sampling, holding and multiplexing phases is negligible.

For data acquisition, effective software is key. In the work described, a commercial data acquisition package and an in-house frequency response analyser software developed at the University of Bath (UK) were employed.

Digital closed-loop control must be implemented in real time, and therefore an environment with a deterministic operating system is required. This can be achieved with an embedded architecture or a PLC. Typically PLCs are used in industrial plants where a vast number of loops need to be controlled or monitored within the framework of a hierarchical (pyramidal type) distributed architecture. A PLC is often interfaced with a PC (often referred to as a HMI, a human–machine interface) for graphic display of the controlled plant parameters. I/O signals can be either hardwired or connected via a network (sometimes referred to as remote I/O). Signals, appropriately routed, can also be transferred to other PCs or PLCs over the network, using a variety of transmission media (*e.g.*, serial links, Ethernet or fibre optics) and appropriate protocols. A typical automotive protocol is the controller area network (CAN bus). Control software can be written in a graphical language (Ladder-like) or in C/C++. Fast dynamics routines can also be written in Assembly language to optimise performance.

Closed-loop control software in the applications described was written in C. Good programming practice suggests to write software in modular form, and develop a user interface that allows the entry of the key configurable parameters of the control algorithm, enabling data to be saved to file for off-line post-processing.

In automotive applications embedded architectures are used and feedback signals are either hardwired using shielded cables (and additional low-pass filtering stages are present in the on-board electronics) or transmitted over the network. Furthermore dedicated circuitries are present to increase the noise rejection.

---

2. The resolution V_0 of a converter can be calculated with the following expression that links range of the converter, resolution and number of bits:

$$[K]\, Log_{10} 2 = Log_{10}\{(V_{max} - V_{min})/V_0 + 1\},$$

where $[K]$ is an integer that represents the number of bits and $V_{max} - V_{min}$ is the range of the converter.

7.2 Car Dynamics Experimental Analysis

A car, thought of as a rigid body unilaterally constrained by the road, is a 6DOF system having three translational (forward and backward motion, side slip and bounce) and three rotational (roll, pitch and yaw) degrees of freedom.

From the viewpoint of assessing the ride performance of a suspension system, bounce, roll and pitch motions are the most pertinent. Side slip, which occurs only for a lack of adhesion, and yaw are of interest mainly to handling studies.

At the outset it is necessary to specify the inputs which excite car dynamic responses of interest and which best represent real conditions on the road. Vehicle driving tests in controlled conditions would appear to be the most natural solution. However it is also possible to emulate the road input with a road simulator. This allows measurements to be made in more controlled conditions; besides, it permits tests which are not feasible on a road to be carried out (*e.g.*, sine wave inputs or the excitation of single suspension units separately to mimic a quarter-car-like behaviour). Moreover on a road simulator the level of external disturbances is reduced or even eliminated (*e.g.*, wind gust) compared to a road test, and hence tests are more repeatable.

Preliminary tests on the car equipped with its original viscous damper are necessary in order to have a set of data to benchmark a semi-active system. In the following developments it will be described how to practically carry out vehicle measurements and how to process data effectively.

7.2.1 The Experimental Set-up

The heart of the vehicle testing equipment is the four-poster road simulator. This is essentially composed of four hydraulic actuators on which the vehicle under testing (a Ford Orion) is placed (Figure 7.1).

The main features of the road simulator available at the University of Bath are now briefly described to identify its potential and its limits. Each hydraulic actuator is position-controlled through a servovalve. The rig is operated via a human–machine interface PC where software manages and supervises all the operations. It is possible to select elementary inputs (sine waves, half-sine waves, square waves) in the 0–25 Hz frequency range and ±125 mm amplitude range. The relative phase shift among the actuators can be controlled as well. In this way pure bounce, roll and pitch inputs, and other customised inputs, can be generated. Furthermore, it is possible to create user-defined waveforms by composing elementary functions via a graphical editor.

In order to gain an understanding of the vehicle dynamic phenomena of interest and to provide a comprehensive set of data to assess suspension performance, tests with the following inputs are essential:

- sinusoidal input to a single wheel
- bounce, roll and pitch tests with sinusoidal input
- pseudo-random input
- bump test

Sinusoidal waveforms do not represent any realistic road condition. However they are simple and well-known inputs and they allow the use of frequency-domain methods for manipulating data. Sinusoidal tests give a first clue to the effectiveness of the suspensions: they permit an assessment of how controlled suspensions behave with respect to passive suspensions.

Fig. 7.1. Test vehicle on the four-poster road simulator

It must be remarked that, because of the non-linear behaviour of the four-poster shaker, it is not possible to obtain a truly sinusoidal motion in spite of a sinusoidal driving input voltage applied to the shaker. In addition the ride dynamic response of the passive car is linear only to a first approximation (the main non-linearities being the tyre characteristics and the hysteretic viscous damper characteristics, including mountings and rubber bushes). As a consequence of the whole system being non-linear, it is not possible to define a proper frequency response (in the control sense). For this reason the most significant plot is the RMS value (or the ratio of two RMS values, for instance the ratio of chassis and wheel accelerations versus the fundamental harmonic of the excitation frequency for a fixed input amplitude). This experimental frequency response should more appropriately be called the acceleration transmissibility ratio, because it includes the effect of all the harmonic content of the signals. However for ease of terminology the term frequency response will be used hereafter.

The passive system frequency responses are ideally expected to be worse (*i.e.*, with larger values) than the semi-active frequency responses in the working frequency range.

This is a first crucial assessment in the frequency domain before further investigations in order to evaluate the properties of the responses in the time domain (harmonic content, peak values, higher-derivative trends, peak values *etc.*).

The vehicle is instrumented with four accelerometers and one displacement transducer mounted on the axle where the semi-active damper is fitted. Two accelerometers are mounted on each rear wheel, connected through small cylinders bolted on the wheels. The connections to the vehicle must be made as compact and stiff as possible in order to reduce unwanted vibration which could affect the measurement. The other two accelerometers are fixed on the chassis of the car, close to each rear suspension. Accelerometers have a principal axis of sensitivity, therefore they need to be mounted vertically in order to measure the vertical component of chassis' and wheel acceleration. The calibration of these devices is based on gravity (by interpolating readings at +1 g, 0 and –1 g). Because of their sensitivity to acceleration due to gravity, it might be necessary to compensate (in hardware or in software) the offset of 1 g that they may introduce.

Signals from the transducers are manipulated in conditioning cards. As mentioned in Section 7.1.1, most conditioning circuitries are essentially composed of an amplification stage followed by a filtering stage. A stabilised power supply provides the required constant voltage to the controller. Pre-amplification is necessary to make the signal more robust, increasing its immunity to noise and making it less interference sensitive. The amplifier gain must be suitably chosen considering the order of magnitude of car vertical accelerations (a good choice could be 1g =10 V for instance). The grounding is done through an earth resistance on the metallic trolley carrying the equipment; the ground is unique to avoid earth loops.

A critical stage of the conditioning chain is filtering. Acceleration is typically a noisy signal: stray effects due to the mounting, rubber bushes, friction *etc.* are picked up by the transducer and therefore it is necessary to low-pass filter the signal. The break frequency of the filter must be chosen trading off among different conflicting requirements: if it is too low it introduces an undesired phase lag and further it smoothes the waveforms by cutting off the high-frequency harmonic content. This only slightly affects integral parameters like the RMS (higher-order harmonics typically have small amplitudes). However it can diminish the actual value of some time-domain information such as peak acceleration values or jerk content which are comfort related. These need to be quantified properly (particularly in a controlled suspension with a switching control logic). On the other hand if the break frequency is too high, a significant level of noise can be superimposed on the signal, affecting not only its readability but also the accurate evaluation of its RMS. In addition, a high level of noise could introduce chattering if acceleration signals were used in a switching logic.

The choice of an appropriate break frequency is a trial-and-error process and can be initially performed with an analogue, manually-adjustable filter. In the applications investigated a second-order Butterworth filter with a break frequency of 40 Hz appeared to be appropriate. With this choice the phase lag introduced is only 10° at 5 Hz. Digital filtering is another option too.

The relative displacement is measured with a position transducer mounted between body and wheel. An externally mounted LVDT is not appropriate as, when the car undergoes testing, tyre torsional modes may turn the wheels, with a risk of damaging the device. Mounting connections through plastic bolts and rose joints are not always sufficient to provide protection. A pull-wire displacement transducer is instead a better choice. This potentiometric device is safe, since the position is measured with a flexible wire, and not a rigid rod. Potentiometers (depending on the materials they are made of, *e.g.*, hybrid conductive plastic) have an extremely high resolution, resulting in a very clean measured signal; no filtering is in fact necessary. In order to calibrate it accurately, a stepper motor can be employed. Finally measurements need to be carried out with tyres inflated at their rated pressures and with wheel brakes on to minimise tyre torsional vibration, which tends to rotate them.

7.2.2 Post-processing and Measurement Results

As far as sinusoidal tests are concerned the most meaningful data are the time responses and the ratio of chassis to wheel RMS acceleration. The ride model of the car is supposed to be linear to a first approximation; this implies that the acceleration ratio is equal to the velocity and displacement ratios:

$$\frac{a_1}{a_2} = \frac{j\omega v_1}{j\omega v_2} = \frac{\omega^2 x_1}{\omega^2 x_2}, \tag{7.1}$$

a_i , v_i , x_i being, respectively, chassis and wheel (subscripts 1 and 2) acceleration, velocity and displacement. Hence transmissibility curves can be readily obtained.

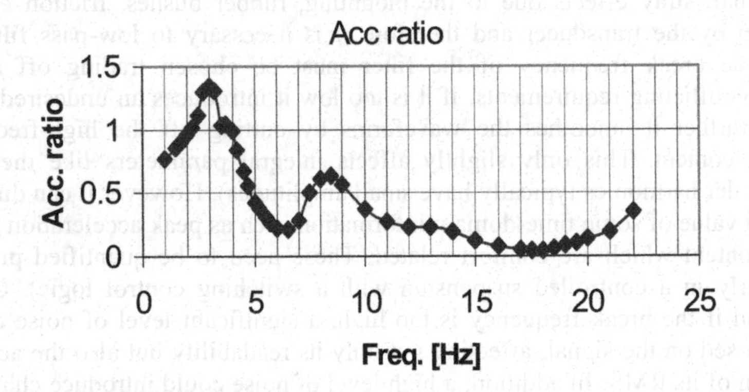

Fig. 7.2. Body-to-wheel acceleration ratio of the rear right corner of the car. Sinusoidal input to one wheel, amplitude: 3 mm

A simple test to verify the amount of linear behaviour of the passive car can be carried out by applying inputs with three different amplitudes. The closer the three responses are to one another the more the system approximates linear behaviour. Figure 7.2 shows the response of the car when only one wheel is excited; the ratio of body over wheel RMS acceleration is plotted.

The four-poster shaker actuator generates a sinusoidal waveform with a peak value of 3 mm in the range 1–22 Hz. In the car response three resonances are evident; the first one is the chassis resonance at about 3 Hz; this is fairly high:

typically this resonance occurs at about 1.5 Hz. This shift is essentially due to the small portion of sprung mass loading the suspension strut, since only the rear right corner of the car is shaken. The second resonance is at about 8.5 Hz; this is due to cross-coupling effects: around that frequency range a compound motion of bounce, roll and pitch is present; an engine mounting resonance is also possible. Eventually the wheel-hop resonance occurs at 18 Hz, although it is not readily recognisable on the graph of Figure 7.2; however if the wheel acceleration response is plotted (Figure 7.3) the resonance becomes clear.

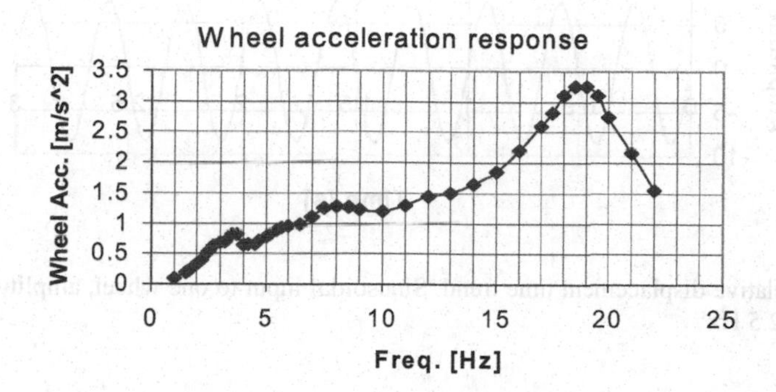

Fig. 7.3. Rear right wheel acceleration response. Sinusoidal input to one wheel, amplitude: 3 mm

Figure 7.4 shows the same response as above when three different input amplitudes (3 mm, 5 mm and 7 mm) are applied (for the purpose of testing linearity). The differences among the three graphs are reasonably small; hence it is possible to state that the non-linear effects, although present, are not so significant and the hypothesis of linearity is fairly realistic within the amplitude range of the signals. Non-linear effects occur typically with large signals (although small signals can also excite non-linear phenomena).

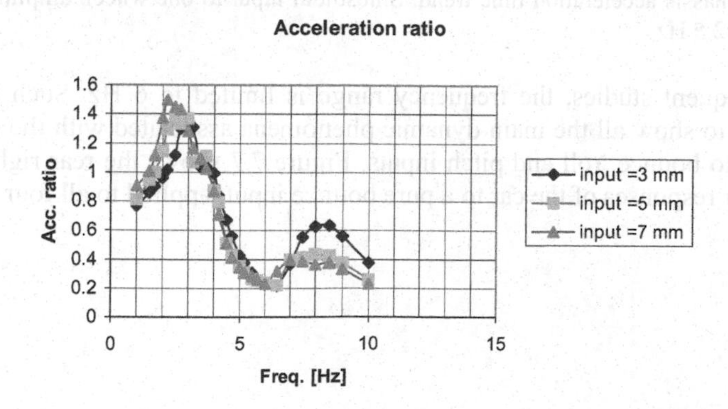

Fig. 7.4. Body-to-wheel acceleration ratio of the rear right corner of the car. Sinusoidal input to one wheel, amplitudes: 3 mm, 5 mm, 7 mm

Figures 7.5 and 7.6 depict two responses in the time domain: relative body-to-wheel displacement and body acceleration (units have been scaled), in response to a 3 mm amplitude sinusoidal signal at a frequency of 2.5 Hz. The traces are nearly sinusoidal. This further confirms the hypothesis of linearity.

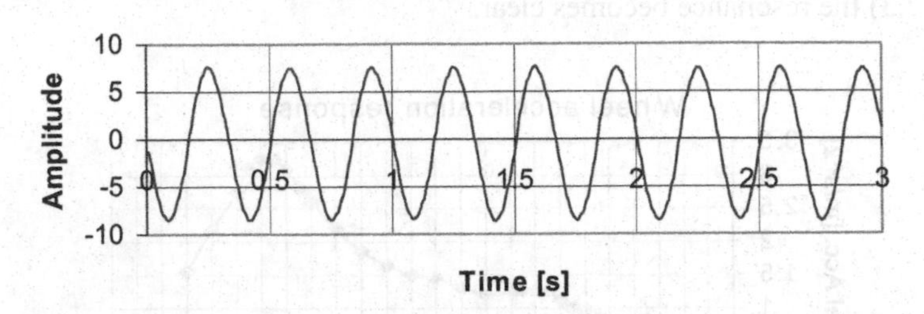

Fig. 7.5. Relative displacement time trend. Sinusoidal input to one wheel, amplitude: 3 mm, frequency: 2.5 Hz .

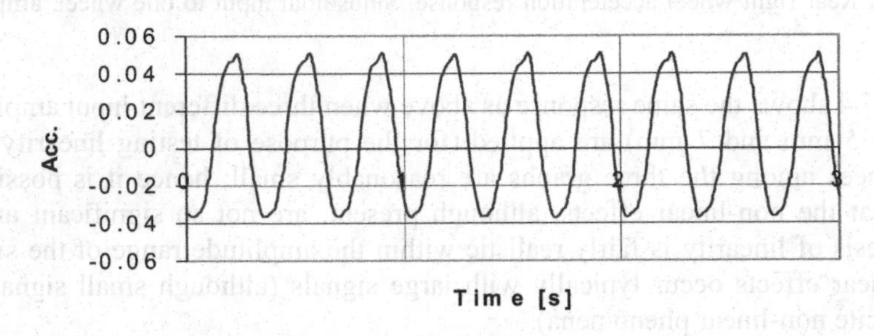

Fig. 7.6. Chassis acceleration time trend. Sinusoidal input to one wheel, amplitude: 3 mm, frequency: 2.5 Hz

For subsequent studies, the frequency range is limited to 6 Hz. Such a range is sufficient to show all the main dynamic phenomena associated with the chassis, in response to bounce, roll and pitch inputs. Figure 7.7 shows the rear right and rear left corner responses of the car to a pure bounce input, applied to all four wheels.

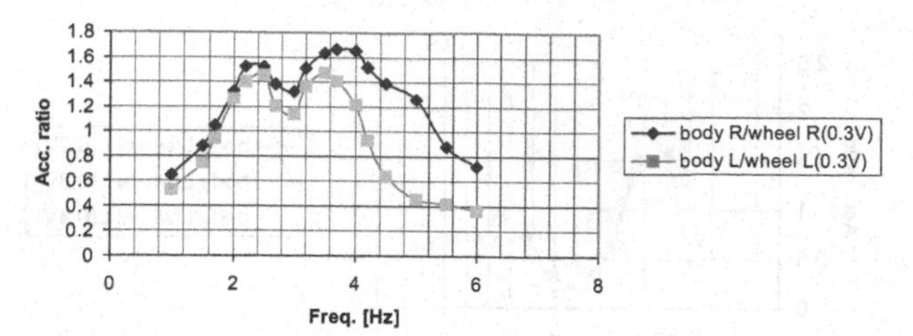

Fig. 7.7. Body-to-wheel acceleration ratio of the rear of the car. Sinusoidal bounce input to all wheels, amplitude: 3 mm

The right and left corner trends exhibit some differences; this is mainly because the centre of gravity of the car does not exactly pass through the vertical plane of symmetry. Two resonances are evident from the figure: the first (around 2.3 Hz) is the bounce resonance while the second resonance (around 3.8 Hz) is due to the pitch–bounce coupled motion. In order to identify which is the dominant parasitic motion (with respect to pure bounce) responsible for that spurious resonance, two more accelerometers can be mounted on the front of the car above each suspension and the relative phase shift recorded. A phase shift of about 180° is associated with roll motion, while a null phase shift corresponds to pitch motion. At the spurious resonance of 3.8 Hz the measured phase shift was about 40°, and hence this resonance is mainly due to the induced pitch motion.

Pitch response is presented in Figure 7.8. Two resonances again are present as before: the pitch resonance at around 2.3 Hz, plus another one at around 3.8 Hz due to cross-coupling effects.

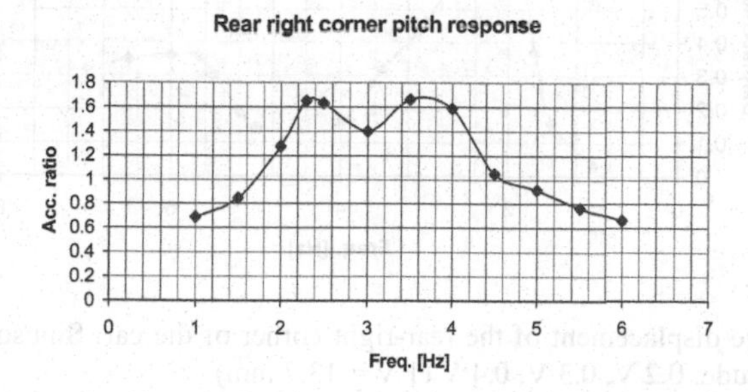

Fig. 7.8. Body-to-wheel acceleration ratio of the rear left corner of the car. Sinusoidal pitch input to all wheels, amplitude: 3 mm

Figure 7.9 shows the roll response for three different inputs applied from an external voltage signal generator (0.2 V, 0.3 V and 0.4 V corresponding to 2.74 mm, 4.11 mm, and 5.48 mm, the calibration factor being 1 V = 13.7 mm). Two resonances are clear from the graph: the roll resonance at around 2.3 Hz and another spurious cross-coupling resonance at about 6 Hz. The closeness of the three responses confirms once again the hypothesis of linearity.

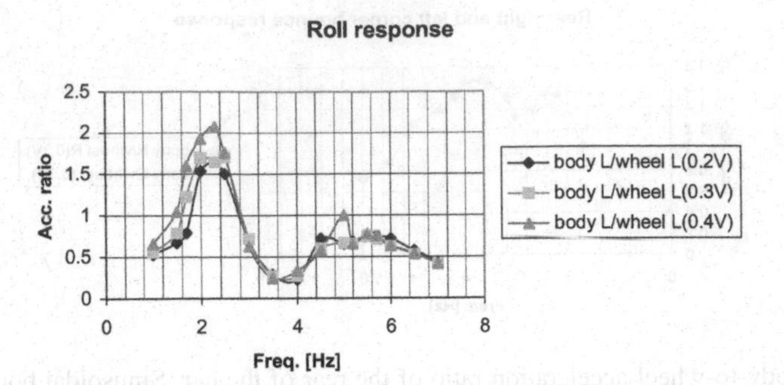

Fig. 7.9. Body-to-wheel acceleration ratio of the rear left corner of the car. Sinusoidal roll input to all wheels, amplitude: 0.2 V, 0.3 V, 0.4 V (1 V = 13.7 mm)

Figure 7.10 depicts the rear right relative displacement (body-to-wheel) response to a roll input. This is proportional to the rolling angle (relative to the wheel). The rolling angle is usually referred to the ground. However in the frequency range of interest the wheel can be considered almost steady so that the rolling angle with respect to the ground is not expected to differ very much from the one measured with respect to the wheel. As expected the resonance measured in terms of acceleration occurs at the same frequency as that of Figure 7.9.

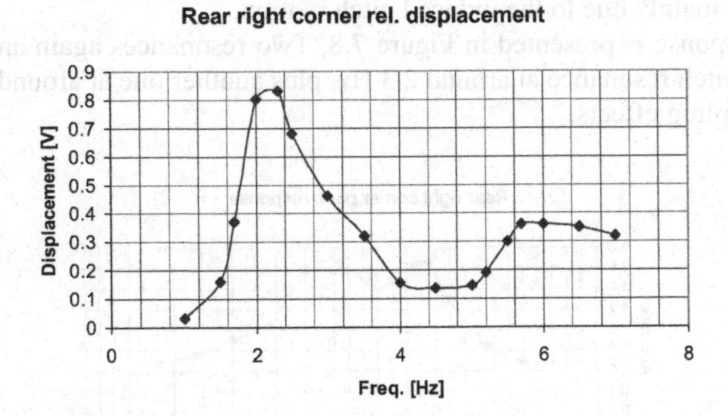

Fig. 7.10. Relative displacement of the rear-right corner of the car. Sinusoidal roll input to all wheels, amplitude: 0.2 V, 0.3 V, 0.4 V (1 V = 13.7 mm)

For the pseudo-random input test, 25-Hz filtered white noise, produced with an external noise generator, is applied to the rear-right wheel. The body acceleration time trend is plotted in Figure 7.11; its RMS value is 0.75 m/s^2.

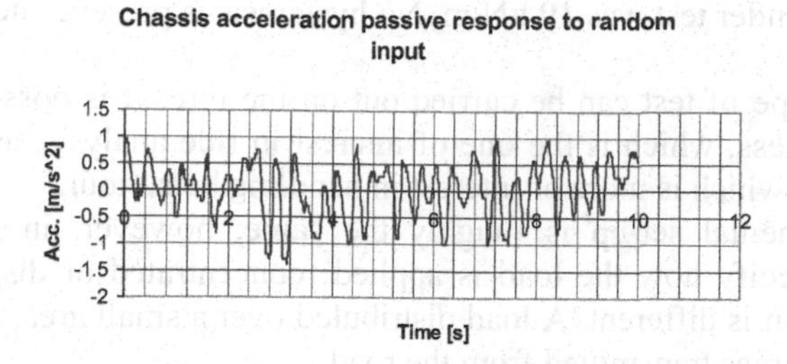

Fig. 7.11. Chassis acceleration time trend. Random input, 25-Hz filtered white noise

The last test is the bump input. A sinusoidal bump with a slight offset has been generated; the demanded input is filtered by actuator dynamics, resulting in a smoother signal. Figure 7.12 shows the response of the car to a bump applied at the rear right wheel.

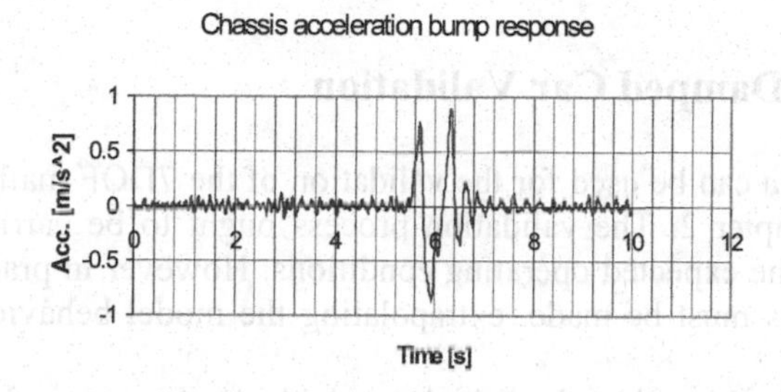

Fig. 7.12. Chassis acceleration time trend. Sinusoidal bump input, amplitude: 30 mm

The most meaningful parameters in a bump test are peak acceleration and number of oscillations. In this case the max overshoot and undershoot are symmetrical with a value of ± 0.85 m/s^2 and the decay of oscillations indicate an equivalent damping ratio of approximately 0.3.

7.2.3 Suspension Spring and Tyre Tests

If accurate figures for spring and tyre stiffness are required, an experimental bench test assessment must be undertaken. This subsection describes a simple method to measure the suspension spring stiffness and the tyre stiffness.

Spring stiffness tests can be carried out by cyclically loading the spring in a material testing machine with a periodic force applied quasi-statically. The compression of the spring under the load can be measured with an extensometer or an LVDT and the force applied with a strain gauge. The spring characteristic is expected to be linear in the measured range. Experiments showed linearity up to a deflection of 200 mm. Such a deflection is far larger than the one experimented in a saloon car under normal conditions. This validates the hypothesis of linearity and allows a linear model to be used for the spring. The spring static stiffness recorded for the vehicle under test was 19 kN/m. No hysteresis is present, the spring being a steel coil.

The same type of test can be carried out on the tyre. It is possible to define a tyre radial stiffness, which is the one of interest in ride analysis, as well as a tyre lateral stiffness, which is more of interest in handling behaviour.

The experimental set-up is roughly the same; however, in this case, it is important to specify how the load is applied (concentrated or distributed) as the stress distribution is different. A load distributed over a small area, however, better represents the forces transmitted from the road.

The typical trend is expected to be slightly non-linear with some hysteresis because rubber is a viscoelastic material. The equivalent linear value of stiffness, measured as the slope of the line interpolating the points corresponding to null force and maximum force, was 74 kN/m. The effective stiffness is twice this value because the tyre as loaded for the test can be thought of as two springs in parallel. If the tyre is loaded from the top the stiffness measured will be half that experienced by a load at the axle, which is what is needed for vehicle dynamics.

7.3 Passively Damped Car Validation

Experimental data can be used for the validation of the 7DOF mathematical model described in Chapter 2. The validation process ought to be carried out over the whole range of the expected operating conditions. However in practice a selection of relevant inputs must be made, extrapolating the model behaviour for all other possible inputs.

The accuracy achieved by the vehicle model has to be assessed both in the time and frequency domains. In the time domain, the simulated trends ought to reproduce the behaviour of the measured quantities, following as closely as possible the trends of the measured variables. In the frequency domain a good match of the frequency response is expected in the range of interest.

Figure 7.13 shows a comparison of measured and predicted rear right body acceleration time history at a frequency of 2.5 Hz. The experimental trend is almost sinusoidal; this confirms again the hypothesis that the behaviour of the car is reasonably linear for typical road profile amplitudes. Under such conditions the simulated behaviour is quite close to the experimental response. The model ought to work in a range of input amplitudes of the order of 10 mm; if the inputs are too large or too small they can excite unmodelled dynamics.

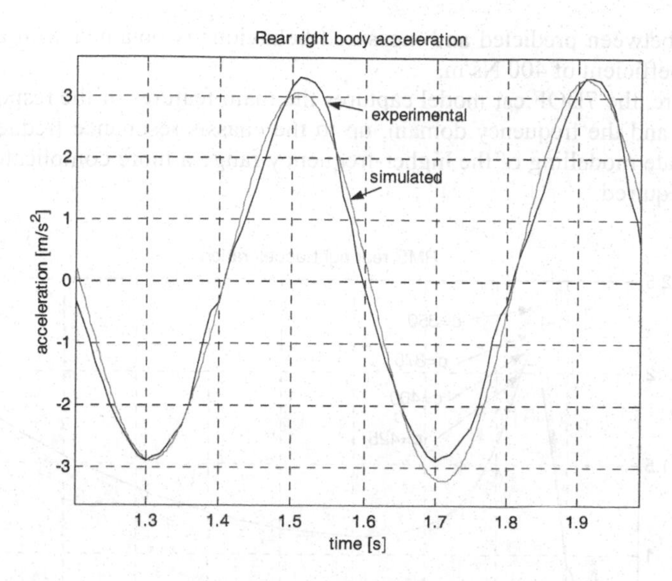

Fig. 7.13. Rear right passive acceleration; sinusoidal input to one wheel, amplitude: 7 mm; frequency: 2.5 Hz

Turning to the frequency-domain analysis, the measured RMS acceleration of the right rear body acceleration for an input excitation of 7 mm is presented in Figure 7.14 (stars on the graph). Predicted behaviour, for four different levels of viscous damping, is also shown. Up to 4 Hz the match is fairly good. At higher frequencies, the simulation overestimates the acceleration. This is presumably due to unmodelled non-linearities, principally associated with the viscous damper characteristic (the viscous damper characteristic employed in the model is obtained by linearising its actual characteristic around the origin, while the real characteristic is non-linear with different trends for bound and rebound strokes). At higher frequencies the local slope of the actual viscous characteristic is smaller than that modelled, resulting in lower damping in practice. Furthermore in a real viscous damper some hysteresis is present, attributable partly to the rubber bushes and parasitic friction as well as to the dissipative internal forces within the hydraulic oil. This explains the mismatch between the experimental data and simulation at frequencies higher than the chassis resonance frequency.

At 8.5 Hz a further resonance is present, due to either the engine mounting or to the induced yaw; such a resonance is not predicted by the model, which would need further degrees of freedom to include these dynamic phenomena. The tyre model is not sophisticated since ride is the main issue here. However highly complicated tyre models are essential in handling studies (as well as in brake/traction analysis) where lateral forces play a major role and not in ride studies where the dynamics of interest are vertical. A small amount of hysteresis is present in the tyre characteristics, but this mainly affects the behaviour around the wheel-hop resonance. The results of a sensitivity analysis, by changing the viscous coefficient in the range 350–425 Ns/m, are also depicted in the figure. Best agreement between predicted and measured behaviour is obtained with a viscous damping coefficient of 400 Ns/m.

Therefore, the 7DOF car model captures the main features of the response both in the time and the frequency domain, up to the chassis resonance frequency. For more accurate modelling of the higher-frequency range a more complicated model would be required.

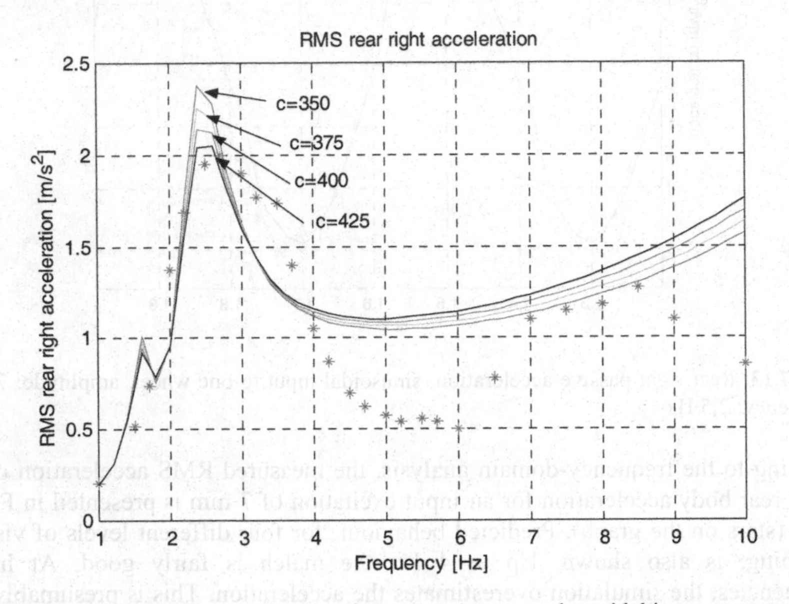

Fig. 7.14. Rear-right passive acceleration frequency response; sinusoidal input to one wheel, amplitude: 7 mm; *c* is the viscous damping coefficient in Ns/m [copyright Elsevier (2003), reproduced from Guglielmino E, Edge KA, Controlled friction damper for vehicle applications, Control Engineering Practice, Vol. 12, N 4, pp 431–443, used by permission]

7.4 Case Study 1: SA Suspension Unit with FD

This section reports the performance of a semi-active suspension unit equipped with a friction damper. After the experimental work on the damper and on its hydraulic drive described in Chapter 5, the friction damper has been installed on a vehicle under test (Figure 7.15).

Fig. 7.15. Friction damper installed on the experimental vehicle

The system supply pressure defined by the relief valve is set to 60 bar for the reasons outlined in Chapter 5. The control valve is connected to the FD, using a hose as short as possible to reduce the volume. The instrumentation is the same employed in the tests on the passively damped car. The results are presented in the following subsections. Firstly the frequency-domain performance of the semi-active FD is discussed. In this domain it is possible to make an initial comparative assessment with the benchmark viscous damper response in terms of RMS values. Subsequently an analysis in the time domain is carried out. Particular care has been given to the issue of ride comfort assessment. This issue is critical in a suspension whose control is based on a switching logic, as time trends can potentially be non-smooth and spiky, causing an uncomfortable ride.

7.4.1 Frequency-domain Analysis

The analysis in the frequency domain is based on RMS values. A comparative analysis of acceleration and working space transmissibility curves in the semi-active and passive case is presented. Figure 7.16 shows the experimentally determined acceleration transmissibility ratio for an input amplitude of 7 mm in the range 1–5 Hz compared with the original (passive) system. The controlled system out-performs the passive system over most of frequency range considered although the passive system response is marginally better up to 1.8 Hz. The controlled response exhibits three peaks: the first, at 1.7 Hz, is the semi-active system chassis resonance; this frequency is lower than the corresponding passive resonance and the peak amplitude is smaller. The inferior behaviour of the semi-active system at low frequency is due to the hydraulic circuit back-pressure, which causes a residual constant-amplitude friction force. The small amplitudes at the lowest frequencies do not produce very significant pressure variations: in this range the constant-amplitude friction force is not negligible. When the frequency increases, the feedback signal is larger and the FD works properly. Hence, the residual friction force deteriorates the performance of the damper if the disturbance is not large, *i.e.*, on smooth roads and at low speed. This confirms that the compensation of the residual friction force via a pre-loaded spring inside the damper would be an effective remedy. Two more peaks are evident in the semi-active curve. This is a non-linear effect of the semi-active system, but in fact the resonances do not create any problem, because they are far lower in size than the corresponding passive values.

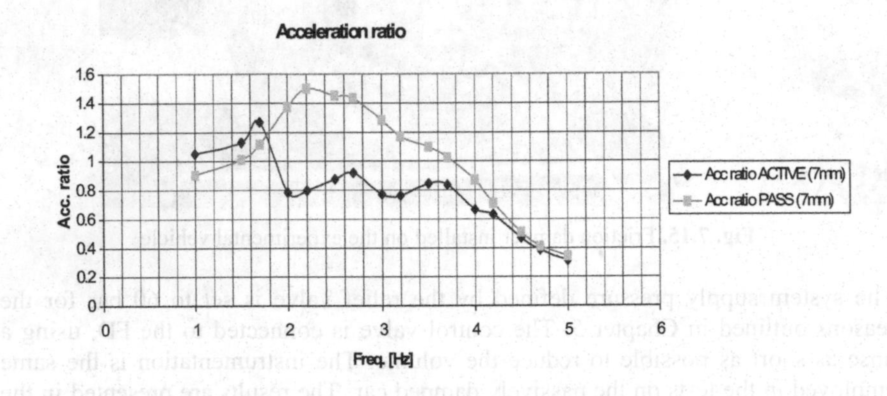

Fig. 7.16. Acceleration ratio transmissibility for passive and semi-active systems. Sinusoidal input to one wheel, amplitude: 7 mm [copyright Elsevier (2003), reproduced with minor modifications from Guglielmino E, Edge KA, Controlled friction damper for vehicle applications, Control Engineering Practice, Vol. 12, N 4, pp 431–443, used by permission]

Figure 7.17 portrays the rear-right suspension working space responses (scaled units) in both cases. Over the frequency range considered the wheel motion can be assumed to be almost steady, hence working space is a good approximation to absolute chassis displacement. The results presented show that the semi-active response is much the same as the passive response, except at the resonant frequency. This is because of the force tracking performed by the controller which inherently promotes higher displacements.

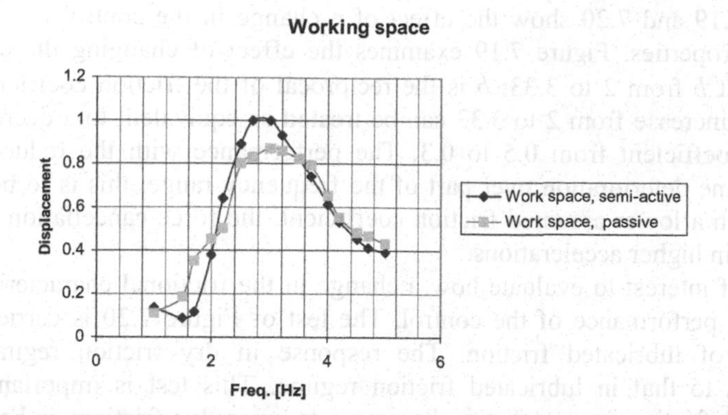

Fig. 7.17. Working space for passive and semi-active systems. Sinusoidal input to one wheel, amplitude: 7 mm [copyright Elsevier (2003), reproduced with minor modifications from Guglielmino E, Edge KA, Controlled friction damper for vehicle applications, Control Engineering Practice, Vol. 12, N 4, pp 431–443, used by permission]

Figure 7.18 depicts the acceleration response for three different input amplitudes. Within reasonable approximations they are fairly similar. This means that nonlinear effects are not very strong. Actually balance logic governing differential equation is piecewise linear and for this particular class of equations the frequency response does not depend on the input amplitude as is the case in nonlinear systems[3].

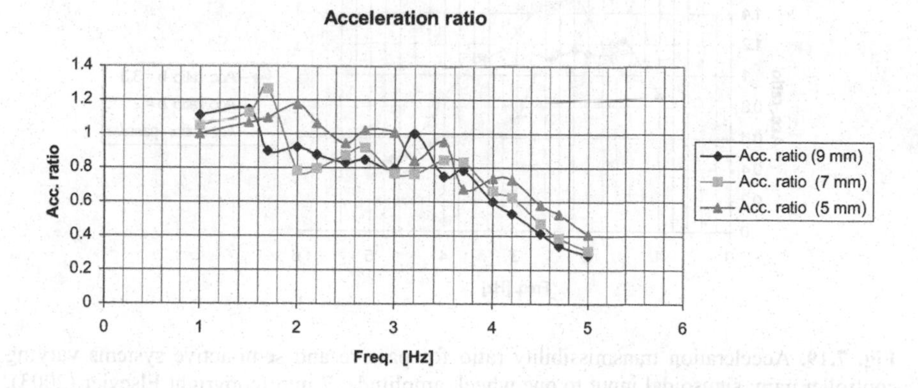

Fig. 7.18. Acceleration ratio transmissibility for semi-active system. Sinusoidal input to one wheel; amplitudes, 5 mm, 7 mm, 9 mm

3. In a linear system the RMS and peak transmissibility ratios (*e.g.*, body over wheel acceleration) are equal and independent of the input signal amplitude. In a piecewise-linear system RMS and peak transmissibility ratios are not equal but still input amplitude independent. In a nonlinear system RMS and peak transmissibility ratios are not equal and both input amplitude dependent.

Figures 7.19 and 7.20 show the effect of a change in the control law and in the friction properties. Figure 7.19 examines the effect of changing the closed-loop coefficient b from 2 to 3.33; b is the reciprocal of the friction coefficient μ, and hence its increase from 2 to 3.33 can be treated as equivalent to a decrease of the friction coefficient from 0.5 to 0.3. The performance with the reduced friction shows some deterioration over part of the frequency range: this is to be expected since, with a lower assumed friction coefficient, the force cancellation is smaller, resulting in higher accelerations.

It is of interest to evaluate how a change in the frictional characteristic would affect the performance of the control. The test of Figure 7.20 is carried out in a situation of lubricated friction. The response in dry friction regime is here compared to that in lubricated friction regime. This test is important, because lubricated friction is a realistic alternative to pure dry friction: it helps reduce stiction and is potentially advantageous in terms of heat dissipation between the friction surfaces.

At low frequencies the dry friction system response is better than the lubricated friction system response although both are worse than the passive response. In the central frequency band the lubricated friction response is better, but at higher frequencies the opposite holds.

Overall, the performance of the controlled suspension is superior to the passive case and is not worsened by lubrication.

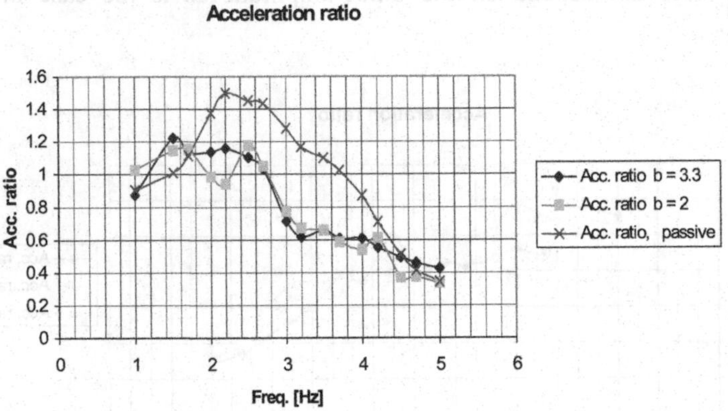

Fig. 7.19. Acceleration transmissibility ratio for passive and semi-active systems varying controller gain; sinusoidal input to one wheel, amplitude: 7 mm [copyright Elsevier (2003), reproduced with minor modifications from Guglielmino E, Edge KA, Controlled friction damper for vehicle applications, Control Engineering Practice, Vol. 12, N 4, pp 431–443, used by permission]

From the foregoing investigations involving changes in the feedback coefficient b (equivalent to a change in dry friction coefficient) and the nature of the lubrication regime, the robustness of the scheme, in a practical sense, has been experimentally verified. Moreover, the semi-active system remains generally superior to the passive system, even in the presence of these changes.

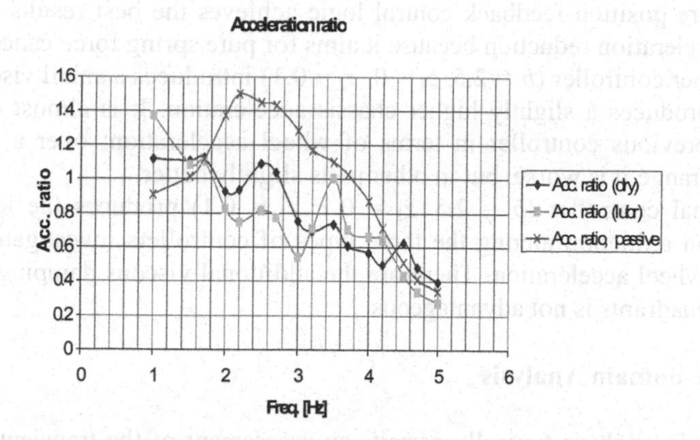

Fig. 7.20. Acceleration ratio transmissibility for passive and semi-active systems; sinusoidal input to one wheel, amplitude: 7 mm [copyright Elsevier (2003), reproduced with minor modifications from Guglielmino E, Edge KA, Controlled friction damper for vehicle applications, Control Engineering Practice, Vol. 12, N 4, pp 431–443, used by permission]

Figure 7.21 shows the effect on RMS chassis acceleration of changing the control law.

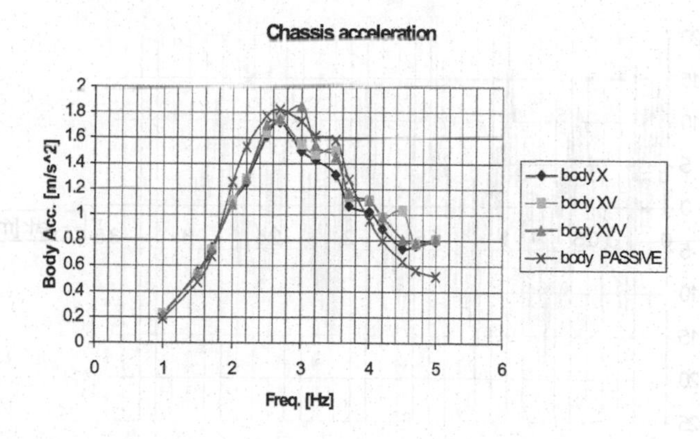

Fig. 7.21. Chassis acceleration transmissibility for passive and semi-active systems; sinusoidal input to one wheel, amplitude 7 mm

The performance of three types of controllers are compared. These are:

- a pure position feedback (b = 2.5, z_1 = 0, z_2 = 0) controller (X in the legend)
- a controller with position feedback in two quadrants and velocity feedback in the other two (b = 2.5, z_1 = 0, z_2 = 0.1; XV in the legend)
- a controller with position plus velocity feedback in two quadrants and velocity feedback in the other two (b = 2.5, z_1 = 0.1, z_2 = 0.1; XVV in the legend)

The pure position feedback control logic achieves the best results in terms of chassis acceleration reduction because it aims for pure spring force cancellation.

The other controller (b = 2.5, z_1 = 0, z_2 = 0.1) introduces a small viscous effect, and thus produces a slightly higher chassis acceleration. It is almost comparable with the previous controller in terms of wheel acceleration: over a part of the frequency range it is worse, but in other parts slightly better.

The final controller (b = 2.5, z_1 = 0.1, z_2 = 0.1) produces the least chassis acceleration reduction among the three types of controllers investigated and also the worst wheel acceleration. Therefore the additional viscous damping in the first and third quadrants is not advantageous.

7.4.2 Time-domain Analysis

Time domain analysis typically permits an assessment of the transient behaviour. In this context it is mainly used to assess ride quality. Figures 7.22 and 7.23 show jerk trends in the passive and semi-active cases. The latter trend presents slightly higher peak values. Jerk is obtained by numerical differentiation of the measured acceleration signal.

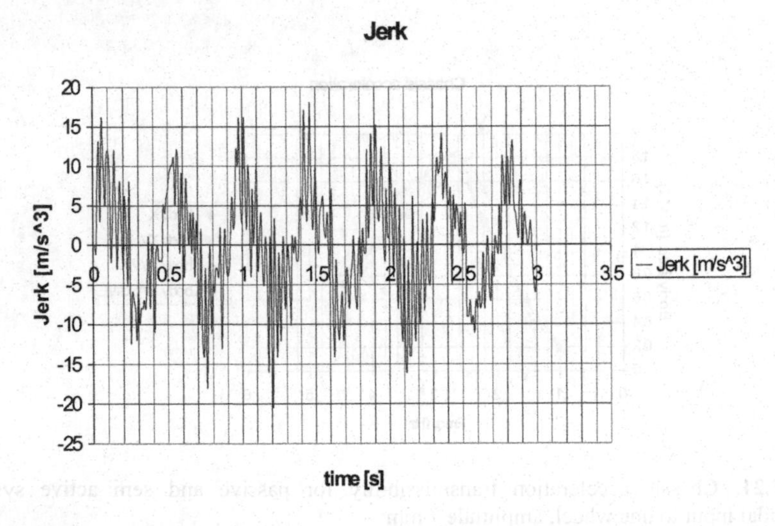

Fig. 7.22. Experimental jerk time trace for the passive system; sinusoidal input, amplitude: 3 mm, frequency: 2.2 Hz

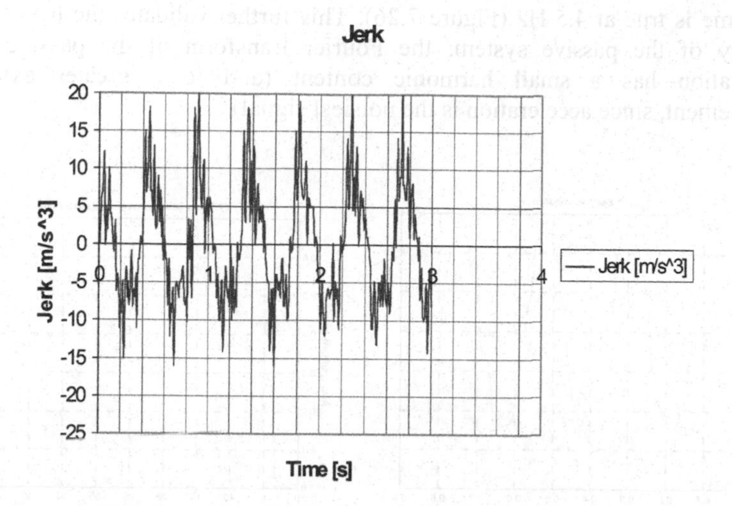

Fig. 7.23. Experimental jerk time trace for the controlled system; sinusoidal input, amplitude: 3 mm, frequency: 2.2 Hz

Next a Fourier analysis of the acceleration waveforms is presented. Although the spectrum is more appropriate to assess the degree of non-linearity rather than comfort, it is possible to establish a qualitative correlation between a high harmonic content and discomfort.

Figures 7.24–7.27 depict experimental time traces and spectra for the passive and semi-active suspension. Figures 7.24 and 7.25 show measured acceleration time histories and spectra for the passive case at frequencies of 2.2 and 4.5 Hz. At 2.2 Hz (Figures 7.24) the behaviour is fairly linear (only a negligible third harmonic is present in the spectrum).

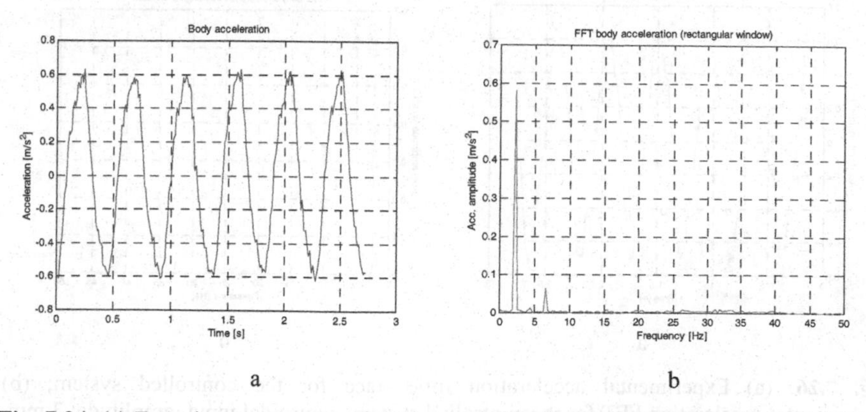

Fig. 7.24. (a) Experimental acceleration time trace for the passive system; (b) experimental acceleration FFT for the passive system; sinusoidal input, amplitude: 3 mm, frequency: 2.2 Hz

The same is true at 4.5 Hz (Figure 7.26). This further validates the hypothesis of linearity of the passive system: the Fourier transform of the passive system acceleration has a small harmonic content (and to a greater extent the displacement, since acceleration is the noisiest signal).

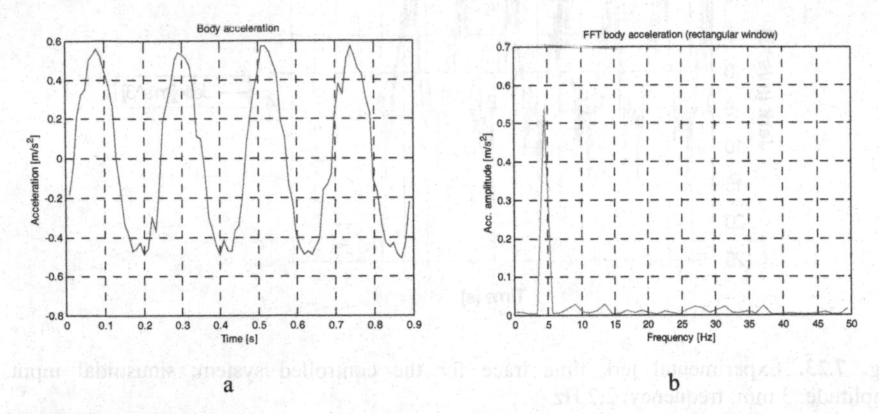

Fig. 7.25. (a) Experimental acceleration time trace for the passive system; (b) experimental acceleration FFT for the passive system; sinusoidal input, amplitude: 3 mm, frequency: 4.5 Hz

Figures 7.26 and 7.27 depict the same graphs for the controlled system. The harmonic content is not very large at 2.2 Hz, but is certainly larger at 4.5 Hz and actually the acceleration time trend is fairly spiky. This spectral analysis of the acceleration has hence confirmed the qualitative correlation between jerk time trends and richer higher-harmonic content.

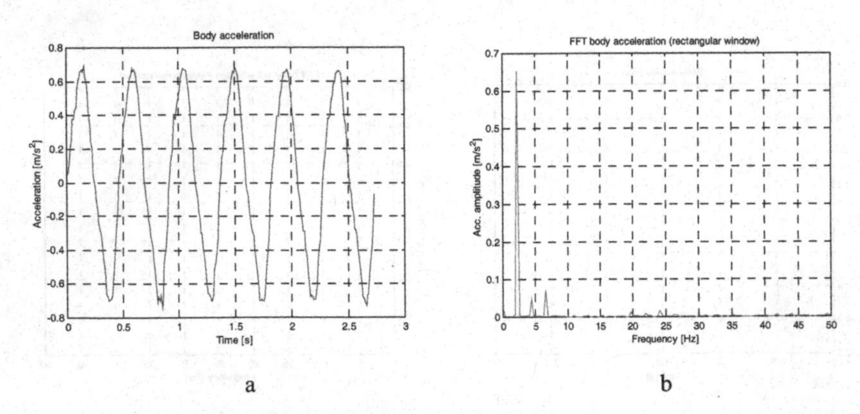

Fig. 7.26. (a) Experimental acceleration time trace for the controlled system; (b) experimental acceleration FFT for the controlled system; sinusoidal input, amplitude: 3 mm, frequency: 2.2 Hz

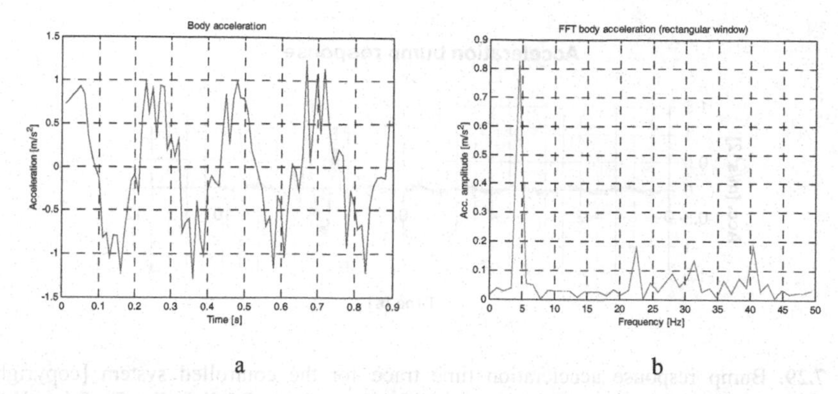

Fig. 7.27. (a) Experimental acceleration time trace for the controlled system; (b) experimental acceleration FFT for the controlled system; sinusoidal input, amplitude: 3 mm, frequency: 4.5 Hz

Figure 7.28 shows the semi-active response to a pseudo-random input. The RMS of the acceleration is 0.58 m/s^2, which is smaller than in the passive case (0.75 m/s^2). Figure 7.29 shows the semi-active system response to a sinusoidal bump. The bump input is the same as in the passive case. The acceleration overshoot and undershoot for the controlled system (in the case $b = 2.5$, $z_1 = 0$, $z_2 = 0$) are 1.1 m/s^2 and –0.8 m/s^2 respectively, whereas for the passive system the values are ±0.85 m/s^2. The number of oscillations is the same for both cases. Thus the controlled system is slightly worse in response to a bump. This is because of the relatively slow response of the pressure control circuit. A bump can be thought as a high-frequency half-wave input (the higher the vehicle velocity, the higher the frequency) and above a certain frequency the servo-system response is not swift enough.

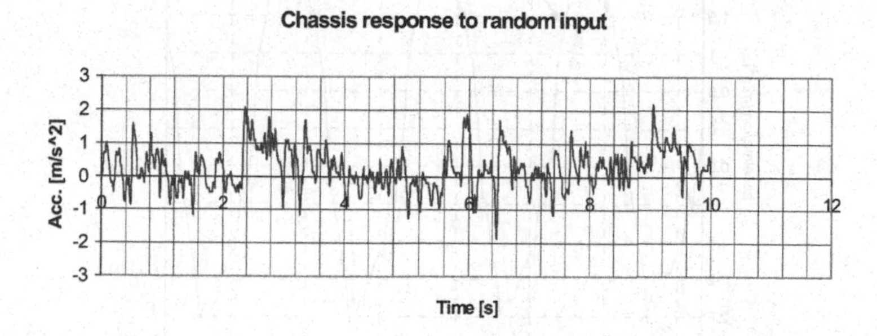

Fig. 7.28. Semi-active chassis acceleration time trend; random input: 25-Hz filtered white noise [copyright Elsevier (2003), reproduced with minor modifications from Guglielmino E, Edge KA, Controlled friction damper for vehicle applications, Control Engineering Practice, Vol. 12, N 4, pp 431–443, used by permission]

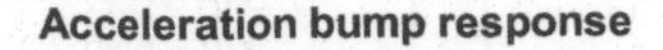

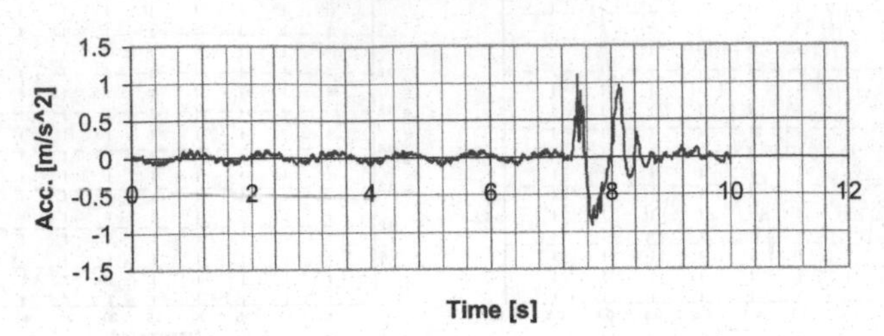

Fig. 7.29. Bump response acceleration time trace for the controlled system [copyright Elsevier (2003), reproduced with minor modifications from Guglielmino E, Edge KA, Controlled friction damper for vehicle applications, Control Engineering Practice, Vol. 12, N 4, pp 431–443, used by permission]

7.4.3 Semi-active System Validation

Having validated separately the electrohydraulic drive and the passive vehicle, the final step is the validation of the whole system. Figure 7.30 shows the predicted and measured sinusoidal time responses after the start-up transient has decayed. The simulation follows the overall trend of the measured acceleration well. The spikes in the experimental results are not captured by the model, but most of them arise from noise present in the measurements. The overall agreement can be considered acceptable.

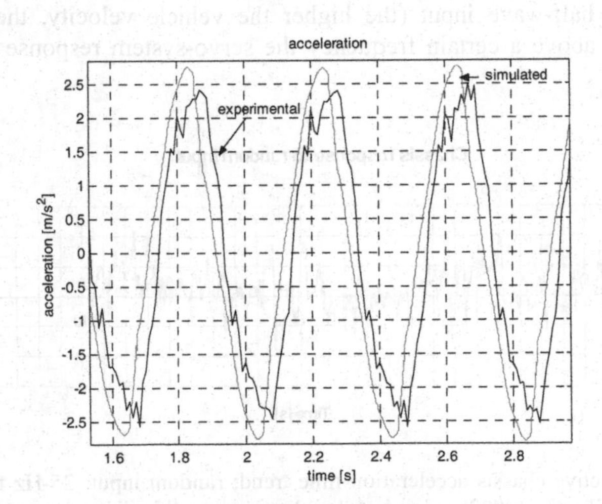

Fig. 7.30. Rear right semi-active acceleration; sinusoidal input to one wheel, amplitude: 7 mm, frequency: 2.5 Hz [copyright Elsevier (2003), reproduced with minor modifications from Guglielmino E, Edge KA, Controlled friction damper for vehicle applications, Control Engineering Practice, Vol. 12, N 4, pp 431–443, used by permission]

Considering now the frequency domain, a sensitivity analysis needs to be carried out to identify the most suitable values for the critical parameters. Figure 7.31 represents the frequency response for different friction coefficients μ between 0.1 and 0.2. At low frequency the trend is virtually independent of the friction coefficient but after the resonance the dependency gets stronger. A smaller friction coefficient produces a larger resonance peak but at higher frequency it tracks the experimental response better, and vice versa for a larger coefficient. The mismatch occurring for frequencies higher than 3.8 Hz is due to the limitations in the car model and the over-simplified model for the other three (passive) dampers, rather than to the hydraulic model of the friction damper drive. The actual value of the nominal friction coefficient of the material was 0.4. The mismatch between simulation and experiments is related to the $\frac{\mu}{\mu_{\text{assumed}}}$ ratio (see Section 5.5).

Figure 7.32 portrays the frequency response for different levels of delay between velocity and friction force created by the frictional memory effect. A change of ±50% does not produce any significant change except at the lowest frequencies.

In Figure 7.33 the effect of a change in the actuator and connecting pipe volume is considered. An increase of an order of magnitude in the volume produces a noticeable effect for frequencies above 2.5 Hz. The impact of the corresponding reduction of the valve-actuator bandwidth, following an increase in volume, is a higher acceleration. This physically occurs because the valve-actuator system cannot catch up with the higher-frequency input; therefore the effect of the residual constant friction force plays the dominant role.

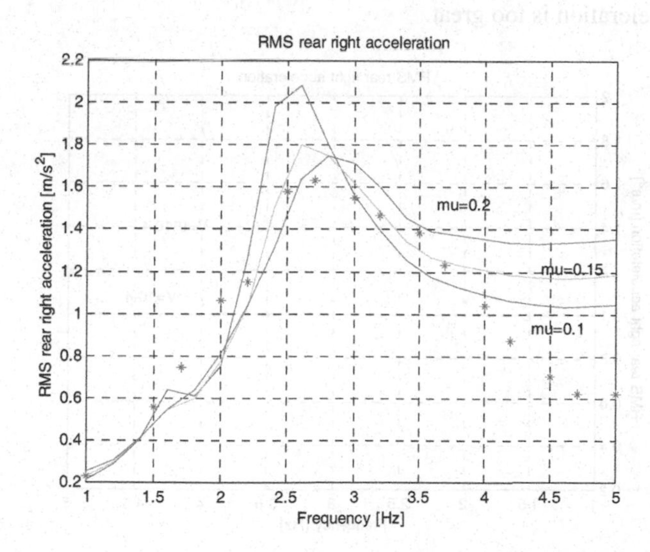

Fig. 7.31. Rear right semi-active acceleration frequency response varying friction coefficient; sinusoidal input to one wheel, amplitude: 7 mm [copyright Elsevier (2003), reproduced with minor modifications from Guglielmino E, Edge KA, Controlled friction damper for vehicle applications, Control Engineering Practice, Vol. 12, N 4, pp 431–443, used by permission]

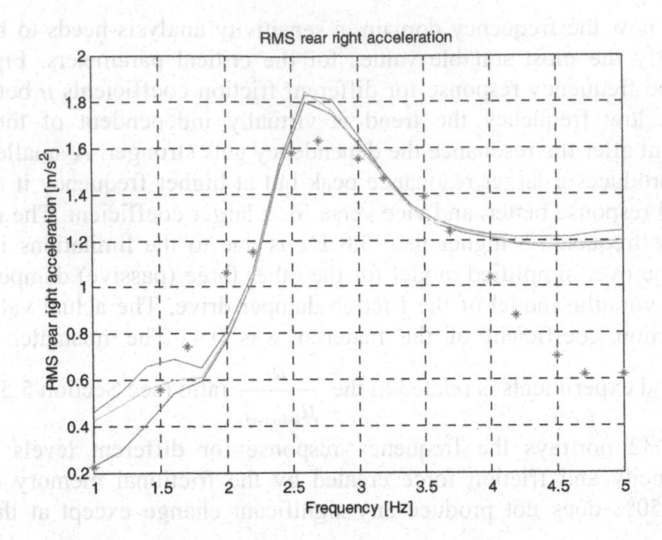

Fig. 7.32. Rear right semi-active acceleration frequency response varying frictional memory; sinusoidal input to one wheel, amplitude: 7 mm [copyright Elsevier (2003), reproduced with minor modifications from Guglielmino E, Edge KA, Controlled friction damper for vehicle applications, Control Engineering Practice, Vol. 12, N 4, pp 431–443, used by permission]

Therefore it can be concluded that simulation provides accurate results in the time domain and in the frequency domain up to almost 4 Hz. Beyond that frequency the predicted acceleration is too great.

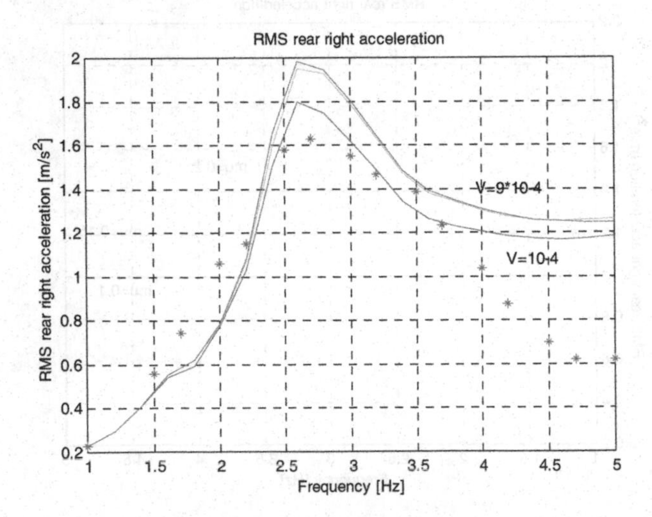

Fig. 7.33. Rear right semi-active acceleration frequency response varying volume; sinusoidal input to one wheel, amplitude: 7 mm [copyright Elsevier (2003), reproduced with minor modifications from Guglielmino E, Edge KA, Controlled friction damper for vehicle applications, Control Engineering Practice, Vol. 12, N 4, pp 431–443, used by permission]

7.5 Case Study 2: MR-based SA Seat Suspension

This case study concerns an investigation of the use of a controlled MR damper for a semi-active seat suspension in vehicles not equipped with primary suspensions (*e.g.*, some types of agricultural, forestry and roadwork vehicles). The control strategy is targeted to improve driver comfort based on a hybrid variable structure fuzzy logic controller. Variable structure control is inherently a switching logic, hence more prone to cause chattering. Fuzzy logic helps reduce chatter without penalising damper dynamic response.

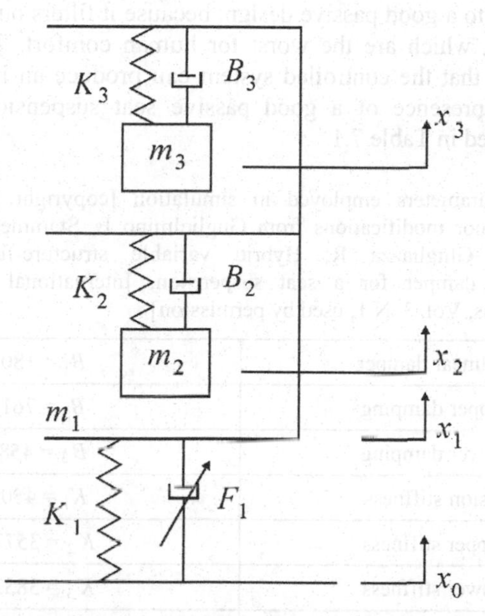

Fig. 7.34. Suspension, seat and driver model [copyright Inderscience (2005), reproduced from Guglielmino E, Stammers CW, Stancioiu D, Sireteanu T and Ghigliazza R, Hybrid variable structure-fuzzy control of a magnetorheological damper for a seat suspension, International Journal of Vehicle Autonomous Systems, Vol. 3, N 1, used by permission]

The whole system, depicted in Figure 7.34, is composed of the suspension, the seat and the driver (*i.e.*, seated body model) and can be modelled as a 3DOF system. The seat is mounted on the semi-active suspension constituted by a linear spring and the MR damper. The seat and driver model (Wei and Griffin, 1998a) consists of a rigid frame m_1 to which two masses, m_2 and m_3 are suspended. The two masses cannot be associated with particular organs of human body (refer to Chapter 3 for further details). For convenience they will be denoted as *upper* mass and *lower* mass. A driver seat suspension usually includes a linkage mechanism between the suspension unit and the seat. Its presence can be taken into account in the model by employing an equivalent value for the suspension stiffness and by appropriate scaling factors for the damping force which take into account the geometry of the linkage. Setting $x = x_1 - x_0$, the equations of motion can be written as

$$
\begin{aligned}
m_3\ddot{x}_3 &= B_3(\dot{x}_1-\dot{x}_3)+K_3(x_1-x_3)=f_3\,,\\
m_2\ddot{x}_2 &= B_2(\dot{x}_1-\dot{x}_2)+K_2(x_1-x_2)=f_2\,,\\
m_1\ddot{x}_1 &= F_1(x,\dot{x},z,u)+K_1x-f_3-f_2\,,
\end{aligned}
\tag{7.2}
$$

where $F_1(x,\dot{x},z,u)$ is the seat suspension damping force (a Bouc–Wen model has been employed), K_1 the suspension stiffness, K_2 and K_3 the lower and upper body stiffness and B_2 and B_3 the lower and upper body damping. The value of the suspension stiffness is chosen so its natural frequency is around 1.5 Hz. Such a value corresponds to a good passive design, because it filters out the frequencies in the range 3–6 Hz, which are the worst for human comfort. The reason for this choice is to show that the controlled system can produce an improvement of the response also in presence of a good passive seat suspension. The simulation parameters are listed in Table 7.1.

Table 7.1. Key parameters employed in simulation [copyright Inderscience (2005), reproduced with minor modifications from Guglielmino E, Stammers CW, Stancioiu D, Sireteanu T and Ghigliazza R, Hybrid variable structure-fuzzy control of a magnetorheological damper for a seat suspension, International Journal of Vehicle Autonomous Systems, Vol. 3, N 1, used by permission]

Passive linear damper	$B_1 = 180$ Ns/m
Body upper damping	$B_2 = 761$ Ns/m
Body lower damping	$B_3 = 458$ Ns/m
Suspension stiffness	$K_1 = 4500$ N/m
Body upper stiffness	$K_2 = 35776$ N/m
Body lower stiffness	$K_3 = 38374$ N/m
Seat mass	$m_1 = 6$ kg
Body upper mass	$m_2 = 33.4$ kg
Body lower mass	$m_3 = 10.7$ kg
Bouc–Wen model coefficients	$A = 2000$; $\beta = \gamma = 2000$ m^{-2}
Bouc–Wen exponent and offset force	$n = 2$; $f_0 = 15$ N

With reference to Figure 7.34, the variable the value of which must be minimised is the arithmetic mean of the RMS of lower and upper mass accelerations, $\ddot{x}_2$ and $\ddot{x}_3$, as this quantity is related to driver vertical acceleration. The frequency range of interest for comfort is up to around 6 Hz (far within the MRD bandwidth).

The aim of the control is to reduce the forces transmitted to the seat by generation of a spring-like (position-dependent damping) control force of sign opposite to that of the spring force.

The logic is implemented by controlling the solenoid current; the control logic can be expressed by the following functional equation:

$$F_1(x,\dot{x},z,u) = F(x,\dot{x},z,u(x,\dot{x})), \tag{7.3}$$

where the current $u(x,\dot{x})$ is expressed by:

$$u(x,\dot{x}) = \begin{cases} a|x| & \text{if } \ x\dot{x} \le 0 \\ 0 & \text{if } \ x\dot{x} > 0, \end{cases} \tag{7.4}$$

a being a gain proportional to suspension spring stiffness. This balance controller is switching type and this may cause chattering problems when the controller switches from one structure to the other.

MRD dynamic response is extremely swift since it is mainly dependent upon electromagnetic dynamics and the time necessary for the oil to reach rheological equilibrium. The fast switching produces periodical acceleration and jerk peaks which degrade ride quality. Chattering problems in fast-dynamics dampers controlled via switched-type algorithms have been reported in work by Choi *et al.* (2000), who pointed out the problem in a study on sliding-mode control of electrorheological dampers. The problem can be tackled at the control level by smoothing the control action by using fuzzy logic. In this way it is possible to soften the fast switching action of the *crisp* balance controller, without the need for low-pass filters which would reduce system bandwidth, which is one the major benefits of using an MRD.

The fuzzy-controlled damping force is expressed as:

$$F_1(x,\dot{x},z,u) = F(x,\dot{x},z,\eta|x|u(x,\dot{x})), \tag{7.5}$$

where $u(x,\dot{x})$ is the fuzzy controller current input and η is a gain.

The variable structure algorithm has been fuzzified by choosing as fuzzy variables the relative displacement and velocity; the linguistic variables are: negative (neg) and positive (pos). The membership functions are depicted in Figure 7.35.

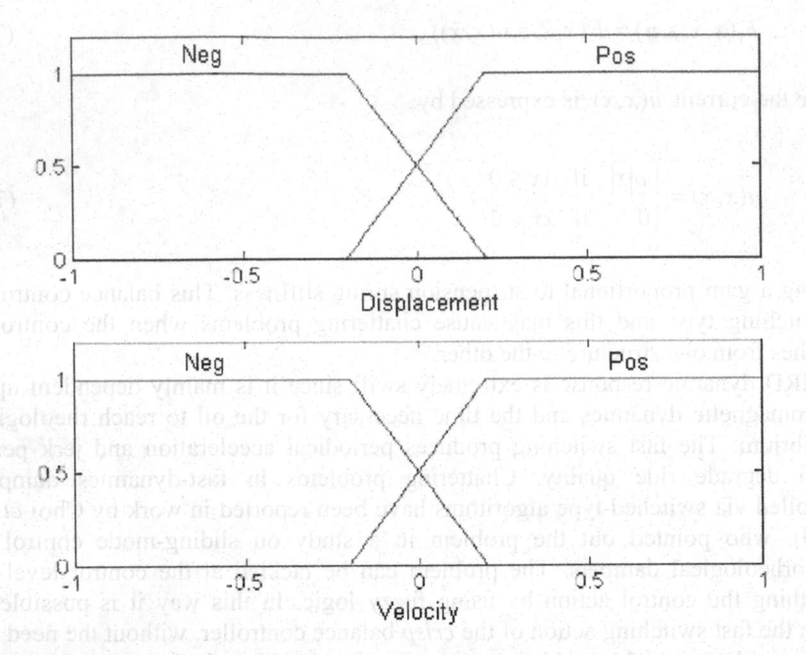

Fig. 7.35. Fuzzy logic membership functions [copyright Inderscience (2005), reproduced from Guglielmino E, Stammers CW, Stancioiu D, Sireteanu T and Ghigliazza R, Hybrid variable structure-fuzzy control of a magnetorheological damper for a seat suspension, International Journal of Vehicle Autonomous Systems, Vol. 3, N 1, used by permission]

The fuzzy controller function is a two-value function $u(x(t),\dot{x}(t)) \in [0,u_{\max}]$ where *Small* = 0 and *Big* = $u_{\max}$. The fuzzy set rules (Table 7.2) are obtained by fuzzifying the control logic defined in Equation 7.4. In this way, the transitions between the two structures are not abrupt.

Table 7.2. Fuzzy logic rules [copyright Inderscience (2005), reproduced with minor modifications from Guglielmino E, Stammers CW, Stancioiu D, Sireteanu T and Ghigliazza R, Hybrid variable structure-fuzzy control of a magnetorheological damper for a seat suspension, International Journal of Vehicle Autonomous Systems, Vol. 3, N 1, used by permission]

Velocity / Displacement	Negative	Positive
Negative	Small	Big
Positive	Big	Small

7.5.1 Numerical Results

The system is tested with a random input $x_0(t)$ obtained from a measured power spectral density of the seat base acceleration in operating conditions. The excitation

RMS value is $\sigma_{x_0} = 0.02$ m. Table 7.3 reports the power spectral density RMS of the controlled variables for the different controllers.

Table 7.3. Performance assessment of passive and semi-active systems [copyright Inderscience (2005), reproduced with minor modifications from Guglielmino E, Stammers CW, Stancioiu D, Sireteanu T and Ghigliazza R, Hybrid variable structure-fuzzy control of a magnetorheological damper for a seat suspension, International Journal of Vehicle Autonomous Systems, Vol. 3, N 1, used by permission]

	$\sigma_{\ddot{x}_2}$ [m/s^2]	$\sigma_{\ddot{x}_3}$ [m/s^2]	$\frac{\sigma_{\ddot{x}_2}+\sigma_{\ddot{x}_3}}{2}$ [m/s^2]
Passive linear viscous damper	3.06	3.23	3.14
Passive MRD	3.68	3.94	3.81
VSC (crisp) with MRD	2.03	2.52	2.27
Fuzzy control with MRD	2.20	2.75	2.48

A reduction of the RMS of the controlled variable of about 21% is achieved with fuzzy control, with respect to a passive system with a traditional viscous damper. The crisp controller performs slightly better (27% reduction). However the merits of the fuzzy controller are in terms of chattering reduction as is evident from Figure 7.36 where the demands for the VSC and fuzzy algorithms are presented.

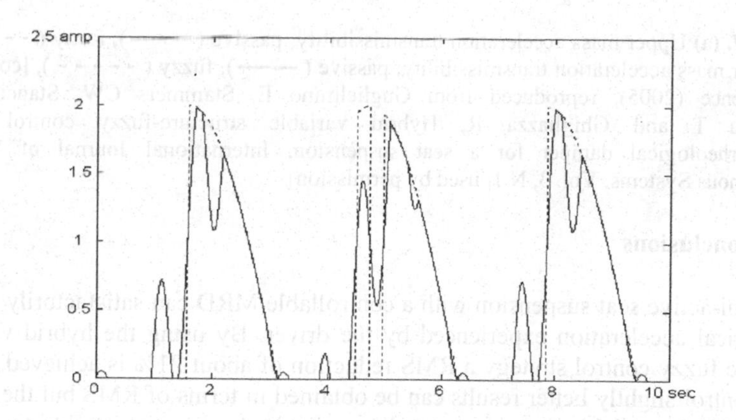

Fig. 7.36. Current demand for VSC (- - - - -) and fuzzy (———) controllers [copyright Inderscience (2005), reproduced from Guglielmino E, Stammers CW, Stancioiu D, Sireteanu T and Ghigliazza R, Hybrid variable structure-fuzzy control of a magnetorheological damper for a seat suspension, International Journal of Vehicle Autonomous Systems, Vol. 3, N 1, used by permission]

The fast switching action of the crisp controller is softened by the fuzzy algorithm and the current transitions between the on and the off state are smoother; this in turn reduces peaks of acceleration arising in the instants of transition, thus causing improvement in the ride quality, although the RMS is slightly higher.

It is also worthwhile noting that an MRD without control performs worse than a linear viscous damper. This is due to the large forces at low velocity as well as to the large hysteresis in its characteristics.

Finally Figure 7.37 shows the transmissibility trends (RMS acceleration versus frequency) for upper and lower mass accelerations, superimposed on the response of the passive system. A reduction in the magnitude of the resonance peak is noticeable, which will produce an increase in comfort. The peak shift to about 1 Hz is beneficial as the body is a little less sensitive to vertical vibration at 1 Hz than at 1.5 Hz.

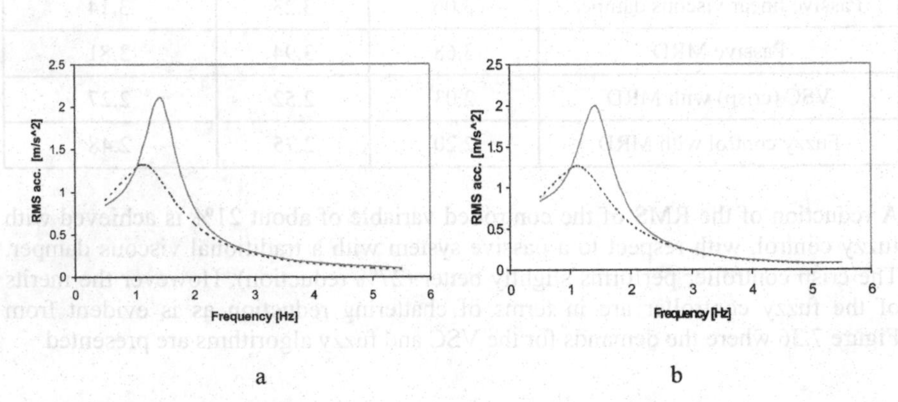

Fig. 7.37. (a) Upper mass acceleration transmissibility; passive (———), fuzzy (- - - - -); (b) lower mass acceleration transmissibility; passive (———), fuzzy (- - - - -), [copyright Inderscience (2005), reproduced from Guglielmino E, Stammers CW, Stancioiu D, Sireteanu T and Ghigliazza R, Hybrid variable structure-fuzzy control of a magnetorheological damper for a seat suspension, International Journal of Vehicle Autonomous Systems, Vol. 3, N 1, used by permission]

7.5.2 Conclusions

The semi-active seat suspension with a controllable MRD can satisfactorily reduce the vertical acceleration experienced by the driver. By using the hybrid variable structure fuzzy control strategy a RMS reduction of about 21% is achieved. Using crisp control slightly better results can be obtained in terms of RMS but the spikes induced by the variable structure controller switching action worsen ride quality. In order to eliminate these spikes filtering would be necessary, but this would reduce the MRD bandwidth; fuzzy logic instead allows full use of the MRD bandwidth.

7.6 Case Study 3: Road Damage Reduction with MRD Truck Suspension

7.6.1 Introduction

Heavy vehicles travel over a variety of road surfaces and experience a wide range of vibration. This affects not only ride quality, but also causes road damage.

Weather conditions as well as vehicle motion are two key causes of road pavement damage. Road damage requires significant investment every year to repair the road surface and causes delays to traffic while it is being done.

Heavy vehicle suspensions ought to be able to isolate the sprung mass from road-induced disturbances as well as improving handling and minimising road damage by reducing dynamic tyre force within the constraint of a set working space.

The reduction of dynamic tyre forces is a challenging field. Cole and Cebon (1996a and 1996b) did extensive work on it, both theoretical and experimental. A thorough analysis of the damage due to dynamic tyre forces and other co-factors is presented by Cebon (1999): an instrumented vehicle was employed to measure the dynamic tyre forces at both low and high excitation frequencies. The study concluded that the wheel dynamic load increases with both vehicle speed and road roughness.

Algorithms are aimed at the reduction of tyre load oscillations, which improves handling on one hand and on the other reduces road damage caused by vehicle wheels. This latter application is particularly important in the case of heavy freight vehicles. Extended groundhook control logic was investigated by Valasek *et al.* (1998) to reduce dynamic tyre forces.

This case study presents a hybrid balance algorithm to reduce road damage and investigate the performance of a heavy articulated vehicle equipped with both MR dampers and passive viscous dampers. A half truck model is employed and system performance investigated via numerical simulation. A variation of the balance logic strategy based on dynamic tyre force tracking has been devised. Algorithm robustness to parametric variations as well as to real-life implementation issues such as feedback signals noise are discussed.

Given the modest cost of MR dampers compared to the overall cost of the vehicle, and the fact that road maintenance requires significant investment worldwide (an estimated of several hundred millions pounds of road damage in the UK was judged to be due to heavy vehicles; Potter *et al.*, 1995) these benefits are cost effective.

A model is developed for the semi-active control of the suspension of a three-axle tractor–trailer combination. Control is applied via an MR damper at either the tractor rear axle or the trailer axle. The load on the tractor front axle is smaller than that on the other two axles and is much less significant from the point of view of road damage.

Control reduces dynamic tyre forces and hence road damage, while improving handling. Trailer sprung mass acceleration (heave and pitch) is also reduced. Robustness of control is established by adding noise to the computed sensor inputs.

7.6.2 Half Truck and MR Damper Model

The half truck model is based on that described in Chapter 2 (Equations 2.31 and 2.32) and pictured in Figure 2.18. Lateral and yaw motions are neglected at this stage, reducing the model complexity. The roll motion is also neglected. Half truck vehicle parameters employed in the simulation are listed in Table 7.4.

The vehicle travels in a straight course with constant speed and is modelled as a three-axle vehicle, the steer axle, the drive tractor axle and the trailer axle, assuming two MR dampers, one fitted on the tractor drive axle and one on the trailer axle of the half truck; the steer axle is equipped with a passive viscous damper.

The objective is to investigate the vehicle performance in terms of ride and road damage using two semi-active dampers controlled by a hybrid logic for various road profiles. The MR damper model employed is based on the work by Lau and Liao (2005), who designed and modelled a prototype damper for a train suspension. Such a damper develops forces of the same order of magnitude as those required in a truck application and in this respect it could be potentially suitable for heavy vehicle applications as well. The schematic diagram of the damper model is shown in Figure 7.38.

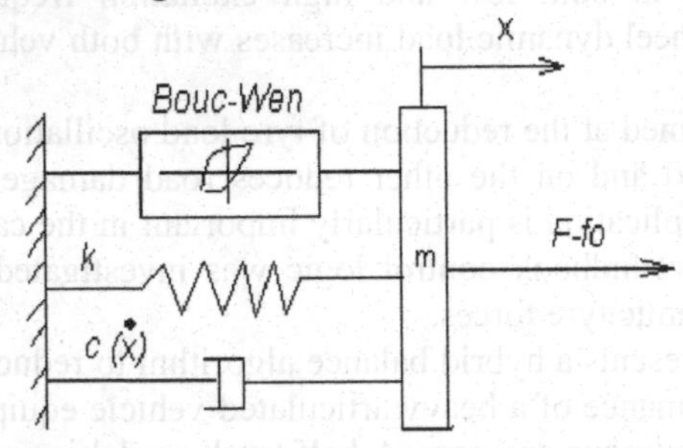

Fig.7.38. MR damper schematic model [copyright IMechE (2008), reproduced from Tsampardoukas G, Stammers CW and Guglielmino E, Semi-active control of a passenger vehicle for improved ride and handling, accepted for publication in Proceedings of the Institution of Mechanical Engineers, Part D: Journal of Automobile Engineering, Publisher: Professional Engineering Publishing, ISSN 0954/4070, Vol. 222, D3/2008, pp 325–352, used by permission]

It is a Bouc–Wen model coupled with a non-linear viscous damper with exponential characteristics and a linear spring term. The governing equations are as follows:

$$\dot{z} = -\gamma\left|\dot{x}\right|\left|z\right|^{n-1} z - \beta\dot{x}\left|z\right|^{n} + A\dot{x}\,, \tag{7.6}$$

$$C(\dot{x}) = -a_1 \exp(-a_2\left|\dot{x}\right|)^{p}\,, \tag{7.7}$$

$$F - f_0 = az + kx + C(\dot{x})\dot{x} + m\ddot{x}\,. \tag{7.8}$$

Table 7.4. Key parameters employed in simulation (copyright Elsevier, reproduced from Tsampardoukas G, Stammers CW and Guglielmino E, Hybrid balance control of a magnetorheological truck suspension, accepted for publication in Journal of Sound and Vibration, used by permission)

Parameter	Value
Tractor chassis mass	$m_c = 4400$ kg
Trailer chassis mass (laden)	$m_t = 12500$ kg
Tractor pitch inertia	$J_c = 18311$ kgm^2
Trailer pitch inertia	$J_t = 251900$ kgm^2
Steer tractor unsprung mass	$m_{u1} = 270$ kg
Drive tractor unsprung mass	$m_{u2} = 520$ kg
Trailer unsprung mass	$m_{u3} = 340$ kg
Distance from steer tractor axle to tractor C.G.	$l_1 = 1.2$ m
Distance from tractor C.G. to drive tractor axle	$l_2 = 4.8$ m
Distance from tractor C.G. to articulation point	$l_4 = 4.134$ m
Distance from articulation point to trailer C.G.	$l_5 = 6.973$ m
Distance from trailer C.G. to trailer axle	$l_6 = 4$ m
Tyre stiffness of steer tractor wheel	$k_{tf} = 847$ kN/m
Tyre stiffness of drive tractor wheel	$k_{tr} = 2$ MN/m
Tyre stiffness of trailer wheel	$k_{tt} = 2$ MN/m
Suspension spring stiffness of steer tractor axle	$k_f = 300$ kN/m
Suspension spring stiffness of drive tractor axle	$k_r = 967430$ N/m
Suspension spring stiffness of trailer axle	$k_t = 155800$ N/m
Suspension damping rate of steer tractor axle	$c_f = 10$ kNs/m
Suspension damping rate of drive tractor axle	$c_r = 27627$ Ns/m
Suspension damping rate of trailer axle	$c_t = 44506$ Ns/m
Stiffness of articulated connection	$k_5 = 20$ MN/m
Damping of articulated connection	$c_5 = 200$ kNs/m

As explained in Chapter 2, the variable z is an evolutionary variable while the parameters β, γ, A and n define the shape of the hysteresis loop. Equation 7.7 models the post-yield plastic damping coefficient, which depends on the relative velocity. This equation is used to describe the MR fluid shear thinning effect, which results in the roll-off of the resisting force of the damper in the low-velocity region. The total exerted force is described by Equation 7.8, which takes into account the evolutionary variable z and the post-yield plastic model, expressed by

(7.7). Table 7.5 lists the numerical value of the constant parameters of the MRD (while the parameters a, a_1, a_2, n and f_o are current dependent).

Table 7.5. Constant parameters for MR damper (copyright Elsevier, reproduced from Tsampardoukas G, Stammers CW and Guglielmino E, Hybrid balance control of a magnetorheological truck suspension, accepted for publication in Journal of Sound and Vibration, used by permission)

Parameter	Value	Parameter	Value
γ	32000 m^{-2}	m	100 kg
β	22 m^{-2}	k	2500 N/m
A	220	p	0.54

The simulated characteristics are depicted in Figure 7.39.

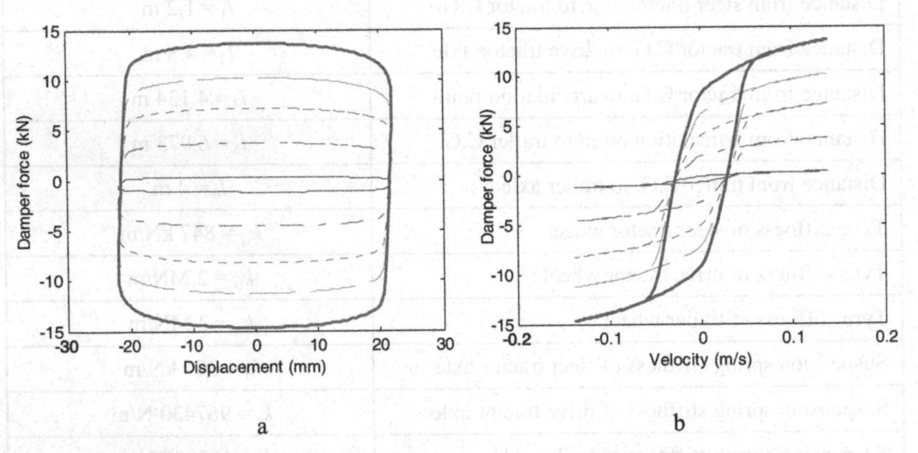

Fig. 7.39. (a) MR damper force velocity and (b) force displacement characteristics for a 22-mm 1-Hz sinusoidal displacement excitation (copyright Elsevier, reproduced from Tsampardoukas G, Stammers CW and Guglielmino E, Hybrid balance control of a magnetorheological truck suspension, accepted for publication in Journal of Sound and Vibration, used by permission)

A benchmark viscous damper having different damping coefficients for the bound and rebound strokes was used (the equivalent damping ratios are, respectively, 0.15 and 0.35).

Appropriate road profile models are required to assess truck performance under realistic operative conditions. Two types of road are considered in this case study: a smooth highway and a highway with gravel. The spectral densities of both road profiles are expressed by $S_g(\Omega)=C_{sp}\Omega^{-n}$, the parameters C_{sp} and n determining the road quality. The smooth highway is described by n = 2.1 and C_{sp}=4.8$\cdot\times 10^{-7}$ whereas the gravel road is determined by setting C_{sp}=4.4$\times 10^{-6}$ and n = 2.1 (Wong, 1993). Heavy goods vehicles are mainly designed to operate on smooth rather than on poor roads. The vehicle operation on highway with gravel is a scenario to examine the performance of the semi-active suspension on damaged roads or in off-road operation.

7.6.3 Road Damage Assessment

It is of paramount importance to establish a quantitative criterion to assess road damage (Cebon, 1989). The most widely employed is the *fourth power law*. This is a result of the experimental work undertaken by the American Association of State Highway and Transportation Officials (AASHO) (Gillespie, 1985). This law shows that the pavement serviceability decreases every time a heavy vehicle axle passes on the road. This reduction is assumed to be related to the fourth power of its static load (Cebon, 1999). Another criterion is known as the *aggregate fourth power force* (Cole *et al.*, 1994) while Potter *et al.* (1995) give a simplified approach to road damage. It is expressed by the following formula:

$$A_k^n = \sum_{j=1}^{Na} P_{jk}^n , \tag{7.9}$$

where k =1, 2, 3,..., Na is the location along the road. The exponent n is chosen depending upon the type of pavement and ranges from $n = 4$ (suitable for fatigue damage) to $n = 1$ (permanent deformation caused by static load). In this work in order to describe the fatigue damage, the aggregate fourth power law with $n = 4$ is used, normalised with respect to the static force *i.e.*,

$$A_k^4 = \sum_{j=1}^{Na} \left(\frac{\text{Static force} + \text{Dynamic force}}{\text{Static force}} \right)_{jk}^4 \tag{7.10}$$

7.6.4 Road Damage Reduction Algorithm

The algorithm outlined here is a variant of the balance logic described in Chapter 4. This hybrid version aims at cancelling the drive tractor and trailer axle tyre forces. The essence of the proposed control algorithm is to cancel the tyre force fluctuations on each axle by ensuring that the wheel follows the road profile closely. The dynamic tyre forces are balanced by applying a controlled damping force in the opposite direction. This is only possible when the control force and the relative velocity have opposite signs and hence energy dissipation takes place.

$$Fc_{\text{rear}} = \begin{cases} b_{\text{r}}\left(-F_{\text{sr}} - m_{u2}\ddot{x}_{\text{wr}}\right) + b_2 F_{\text{dr}} & \text{if } Fc_{\text{rear}} \cdot \text{rear\_rel\_vel} < 0 \\ b_3 F_{\text{dr}} & \text{if } Fcr_{\text{rear}} \cdot \text{rear\_rel\_vel} > 0 \end{cases} \tag{7.11}$$

$$Fc_{\text{trailer}} = \begin{cases} b_{\text{t}}\left(-F_{\text{st}} - m_{u3}\ddot{x}_{\text{wt}}\right) + b_2 F_{\text{dt}} & \text{if } Fc_{\text{trailer}} \cdot \text{trailer\_rel\_vel} < 0 \\ b_3 F_{\text{sd}} & \text{if } Fc_{\text{trailer}} \cdot \text{trailer\_rel\_vel} > 0 \end{cases} \tag{7.12}$$

A pseudo-viscous damping term is added to the control forces to reduce transients, particularly when inputs are near to the wheel-hop frequency. Studies (not reported here) have indicated that the optimal values of b_2 and b_3 (to minimise road damage) should be 20% of critical passive damping when the vehicle travels on smooth or gravel roads. Smaller or higher values of b_2 and b_3 result in larger dynamic tyre forces and higher vibration levels. However, the optimum values of those parameters (in terms of road damage) alter when the vehicle wheels come into contact with bumps or potholes.

7.6.5 Time Response

In this section, and in the following ones, the response of the semi-active controlled MRD system is benchmarked against the passive system. The performance of the controlled system is assessed numerically both in the time domain and in the frequency domain.

The time response to a sinusoidal road surface is firstly investigated. Figures 7.40–7.42 depict the time histories for a 5-mm 2-Hz sinusoidal road input. The hybrid control logic significantly reduces the dynamic tyre forces of the tractor drive and trailer axles. The same trend is also observed for the trailer chassis acceleration. In contrast, the controlled suspension increases the chassis acceleration of the tractor unit. As expected a higher harmonic content is present in the semi-active time trends.

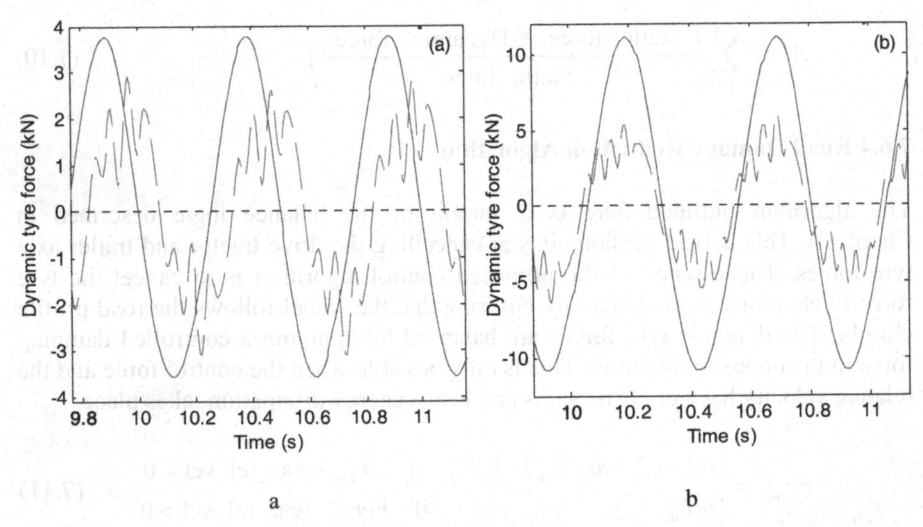

Fig. 7.40. Dynamic tyre forces on tractor drive and trailer axles: (a) dynamic tyre force of tractor drive axle; (b) dynamic tyre forces of trailer axle; (- - - - -) semi-active suspension, (——) passive suspension (copyright Elsevier, reproduced from Tsampardoukas G, Stammers CW and Guglielmino E, Hybrid balance control of a magnetorheological truck suspension, accepted for publication in Journal of Sound and Vibration, used by permission)

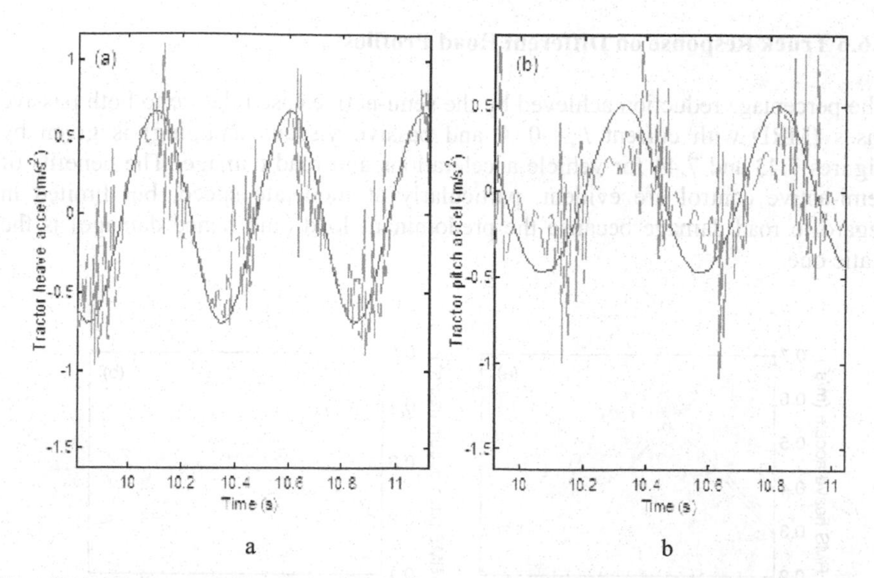

Fig. 7.41. Tractor and trailer chassis heave acceleration: (a) tractor chassis heave acceleration; (b) tractor chassis pitch acceleration; (- - - - -) semi-active suspension, (——) passive suspension (copyright Elsevier, reproduced from Tsampardoukas G, Stammers CW and Guglielmino E, Hybrid balance control of a magnetorheological truck suspension, accepted for publication in Journal of Sound and Vibration, used by permission)

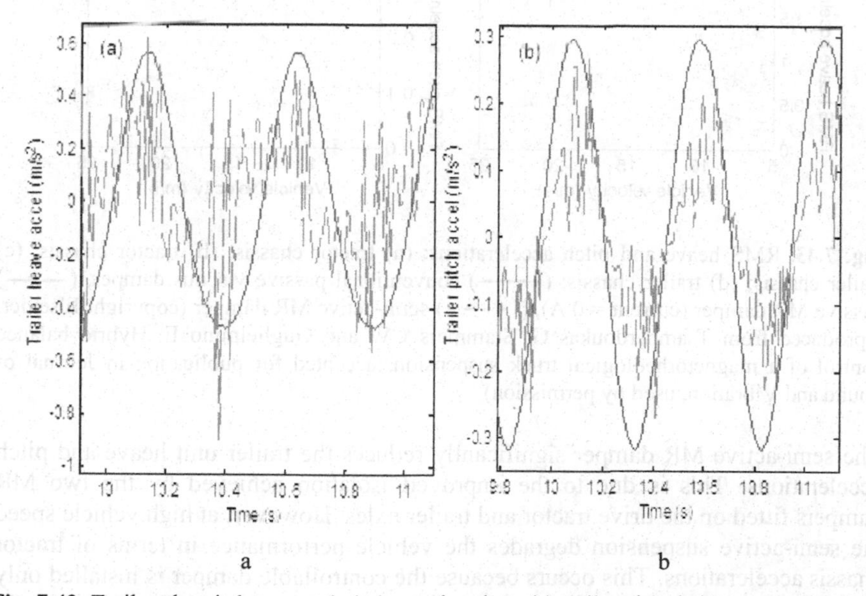

Fig. 7.42. Trailer chassis heave and pitch accelerations (a) trailer chassis heave acceleration; (b) trailer chassis pitch acceleration; (- - - - -) semi-active suspension, (——) passive suspension (copyright Elsevier, reproduced from Tsampardoukas G, Stammers CW and Guglielmino E, Hybrid balance control of a magnetorheological truck suspension, accepted for publication in Journal of Sound and Vibration, used by permission)

7.6.6 Truck Response on Different Road Profiles

The percentage reduction achieved by the semi-active case relative to both passive cases (MRD with current I = 0 A and passive viscous damping) is given by Figures 7.43 and 7.44 for vehicle accelerations and road damage. The benefits of semi-active control are evident, particularly at moderate speed, but limited in regard to road damage because the predominant load (and hence damage) is the static one.

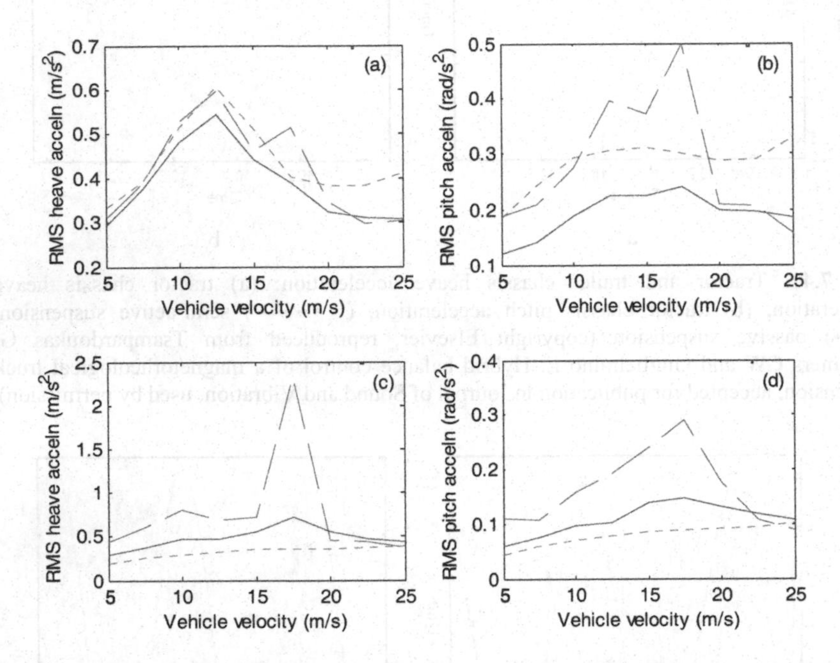

Fig. 7.43. RMS heave and pitch accelerations: (a) tractor chassis; (b) tractor chassis; (c) trailer chassis; (d) trailer chassis; (——) conventional passive viscous damper, (— —) passive MR damper (current = 0 A), (- - - -) semi-active MR damper (copyright Elsevier, reproduced from Tsampardoukas G, Stammers CW and Guglielmino E, Hybrid balance control of a magnetorheological truck suspension, accepted for publication in Journal of Sound and Vibration, used by permission)

The semi-active MR damper significantly reduces the trailer unit heave and pitch accelerations. This is due to the improved isolation achieved by the two MR dampers fitted on the drive tractor and trailer axles. However, at high vehicle speed the semi-active suspension degrades the vehicle performance in terms of tractor chassis accelerations. This occurs because the controllable damper is installed only on the drive tractor axle. It has been verified that the tractor performance would improve if a third MR damper was fitted on the steer axle as well. However truck driver seat are often equipped with a seat damper (ideally semi-active; see case study 2), which in practice reduces the vibration level experienced by the driver.

The application of the road damage criterion given in Figure 7.44 reveals that the damage caused by the dynamic tyre forces is significantly reduced when the balance control cancels them by 100%, while the coefficients b_2 and b_3 are equal to 0.2 of the critical damping force. Figure 7.44 shows that the semi-active suspension reduces significantly the road damage caused by each individual axle as well as the total vehicle damage. It is important to note that the vehicle suspension employing passive MR dampers (*i.e.*, $I = 0$ A) degrades the vehicle response due to its low damping.

The amount of dynamic tyre force cancellation is a critical parameter which affects the system response as Figure 7.45 shows. At low and medium vehicle velocities 100% cancellation is the best option because both RMS and max dynamic tyre forces at each axle are significantly reduced relative to the passive system, resulting in lower road damage with respect to the damage criterion used. On the other hand, it is beneficial to reduce the amount of cancellation for the damper fitted on the drive tractor axle in order to improve tractor unit comfort, particularly at high vehicle velocities, but such a reduction adversely affects road damage at high velocities. Consequently, the optimal choice for the amount of spring force cancellation depends on the control objective.

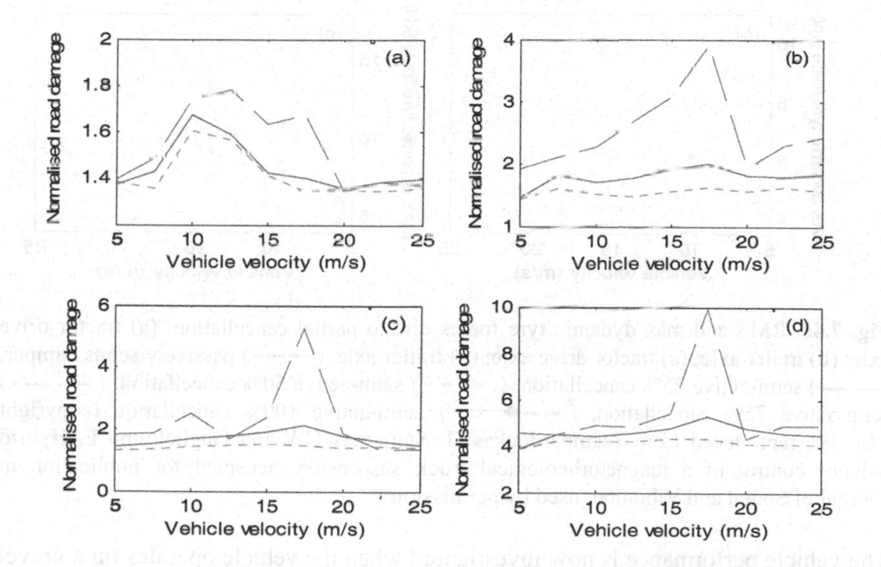

Fig. 7.44. Maximum normalised road damage: (a) steer axle; (b) drive axle; (c) trailer axle; (d) total vehicle: (———) passive viscous damper, (— —) passive MR damper (current = 0 A), (- - - -) semi-active MR damper (copyright Elsevier, reproduced from Tsampardoukas G, Stammers CW and Guglielmino E, Hybrid balance control of a magnetorheological truck suspension, accepted for publication in Journal of Sound and Vibration, used by permission)

A design solution which achieves a compromise between these two requirements entails the use of a suspended driver cab and seat to help reduce the vibration levels transmitted to the human body. In that case, 100% cancellation is the best solution in terms of lower maximum dynamic tyre forces for the semi-active device located at the tractor drive axle, while 50% cancellation is the best solution overall for the same device at the trailer axle. The variation between the amount of cancellation between the units is mainly affected by the coupled vehicle motion and the model (half truck) used. Consequently, a full truck model should be developed in order to evaluate the performance of the hybrid control logic not only for ride but also for handling manoeuvres.

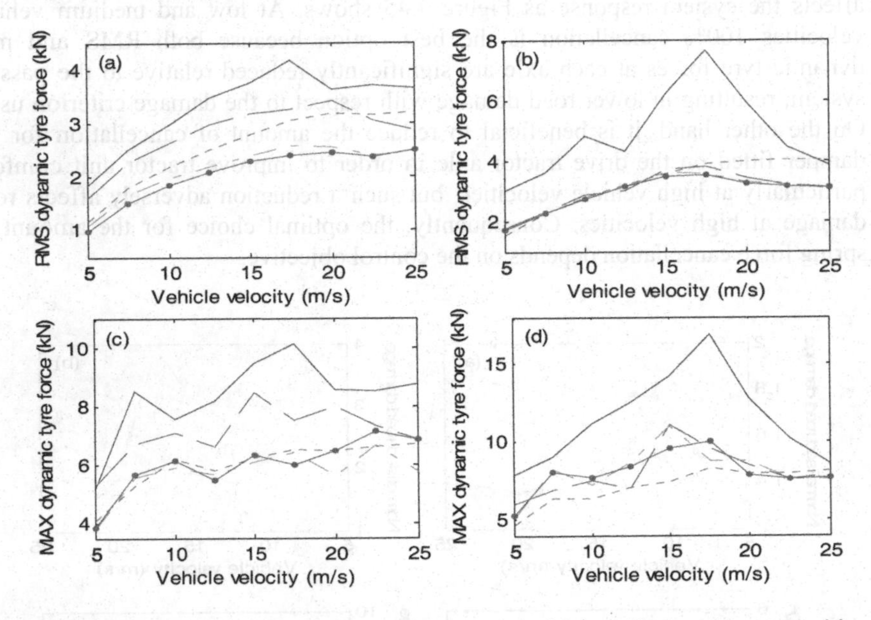

Fig. 7.45. RMS and max dynamic tyre forces due to partial cancellation: (a) tractor drive axle; (b) trailer axle; (c) tractor drive axle; (d) trailer axle. (———) passive viscous damper, (— —) semi-active 25% cancellation, (- - - -) semi-active 50% cancellation, (—·—·) semi-active 75% cancellation, (——◆——) semi-active 100% cancellation (copyright Elsevier, reproduced from Tsampardoukas G, Stammers CW and Guglielmino E, Hybrid balance control of a magnetorheological truck suspension, accepted for publication in Journal of Sound and Vibration, used by permission)

The vehicle performance is now investigated when the vehicle operates on a gravel road. A smooth road surface may have sections where the surface is rough due to maintenance or resurfacing work. It is essential to assess the vehicle and the control logic behaviour under these conditions. Heavy good vehicles are mainly designed to operate on smooth highways rather than on poor roads. Consequently, the vehicle operation on highway with gravel is a scenario to assess the performance of the semi-active suspension either off-road or under large amplitude-road inputs.

Simulation results show that the heave and pitch accelerations of the tractor and trailer units are reduced by control. Figure 7.46 presents the dynamic tyre forces of the vehicle in three different cases. The semi-active suspension performs better than the passive system over the velocity range investigated; however, the control logic becomes ineffective when the vehicle velocity is higher than 25 m/s. The vehicle performance is similar to that on smooth road in terms of normalised road damage.

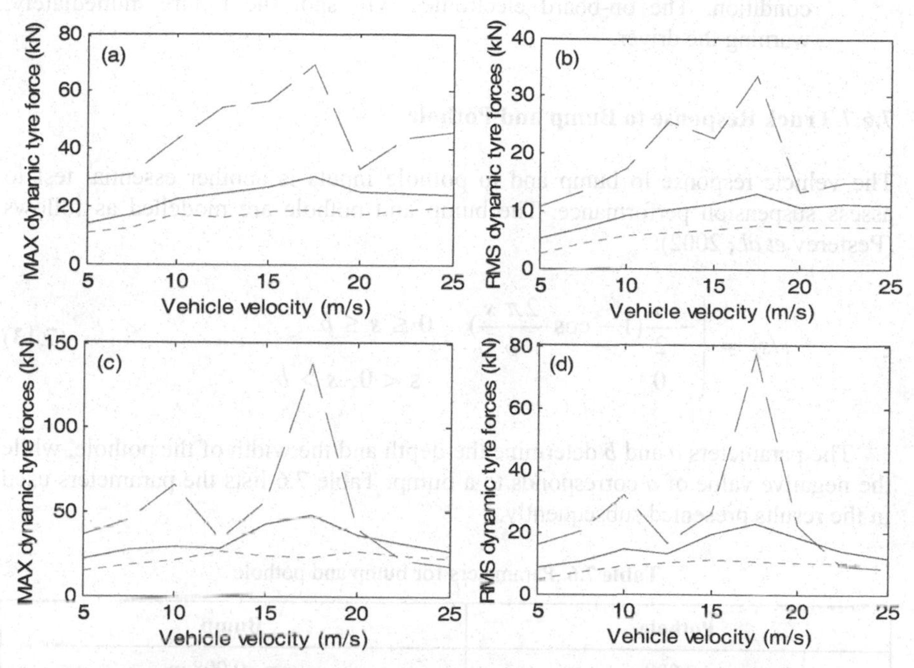

Fig. 7.46. Dynamic tyre forces: (a) tractor drive axle; (b) tractor drive axle; (c) trailer axle; (d) trailer axle; (——) passive viscous damper, (— —) passive MR damper ($I = 0$ A), (- - - -) semi-active MR damper (copyright Elsevier, reproduced from Tsampardoukas G, Stammers CW and Guglielmino E, Hybrid balance control of a magnetorheological truck suspension, accepted for publication in Journal of Sound and Vibration, used by permission)

The results presented so far can be summarised as follows:

- Semi-active hybrid balance control is beneficial to heavy vehicle performance: chassis accelerations are significantly reduced and road damage moderately so,
- The semi-active MR damper significantly reduces axle loads at all speeds on both harsh and smooth roads,
- By employing balance logic on the drive tractor and trailer controlled dampers, trailer chassis acceleration is substantially reduced. The tractor chassis acceleration slightly increases because a passive viscous damper is used on the steer tractor axle,
- The partial cancellation of the dynamic tyre forces is mainly affected by the vehicle speed; 100% cancellation is the optimal solution in terms of lower road damage in the moderate vehicle speed range (from 12.5 m/s to 20 m/s) while 75% cancellation is the best solution in the low (from 5m/s to 10m/s) and high vehicle speed ranges (from 20 m/s to 25 m/s),

- The passive MR damper is not able to produce high forces, causing excessive load on axles. However an MR damper operates in this mode only because of an electrical fault (*e.g.*, if the control system power supply fails). The system performs poorly but this is a provisional fail-safe condition. The on-board electronics will spot the failure immediately, warning the driver.

7.6.7 Truck Response to Bump and Pothole

The vehicle response to bump and to pothole inputs is another essential test to assess suspension performance. The bump and pothole are modelled as follows (Pesterev *et al.*, 2002):

$$r(s) = \begin{cases} -\frac{a}{2}(1-\cos\frac{2\pi s}{b}) & 0 \le s \le b \\ 0 & s<0,\ s>b \end{cases} \tag{7.13}$$

The parameters a and b determine the depth and the width of the pothole, while the negative value of a corresponds to a bump. Table 7.6 lists the parameters used in the results presented subsequently.

Table 7.6. Parameters for bump and pothole

Pothole	Bump
$a = 0.008$ m	$a = -0.008$ m
$b = 1$ m	$b = 1$ m
Vehicle velocity = 10 m/s	
100% cancellation of dynamic tyre forces $b_r = 1$ and $b_t = 1$	
$b_2 = 0.5$, amount of additional critical damper force when damper is on $b_3 = 0.2$, amount of critical damper force when damper is off	

Figures 7.47 and 7.48 show that the semi-active control scheme reduces the peak values of the dynamic tyre forces on both drive tractor and trailer axles. The controlled suspension reduces the number of oscillations with respect to the passive suspension. Similar trends can be observed for vehicle speeds ranging from 5 to 25 m/s (results not shown here). Control produces high damping during the settling time but more or less the same level of damping as the passive case during the impact stage.

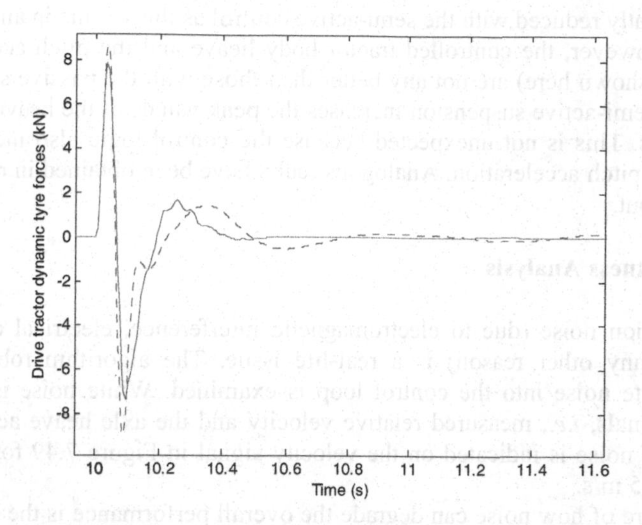

Fig. 7.47. Tractor drive dynamic tyre forces due to bump; (- - - - -) passive suspension, (——) semi-active suspension (copyright Elsevier, reproduced from Tsampardoukas G, Stammers CW and Guglielmino E, Hybrid balance control of a magnetorheological truck suspension, accepted for publication in Journal of Sound and Vibration, used by permission)

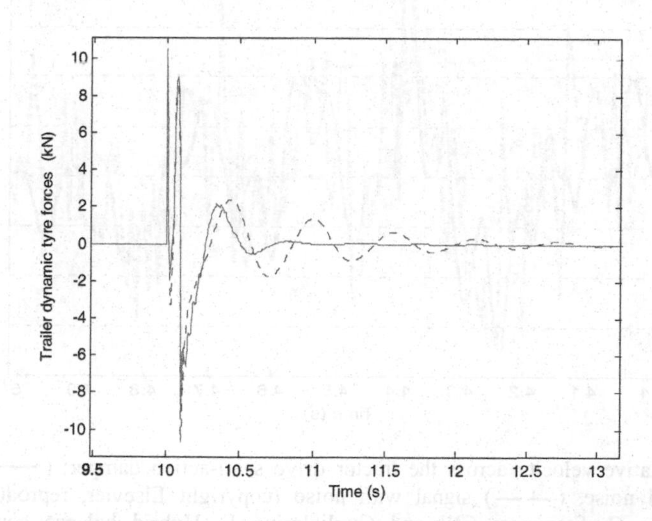

Fig. 7.48. Trailer drive dynamic tyre forces due to bump; (- - - - -) passive suspension, (——) semi-active suspension (copyright Elsevier, reproduced from Tsampardoukas G, Stammers CW and Guglielmino E, Hybrid balance control of a magnetorheological truck suspension, accepted for publication in Journal of Sound and Vibration, used by permission)

Simulation has also shown that an improvement is present on the trailer body heave and pitch accelerations too (graphs not depicted here): the free oscillations are significantly reduced with the semi-active control as this results in an additional damping. However, the controlled tractor body heave and the pitch accelerations (graphs not shown here) are not any better than those with the passive suspension. In fact, the semi-active suspension increases the peak values of the heave and pitch accelerations. This is not unexpected because the control logic also increases the tractor body pitch acceleration. Analogous reults have been obtained in response to a pothole input.

7.6.8 Robustness Analysis

Instrumentation noise (due to electromagnetic interference, electrical component damage or any other reason) is a real-life issue. The algorithm robustness to injected white noise into the control loop is examined. White noise is added to feedback signals, *i.e.*, measured relative velocity and the axle heave acceleration. The level of noise is indicated on the velocity signal in Figure 7.49 for a vehicle velocity of 15 m/s.

A measure of how noise can degrade the overall performance is the increase in the number of switches of the controller (*i.e.*, increased chattering)

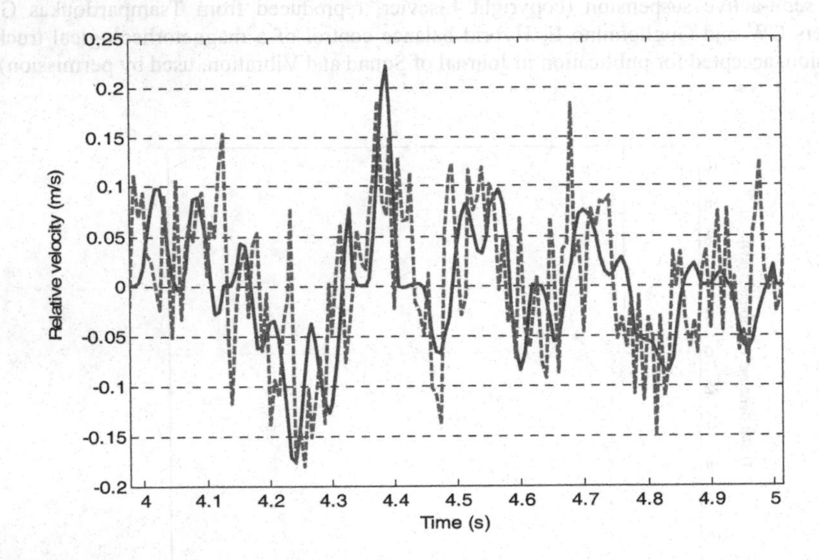

Fig. 7.49. Relative velocity across the tractor drive semi-active damper; (——) signal without added noise, (— —) signal with noise (copyright Elsevier, reproduced from Tsampardoukas G, Stammers CW and Guglielmino E, Hybrid balance control of a magnetorheological truck suspension, accepted for publication in Journal of Sound and Vibration, used by permission)

A moderate penalty due to noise is observed in the heave tractor chassis acceleration.

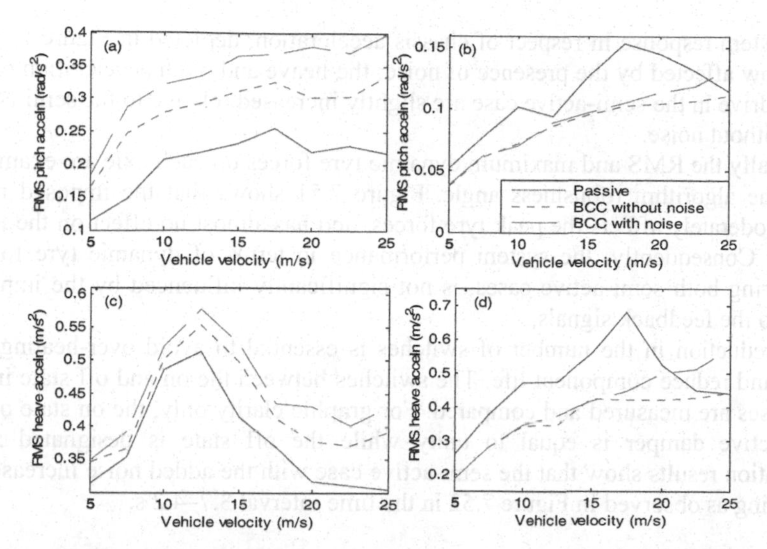

Fig. 7.50. Chassis RMS heave and pitch acceleration on both vehicle units: (a) tractor chassis; (b) trailer chassis; (c) tractor chassis; (d) trailer chassis; (——) passive, (- - - -) semi-active (BCC) without imposed noise, (— —) semi-active (BCC) with noise (copyright Elsevier, reproduced from Tsampardoukas G, Stammers CW and Guglielmino E, Hybrid balance control of a magnetorheological truck suspension, accepted for publication in Journal of Sound and Vibration, used by permission)

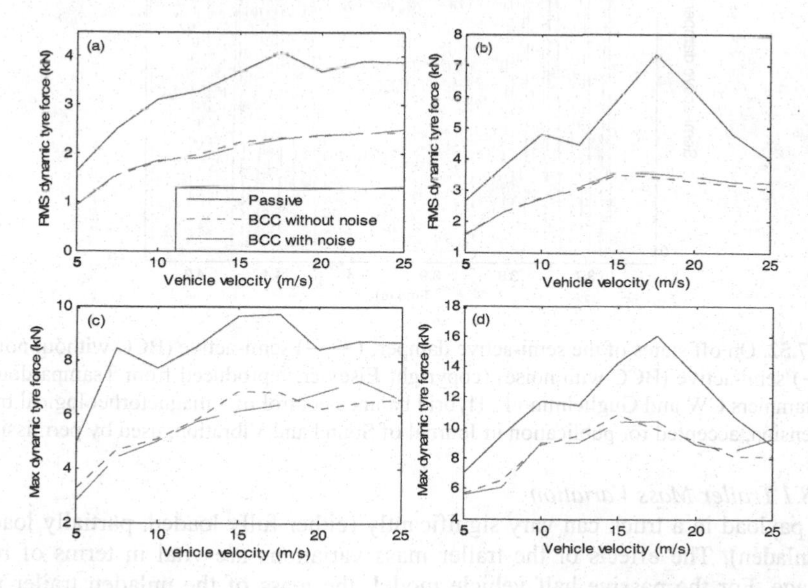

Fig. 7.51. RMS and max dynamic tyre forces on tractor drive and trailer axles: (a) tractor drive axle; (b) trailer axle; (c) tractor drive axle; (d) trailer axle; (——) passive, (- - - -) semi-active (BCC) without imposed noise, (— —) semi-active (BCC) with noise (copyright Elsevier, reproduced from Tsampardoukas G, Stammers CW and Guglielmino E, Hybrid balance control of a magnetorheological truck suspension, accepted for publication in Journal of Sound and Vibration, used by pemission)

The system response in respect of chassis acceleration, depicted in Figure 7.50, is somehow affected by the presence of noise: the heave and pitch acceleration of the trailer drive in the semi-active case are slightly increased relative to the semi-active case without noise.

Finally the RMS and maximum dynamic tyre forces on each axle are examined from the algorithm robustness angle. Figure 7.51 shows that the imposed noise only moderately affects the peak tyre forces, and has almost no effect on the RMS values. Consequently, the system performance in terms of dynamic tyre forces, comparing both semi-active cases, is not significantly influenced by the imposed noise to the feedback signals.

A reduction in the number of switches is essential to avoid over-heating and wear, and reduce component life. The switches between the on and off state in the two cases are measured and compared. For graphic clarity only, the on state of the semi-active damper is equal to unity while the off state is designated zero. Simulation results show that the semi-active case with the added noise increase the chattering as observed in Figure 7.52 in the time interval 3.7–4.2 s.

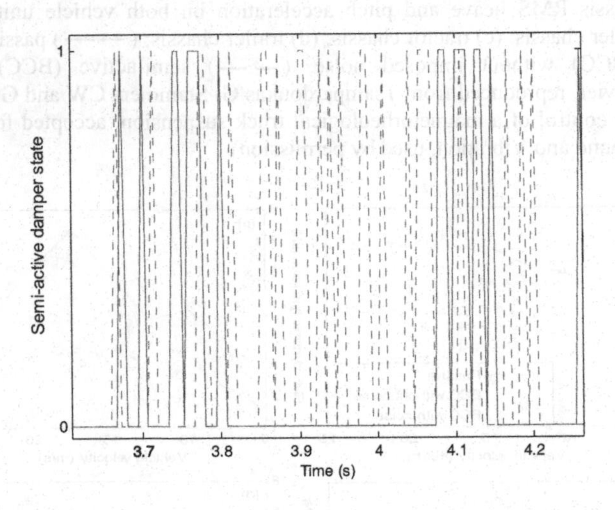

Fig. 7.52. On/off states of the semi-active damper; (——) semi-active (BCC without noise), (----) semi-active (BCC with noise) (copyright Elsevier, reproduced from Tsampardoukas G, Stammers CW and Guglielmino E, Hybrid balance control of a magnetorheological truck suspension, accepted for publication in Journal of Sound and Vibration, used by permission)

7.6.8.1 Trailer Mass Variation

The payload in a truck can vary significantly (either fully loaded, partially loaded or unladen). The effects of the trailer mass variations are vital in terms of road damage. For the passive half vehicle model, the mass of the unladen trailer was taken as 5000 kg, the fully loaded trailer 12500 kg and a partly-loaded trailer 8500 kg.

The variations of the trailer mass are considered in the simulation process to examine the response of the system in the three cases. Figure 7.53 shows the reduction of the dynamic forces at a specified vehicle speed when the three different trailer masses A convergence in the trends is observed at high vehicle speeds.

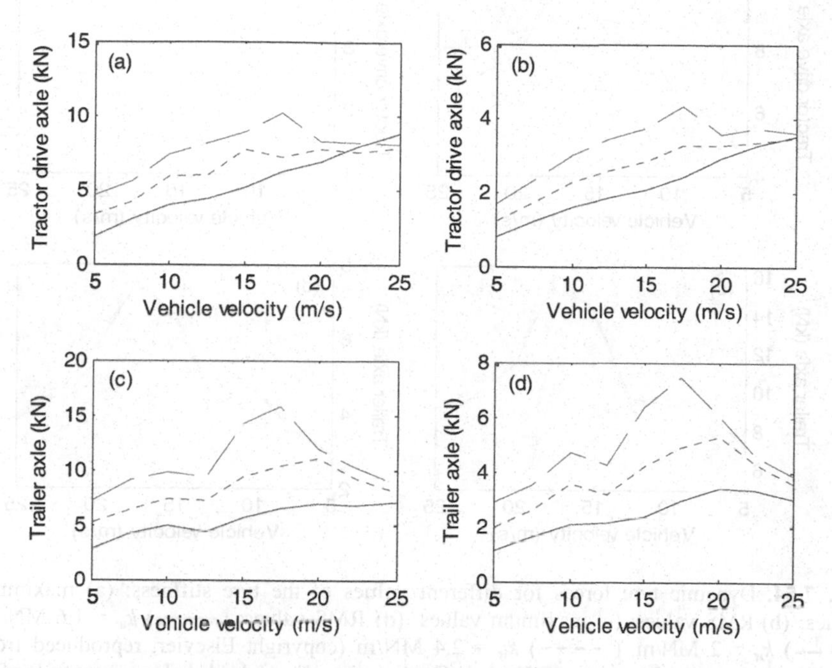

Fig. 7.53. Dynamic tyre forces for different values of the trailer mass: (a) maximum values; (b) RMS values; (c) maximum values; (d) RMS values; (———) m_t = 5000 kg, (- - - -) m_t = 8500 kg, (— —) m_t = 12500 kg (copyright Elsevier, reproduced from Tsampardoukas G, Stammers CW and Guglielmino E, Hybrid balance control of a magnetorheological truck suspension, accepted for publication in Journal of Sound and Vibration, used by permission)

7.6.8.2 Tyre Stiffness Variation

In order to simulate the vertical tyre motion due to road irregularities the tyre is modelled as a spring with high stiffness. In real operating conditions the tyre pressure is not constant and slowly changes over time, resulting in variations of the tyre stiffness. A tyre stiffness of 2 MN/m has normally been used in the simulation, while lower or higher values result in different tyre pressures.

The variation of the tyre stiffness of the drive tractor wheel shows that the dynamic tyre forces are slightly affected at low and moderate vehicle speeds (Figure 7.54). The system response alters at high vehicle velocities, producing larger tyre forces, resulting in higher road damage (as expressed by Equation 7.9). The simulation results plotted in Figure 7.54 show that the dynamic tyre forces at both axles are little affected by this variation. Consequently, the road damage (see road damage criterion) is also unaffected at low vehicle speeds with a moderate increase at high vehicle velocities.

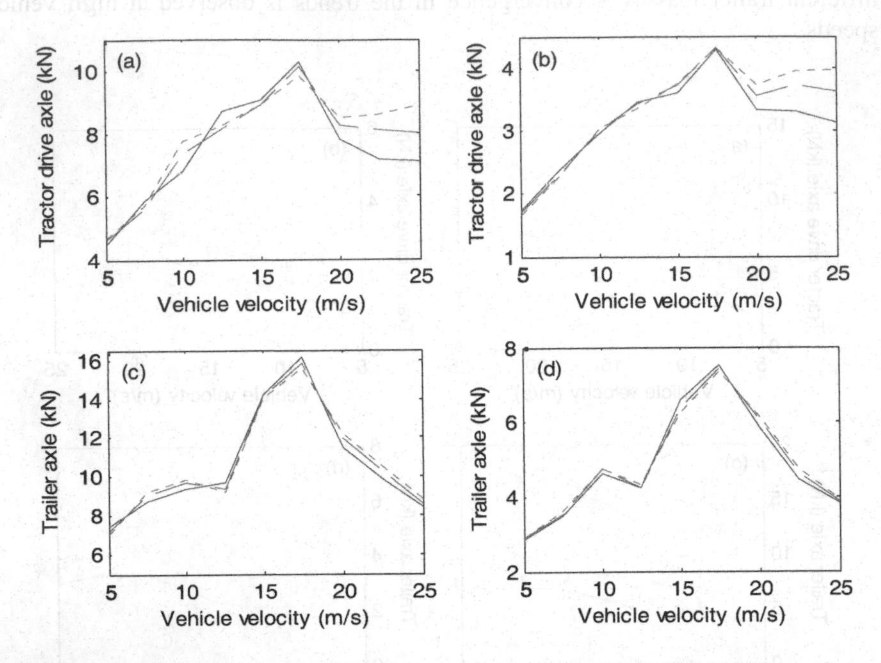

Fig. 7.54. Dynamic tyre forces for different values of the tyre stiffness: (a) maximum values; (b) RMS values; (c) maximum values: (d) RMS values; (——) k_{tr} = 1.6 MN/m, (— —) k_{tr} = 2 MN/m, (- - - -) k_{tr} = 2.4 MN/m (copyright Elsevier, reproduced from Tsampardoukas G, Stammers CW and Guglielmino E, Hybrid balance control of a magnetorheological truck suspension, accepted for publication in Journal of Sound and Vibration, used by permission)

7.6.8.3 MRD Response Time

The response time of MR dampers for vehicle applications is a critical factor because it determines the effectiveness of the MR damper used. The aim is now to assess the effect of the damper response to the vehicle performance.

The response time is defined as the time required for the MR damper to reach 64% or 95% (Goncalves *et al.*, 2003) of the final exerted force, starting from the initial state. The time response depends both on the fluid transformation from a mineral-oil-like consistency (but not majorly as this time is less than 1 ms) and on the inductance of the electromagnetic circuit as well as the output impedance of the driving electronics.

Extensive experimental work by Koo *et al.* (2004) showed that the time response of the MR damper is affected by several parameters such as the response of the driving electronics, the applied current, the piston velocity and the system compliance. The time response of MR damper for commercial vehicle application is less than 25 ms (effective time response to reach 95% of its final value).

In the simulation work a first-order lag is employed in the semi-active control schemes in order to model the MR damper dynamics. The time constant of the first-order lag is chosen so that the system reaches 64% of the final value in 11 ms (one time constant), which corresponds to 25 ms to reach 95% of its final value (effective time constant). In the current study the time constant quoted is that to reach 64% of the final state.

Simulation results indicate that a fast-response MR damper (5 ms time constant) is the preferable device to improve the vehicle response in terms of lower dynamic tyre forces. Figure 7.55 shows that the maximum values rather than the RMS values of the dynamic tyre forces are more sensitive to the time response of the semi-active damper.

A damper with time constant equal to 20 ms produces larger maximum tyre forces because it cannot respond fast enough to exert the required control force in order to cancel the dynamic tyre forces.

However, the latter value of the time constant is high and is the typical of a bigger damper, more suitable for controlling structural vibration. A time constant of 15 ms, which corresponds to a 37.5 ms effective time constant, might be a more realistic value for semi-active dampers used in heavy-vehicle applications.

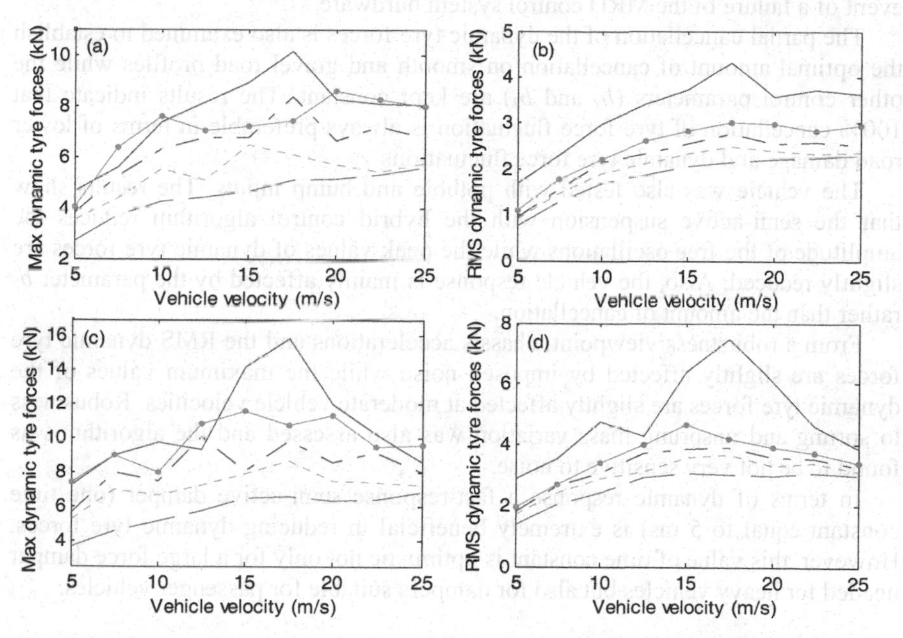

Fig. 7.55. Max dynamic tyre forces for different time constants: (a) tractor drive axle; (b) tractor drive axle; (c) trailer axle; (d) trailer axle; (———) passive damper, (— —) *Tc* = 5 ms (- - - -), *Tc* = 10 ms (— - — -), *Tc* = 15 ms (——◆——), *Tc* = 20 ms (copyright Elsevier, reproduced from Tsampardoukas G, Stammers CW and Guglielmino E, Hybrid balance control of a magnetorheological truck suspension, accepted for publication in Journal of Sound and Vibration, used by permission)

7.7 Conclusions

This case study centred on road damage reduction has shown that the semi-active truck suspension response is superior overall to the passive response on both road profiles (smooth and gravel). The maximum and RMS values of the dynamic tyre forces are substantially reduced by the control logic and the road damage follows the same pattern as the dynamic tyre forces. Similarly, a reduction of the trailer chassis acceleration is obtained because the trailer unit is well isolated from the ground irregularities. Conversely, the tractor chassis acceleration slightly increases because the steer axle is assumed to be equipped with conventional viscous dampers.

Additionally, the simulation results show that the MRD has poor performance when the applied current is zero (passive operation) due to the very low damping provided by the semi-active device. However, this scenario occurs only in the event of a failure of the MRD control system hardware.

The partial cancellation of the dynamic tyre forces is also examined to establish the optimal amount of cancellation on smooth and gravel road profiles while the other control parameters (b_2 and b_3) are kept constant. The results indicate that 100% cancellation of tyre force fluctuation is always preferable in terms of lower road damage and dynamic tyre force fluctuations.

The vehicle was also tested with pothole and bump inputs. The results show that the semi-active suspension with the hybrid control algorithm reduces the amplitude of the free oscillations while the peak values of dynamic tyre forces are slightly reduced. Also, the vehicle response is mainly affected by the parameter b_2 rather than the amount of cancellation.

From a robustness viewpoint, chassis accelerations and the RMS dynamic tyre forces are slightly affected by imposed noise while the maximum values of the dynamic tyre forces are slightly affected at moderate vehicle velocities. Robustness to sprung and unsprung mass variation was also assessed and the algorithm was found to be not very sensitive to noise.

In terms of dynamic response a fast-response semi-active damper (one time constant equal to 5 ms) is extremely beneficial in reducing dynamic tyre forces. However, this value of time constant is optimistic not only for a large force damper needed for heavy vehicles but also for dampers suitable for passenger vehicles.

References

Abdi H, Valentin D, Edelman BE (1999). Neural networks. Sage, Thousand Oaks

Abramowitz M, Stegun AI (1970) Handbook of mathematical function with formulas, graphs, and mathematical tables. Applied Mathematical Series, N.B.S.

Agarwal M (1997) A systemic classification of neural-network-based control. IEEE Control Syst Mag 17, 2:75–93

Agrawal A, Kulkarni P, Vieira SL, Naganathan NG (2001) An overview of magneto- and electro-rheological fluids and their applications in fluid power systems. Int J Fluid Power 2:5–36

Ahmadian M, Marjoram RH (1989) Effects of passive and semi-active suspension on body and wheel-hop control. SAE paper 892487

Al-Houlu N, Weaver J, Lahdhiri T, Joo DS (1999) Sliding mode-based fuzzy logic controller for a vehicle suspension system. American Control Conference, San Diego, USA, pp 4188–4192

Alanoly J, Sankar S (1987) A new concept in semi-active vibration isolation. J Mech Transmissions Autom Des 109:242–247

Alleyne A, Neuhaus PE, Hedrick JK (1993)Application of non-linear control theory to electronically controlled suspensions. Vehicle Syst Dyn 22:309–320

Amontons G (1699) On the resistance originating in machines. French R Acad Sci A12:206–22

Anderson J, Ferri AA (1990) Behavior of a single-degree-of-freedom system with a generalized friction law. J Sound Vib 140; 2:287–304

Armstrong-Helouvry B, (1990) Stick-slip arising from Stribeck friction. Proc IEEE Int Conf on Robotics and Automation, Cincinnati, USA, pp 1377–1382

Armstrong-Helouvry B, Dupont P, Canudas de Wit , (1994) A survey of models, analysis tools and compensation methods for the control of machines with friction. Automatica 30; 7:1083–1138.

Åström KJ (1998) Control of systems with friction. MOVIC '98, Zurich, Switzerland 1:25–32

Åström KJ, Hägglund T, Hang CC, Ho WK (1993) Automatic tuning and adaptation for PID controllers — a survey. Control Eng Practice 1; 4:699–714

Baker A (1984) Lotus active suspension. Automotive Engineer, pp 56–57

Baker GA, Graves-Morris P, (1996) Padé Approximants. Cambridge University Press

Barak P, Hrovat H (1988) Application of the LQG approach to the design of an automotive suspension for three-dimensional vehicle model. Proc IMechE Advanced Suspensions Conference, London, UK, pp 11–26

Bastow D (1993) Car suspension and handling. Pentech, London

Bellizzi S, Bouc R (1989) Adaptive control for semi-active isolators. Proc ASME-DED Conf on Machinery Dynamics — Applications and Vibration Control Problems, Montreal, Canada 18; 2: 317–323

Bendat JS, Piersol AG (1971) Random data: analysis and measurement procedures. Wiley, New York, NY

Bendat JS, Piersol AG (1980) Engineering applications of correlation and spectral analysis. Wiley, New York, NY

Bernard J, Vanderploeg M, Jane R (1981) Tire models for the determination of vehicle structural loads. Vehicle Syst Dyn 10:169–173

Berkovitz LD (1974) Optimal control theory. Springer Verlag, New York, NY

Besinger FH, Cebon D, Cole DJ (1991) An experimental investigation into the use of semi-active dampers on heavy lorries. Proc 12th IAVSD Symposium on Dynamics of Vehicles on Roads and on Railways Tracks, pp 26–30

Bliman PA, Sorine M (1991) Friction modelling by hysteresis operators. Application to Dahl, stiction and Stribeck effects. Proc Conf on Model of Hysteresis, Pitman Research Notes in Mathematics, Trento, Italy

Bliman PA, Sorine M (1995) Easy-to-use realistic dry friction models for automatic control. Proc 3rd European Control Conference, Rome, Italy, pp 3788–3794

Bosso N, Gugliotta A, Somà A (2000) Simulation of a freight bogie with friction dampers. 5th ADAMS Rail Users' Conference, Haarlem, The Netherlands

British Standards Institution (1996) BS 7853. Mechanical vibration — road surface profiles — reporting of measured data

British Standards Institution (1987) BS 6841. Measurement and evaluation of human exposure to whole-body mechanical vibration

Brokate M, Visintin A (1989) Properties of the Preisach model for hysteresis. J Reine und Angewandte Math 402:1–40

Bouc R (1971) Modèle mathèmatique d'hystèrèsis. Acustica 24:16–25

Blundell M, Harty D (2004) The multi-body systems approach to vehicle dynamics. Elsevier Butterworth-Heinemann, UK

Brun X, Sesmat S, Scavarda S, Thomasset D (1999) Simulation and experimental study of the partial equilibrium of an electropneumatic positioning system, cause of the sticking and restarting phenomenon. Fourth Japan Hydraulics and Pneumatics Society Int Symposium on Fluid Power, Tokyo, Japan, pp 125–130

Burns RS (2001) Advanced control engineering. Elsevier Butterworth-Heinemann, UK

Butz T, von Stryk O (1999) Modelling and simulation of rheological fluid devices, Preprint SFB-438-9911, Technische Universität München, Germany

Canudas de Wit C, Olsson HJ, Åström KJ, Lischinsky P (1993) Dynamic friction models and control design. American Control Conference, San Francisco, USA, pp 1920–1926

Canudas de Wit C, Olsson HJ, Åström KJ, Lischinsky P (1995) A new model for control of systems with friction. IEEE Trans Autom Control 40; 5:419–425

Carlson JD, Spencer BF (1996) Magneto-rheological fluid dampers for semi-active seismic control. Proc 3rd Int Conf on Motion and Vibration Control, III, pp 35–40

Chalasani RM (1987) Ride performance potential of active suspension systems-part II: comprehensive analysis based on a full-car model. ASME Symposium on Simulation and Control of Ground Vehicles and Transportation Systems 80:205–234

Chang SL, Wu CH (1997) Design of an active suspension system based on a biological model. American Control Conference, Albuquerque, USA, pp 2915–2919

Chantranuwathana S, Peng H (1999) Adaptive robust control for active suspensions. Proc American Control Conference, San Diego, USA, pp 1702–1706

Chen BM (2000) Robust and H-infinity control. Springer, London, UK

Choi SB, Choi YT, Park DW (2000) A sliding mode control of a full-car electrorheological suspension via hardware-in-the-loop simulation. J Dyn Syst Measure Control 122:114–121

Choi SB, Lee K, Parrk P (2001) A hysteresis model for the field-dependent damping force of a magnetorheological damper. J Sound Vib 245; 2:375–383

Chua LO, Bass SC (1972) A generalized hysteresis model. IEEE Trans Circuit Theory CT 19; 1:36–48

Cohen GH, Coon G (1953) Theoretical consideration of retarded control. Trans ASME 75:827–834

Cole DJ, Cebon D, Besinger FH (1994) Optimization of passive and semi-active heavy vehicle suspensions. SAE paper 942309

Cole DJ, Cebon D (1996a) Influence of tractor-trailer interaction on assessment of road performance. Proc IMechE, J Automobile Eng, D1; 212:1–10

Cole DJ, Cebon D (1996b) Truck suspension design to minimize road damage. Proc IMechE, J Automobile Eng, D2; 210:95–97

Cebon D (1989) Vehicle-generated road damage: a review.Vehicle Syst Dyn 18; 1–3

Cebon D (1999) Handbook of vehicle-road interaction. Swets and Zeitlinger, Lisse, The Netherlands

Cole DJ, (2001) Fundamental issues in suspension design for heavy road vehicles. Vehicle Syst Dyn 35; 4/5:319–360

Constantinescu VN, (1995) Laminar viscous flow. Springer, Berlin Heidelberg New York

Coulomb CA (1785) Theorie des machines simples, en ayant egard au frottement de leurs parties, et a la roideur des cordages. Memoires de Mathematiques et de Physique de l'Academie des Sciences, pp 161–342

Council of the European Union (1999) Amended proposal for a council directive on the minimum health and safety requirements regarding the exposure of workers to the risks arising from physical agents (vibration) — individual directive within the meaning of article 16 of the directive 89/391/EEC, 1999

Crolla DA, (1995) Vehicle dynamics — theory into practice. J Automobile Eng 209:1–12

Crolla DA, Abdel Hady MBA (1988) Semi-active suspension control for a full vehicle model. SAE paper 911904

Crolla DA, Aboul Nour AMA (1988) Theoretical comparisons of various active suspension systems in terms of performance and power requirements. Proc IMechE Advanced Suspensions Conference, London, UK, pp 1–9

Crolla DA, Horton DNL, Dale AK (1984) Off road vehicle ride vibration. Proc IMechE, C131/8, part C, pp 55–64

Crolla DA, Horton DNL, Pitcher RH, Lines JA (1987) Active suspension control for an off-road vehicle. Proc ImechE D1; 201:1–10

Crosby MJ, Karnopp DC (1973) The active damper - a new concept for shock and vibration control. Shock Vib Bull 43; 4:119–133

Cross J (1999) Farewell to rock and roll. Automotive Eng 24; 7:42–43, Professional Engineering, UK

Crossley TR, Porter B (1979) Modern approaches to control system design. Peregrinus, ed. Munro N

Curtis A (1991) Ride revolution. Car Design and Technology, June/July, pp 30–36

Dahl P (1968) A solid friction model. Technical Report TOR–0158(3107–18)–1. The Aerospace Corporation, El Segundo, USA

Dahl P (1976) Solid friction damping of mechanical vibrations. AIAA 14; 10:1675–1682

Dahlin EG (1968) Designing and tuning digital controllers. Instrum Control Syst 41:6

Davis DC (1974) A radial spring terrain enveloping tire model. Vehicle Syst Dyn 3:55–69

De Carlo RA, Zak SH, Matthews GS (1988) Variable structure control of nonlinear multivariable systems: a tutorial. Proc IEEE 76; 3:212–232

Demic M (1984) Assessment of random vertical vibration on human body fatigue, using a physiological approach. Proc IMechE Conf Vehicle Noise and Vibration, London, UK, pp 91–95

Dieckmann D (1958) A study of the influence of vibration in man. Ergonomics 4:347

Dinca F, Teodosiu C (1973) Nonlinear and random vibration. Academic, New York, London

Dorf RC, Bishop RH (1995) Modern control systems. Addison-Wesley, Reading, MA

Dukkipati RV (2000) Vehicle dynamics. CRC Press, Boca Raton, FL

Dupont P, Kasturi P, Stokes A (1997) Semi-active control of friction dampers. J Sound Vib 202, 2:203–218

Dupont P, Armstrong B, Hayward V (2000) Elasto-plastic friction model: contact compliance and stiction. Proc American Control Conference, Boston USA

Dyke SJ, Spencer BF, Sain MK, Carlson JD (1996) Modelling and control of magnetorheological dampers for seismic response reduction. Smart Mater Struct 5:565–575

Eryilmaz B, Wilson BH, (2000) Modeling the internal leakage of hydraulic servovalves. J Dyn Syst, Measure Control 122; 3:576–579

Fard M, Ishihara T, Inooka H (2003) Dynamics of the head-neck complex in response to trunk horizontal vibration. J Biomech Eng 125; 4:533–539

Federspiel-Labrosse GM (1954) Contribution a l'etude et au perfectionement de la suspension des vehicules. J de la SIA, FISITA, pp 427–436

Federspiel JM (1976) Quelques aspects modernes de probleme de dynamique automobile. Ingeneures de l'Automobile 5:33–140

Gao W, Hung JC (1993) Variable structure control of nonlinear systems: a new approach. IEEE Trans Ind Electron 40; 1:45–55

Gawthrop PJ, Smith L (1996) Metamodelling: bond graphs and dynamics systems. Prentice Hall, Hemel Hempstead, Herts, UK

Genta G (1993) Meccanica dell' autoveicolo. Levrotto & Bella, Torino, Italy

Gerdes JC, Hedrick JK (1999) Hysteresis control of nonlinear single-acting actuators as applied to brake/throttle switching. American Control Conference, San Diego, USA, pp 1692–1696

Ghazi Zadeh A, Fahim A, El–Gindy M (1997) Neural network and fuzzy logic applications to vehicle systems: literature survey. Int J Vehicle Des 18; 2:132–193

Ghita G, Giuclea M (2004) Magnetorheological fluids and dampers . In: Topics in applied mechanics (2004) edited by Chiroiu L and Sireteanu T, Vol II, Ch 5, pp 118-161 ed. Chiroiu L and Sireteanu T, Publishing House of Romanian Academy

Gillespie TD (1985), Heavy Truck Ride, Society of Automotive Engineers, 85001

Gillespie TD (1992) Fundamentals of Vehicle Dynamics. SAE

Giuclea M, Sireteanu T, Mita AM, Ghita G (2002) Genetic algorithm for parameter identification of Bouc-Wen model. Rev Roum Sci Techn Mec Appl 51; 2:79–188

Giuclea M, Sireteanu T, Stancioiu D, Stammers CW, (2004) Model parameter identification for vehicle vibration control with magnetorheological dampers using computational intelligence methods. Proc IMechE 218 Part I, J Syst Control Eng 17:569–581

Goldberg DE (1989) Genetic algorithms in search, optimization and machine learning. Addison-Wesley, Reading, USA

Goncalves FD, Koo JH, Ahmadian M (2003) Experimental approach for finding the response time of MR dampers for vehicle applications. Proc ASME DETC'03, Chicago USA

Goodall RM, Kortüm W (1983) Active control in ground transportation — a review of the state-of-the-art and future potential. Vehicle Syst Dyn 12:225–257

Goodall RM, Williams RA, Lawton A, Harbrough PR (1981) Railway vehicle active suspensions in theory and practice. Proc 7th IAVSD Symposium on the Dynamics of the Vehicles on Roads and Railway Tracks, pp 301–306
Griffin M (1984) Vibration dose values for whole body vibration: some examples. Human Factor Research Unit, ISVR, Southampton, UK
Griffin M (1998) A comparison of standardized methods for predicting the hazards of whole body vibration and repeated shock. J Sound Vib 215:883–914
Groenewald ML, Gouws J (1996) In-motion tyre pressure control system for vehicles. Proc IEEE Mediterranean Electrotech Conf 3:1465–1468
Guglielmino E (2001) Robust control of a hydraulically actuated friction damper for vehicle applications, PhD Thesis, University of Bath, UK
Guglielmino E, Stammers CW, Stancioiu D, Sireteanu T (2002) Experimental assessment of suspension with controlled dry friction damper. The National Conference with International Participation, Automobile, Environment and Agricultural Machinery, Cluj-Napoca, Romania
Guglielmino E, Edge KA (2004) Controlled friction damper for vehicle applications. Control Eng Practice 12, 4:431–443
Guglielmino E, Edge KA, Ghigliazza R (2004) On the control of the friction force. Meccanica 39; 5:395–406
Guglielmino E, Stammers CW, Stancioiu D, Sireteanu T, Ghigliazza R (2005) Hybrid variable structure-fuzzy control of a magnetorheological damper for a seat suspension. Int J Vehicle Auton Syst 3; 1:34–46
Guglielmino E, Stammers CW, Stancioiu D, Sireteanu T (2005) Conventional and non-conventional smart damping systems for ride control. Int J Vehicle Auton Syst 3; 2/3/4:216–229
Guglielmino E, Stammers CW, Edge KA (2000) Robust force control in electrohydraulic friction damper systems using a variable structure scheme with non linear state feedback. Internationales Fluidtechnisches Kolloquium, Dresden, Germany
Guglielmino E, Edge KA (2000) Robust control of electrohydraulically actuated friction damper. Proc ASME IMECE 2000, Orlando, USA
Guglielmino E, Edge KA (2001) Modelling of an electrohydraulically-activated friction damper in a vehicle application. Proc ASME IMECE 2001, New York, USA
Guglielmino E, Stammers CW, Edge KA, Stancioiu D, Sireteanu T (2004) Damp-by-wire: magnetorheological vs. friction dampers. IFAC 16th World Congress, Prague, Czech Republic
Gunston T, Rebelle J, Griffin MJ (2004) A comparison of two methods of simulating seat suspension dynamic performance. J Sound Vib 278:117–134
Haalman A (1965) Adjusting controllers for deadtime process. Control Eng, pp 71–73
Hac A (1987) Adaptive control of vehicle suspension. Vehicle Syst Dyn 16:57–74
Hac A (1992) Optimal linear preview control of active vehicle suspension. Vehicle Syst 21; 3:167–195
Hac A, Youn I (1992) Optimal semiactive suspension with preview based on a quarter car model. J Vib Acoust 114; 1:84–92
Haessig DA, Friedland B (1991) On the modelling and simulation of friction. ASME J Dyn Syst Measure Control 113; 3:354–362
Hayward VB, Armstrong B (2000) A New computational model of friction applied to haptic rendering. Exp Rob VI: 404–412
Hedrick JK, Butsuen T (1988) Invariant properties of automotive suspensions. Proc IMechE, pp 35–42
Hedrick JK, Wormely DN (1975) Active suspension for ground support vehicles - a state-of-the-art review. ASME-AMD 15:2–40
Hertz H (1881) On the contact of elastic solids. J Reine und Anges Math 92:156–171

Hess DP, Soom A (1990) Friction at a lubricated line contact operating at a oscillating sliding velocities. J Tribol 112, 1:147–152

Hillebrecht P, Konik D, Pfeil D, Wallentowitz F, Zieglmeier F (1992) The active suspension between customer benefit and technological competition. Proc XXIV FISITA Congress, Total Vehicle Dynamics 2:221–230

Hocking LM (1991) Optimal control — an introduction to the theory with applications. Oxford University Press

Hollerbach JM, Nahvi A, Freier R, Nelson DD (1998) Display of friction in virtual environments based on human finger pad characteristics. Proc ASME Dynamic Systems and Control Division - DSC, Anaheim, USA, 64:179–184

Horrocks BS, Stammers CW, Gartner M (1997) The performance of a hydraulically activated semi-active friction damper. Fifth Scandinavian Int Conf on Fluid Power, Linköping, Sweden 3:47–55

Hrovat D, Hubbard M (1987) A comparison between jerk optimal and acceleration optimal vibration isolation. J Sound Vib 112; 2:201–210

Hung YJ, Gao W, Hung JC (1993) Variable structure control: a survey. IEEE Trans Ind Electron 40; 1:1–22

International Standard Organization for Standardization (1978) Mechanical Vibration and Shock - Evaluation of Human Exposure to Whole-Body Vibration. Part 1: General Requirements. Geneva, Switzerland

Itkis Y (1976) Control systems of variable structure. Wiley, New York, NY

Jiang Z, Streit DA, El-Gindy M (2001) Heavy vehicle ride comfort: literature survey. Int J Heavy Vehicle Syst 8; 3/4:258–284

Jolly MR, Bender JW, Carlson JD (1998) Properties and applications of commercial magnetorheological fluids. SPIE 5th Annual Int Symposium on Smart Structures and Materials, San Diego, USA

Johnson KL (1987) Contact mechanics. Cambridge University Press

Karnopp DC, Crosby MJ, Harwood RA (1974) Vibration control using semi-active force generators. J Eng Ind 97:619–626

Karnopp DC (1985) Computer simulation of stick-slip friction in mechanical dynamic systems. ASME Trans 107:100–103

Karnopp DC, Margolis DL, Rosenberg RC (1990) System dynamics: a unified approach. Wiley, New York

Kato S, Sato N, Matsubayashi T (1972) Some considerations of characteristics of static friction of machine tool slideway. J Lubrication Technol 94; 3:234–237

Kyle CS, Roschke PN (2000) Fuzzy modelling of a magnetorheological damper using Anfis. Proc IEEE Fuzzy Conf, San Antonio, Texas, USA

Kim C, Ro PI (1998) A sliding mode controller for vehicle active suspension systems with non-linearities. Proc IMechE J Automobile Eng 212; 2:79–92

Kirk DE (2004) Optimal control theory: an introduction. Dover, New York, NY

Koo JH, Ahmadian M, Setareh M, Murray T (2004) In search of suitable control methods for semi-active tuned vibration dampers. J Vib Control 10; 2:163–174

Krasnoel´skii MA, Pokrovskii AV (1989) Systems with hysteresis. Springer, Berlin Heidelberg New York

Lambert JD (1973) Computational methods in ordinary differential equations. Wiley, NY

Lampaert V, Swevers J, AI-Bender F (2004) Comparison of model and non-model based friction compensation techniques in the neighbourhood of pre-sliding friction. American Control Conf, Boston, USA

Lampaert V, AI-Bender Fm Swevers J, (2003) A generalized Maxwell-slip friction model appropriate for control purposes. Proc IEEE Int Conf Physics and Control, Saint-Petersburg, Russia

Lakie M , Walsh EG, Wright GW (1984) Resonance at the wrist demonstrated by the use of a torque motor: an instrumental analysis of muscle tone in man. J Physiol 353:265–285
Lanchester FW (1936) Motor car suspension and independent springing. Proc IMechE J Automobile Eng D1; 30:668–762
Lau YK, Liao WH (2005) Design and analysis of magnetorheological dampers for train suspension. IMechE, Part F, J Rail Rapid Transit 219:261–276
Leatherwood J, Dempsey T, Clevenson S (1980) A design tool for estimating passenger ride discomfort within complex ride environments. Human Factors 22, 3:291–312
Lee R, Pradko F (1969) Analytical analysis of human vibration. SAE paper 680091
Leigh JR (1992) Applied digital control. Theory, design & implementation. Prentice Hall, UK
Leonardo da Vinci (1519) The Notebooks. Ed. Dover, New York, NY
Liu Y, Waters TP, Brennan MJ (2005) A comparison of semi-active damping control strategies for vibration isolation of harmonic disturbances. J Sound Vib 208:21–39
Luenberger DG (1964) Observing the state of a linear system. IEEE Trans Mil Electron 8:74–80
Mansfield NJ, Griffin MJ (2000) Non-linearities in apparent mass and transmissibility during exposure to whole-body vertical vibration. J Biomech 3:933–941
Margolis DL (1982) The response of active and semi-active suspensions to realistic feedback signals. Vehicle Syst Dyn 11; 5–6:267–282
Margolis D, Nobles CM (1991) Semi-active heave and roll control for large off-road vehicles. SAE paper 912672
Majjad R (1997) Estimation of suspension parameters. Proc IEEE Int Conf on Control Applications, Hartford, USA, pp 522–527
Martins I, Esteves J, Da Silva FP, Verdelho P (1999) Electromagnetic hybrid active-passive vehicle suspension system. Proc IEEE 49th Vehicular Technology Conference, Houston, USA 3:2273–2277
Mayergoyz ID (1991) Mathematical models of hysteresis. Springer Verlag, New York, NY
McCloy D, Martin HR (1980) Control of fluid power: analysis and design. Ellis Horwood, Chichester, UK
Melzner K, Fleischer J, Odenbach S (2001) New developments in the investigation of magnetoviscous and viscoelastic effects in magnetic fluids. Magnetohydrodynamics 37; 3:285–290
Miller LR, Nobes CM (1988) The design and development of a semi-active suspension for a military tank. SAE paper 881133
Milliken WF (1987) Lotus active suspension system. Proc 11th Int EVS Conference
Milliken WF, Milliken DL (1995) Race car vehicle dynamics. SAE, Warrendale
Mohan B, Phadke SB (1996) Variable structure active suspension system. Proc IEEE 22nd Int Conf on Industrial Electronics, Control, and Instrumentation, IECON, Taipei, Taiwan, pp 1945–1948
Moran A, Nagai M (1992) Analysis and design of active suspensions by H_∞ robust control theory. JSME Int J, Series III, 35, 3:427–437
Moran A, Nagai M (1994) Optimal active control of nonlinear vehicle suspensions using neural networks. JSME Int J, Series C, 37; 1:707–717
Morin AJ (1833) New friction experiments carried out at Metz in 1831-1833. Proc French Acad Sci 4:1–128
Moulton AE, Best A (1979a) From hydrolastic to hydragas suspension. Proc IMechE 193; 9:15–25
Moulton AE, Best A (1979b) Hydragas suspension. SAE paper 790374
Nishitani A, Nitta Y, Ishibashi Y, Itoh A (1999) Semi-active structural control with variable friction dampers, American Control Conference, San Diego, USA, pp 1017–1021

Ngwompo RF, Guglielmino E, Edge KA (2004) Performance enhancement of a friction damper system using bond graphs. 7th Biennial ASME Conference on Engineering Systems Design and Analysis, Manchester, UK

Novak M, Valasek M (1996), A new concept of semi-active control of truck's suspension. AVEC 96, Aachen, Germany

Ogawa Y, Kawasaki H, Arai J, Nakazato M (1999) Application of semi-active suspension system to railway vehicles. Fourth Japan Hydraulics and Pneumatics Society Int Symposium on Fluid Power, Tokyo, Japan, pp 267–272

O'Neill H, Wale GD (1994) Semi-active suspension improves rail vehicles ride. Comput Control Eng 5; 4:183–188

Pacejka HB, Bakker E (1991). The magic formula tyre model. Proc 1st Tyre Colloquium, Delft, The Netherlands

Paddan GS, Griffin MJ (2002) Effect of seating on exposures to whole-body vibration in vehicles. J Sound Vib 253; 1:215–241

Park JH, Kim YS (1998) Decentralized variable structure control for active suspensions based on a full-car model. Proc IEEE Int Conf on Control Applications, Trieste, Italy, pp 383–387

Passino KM, Yurkovich S (1998) Fuzzy control. Addison-Wesley-Longman, Menlo Park, CA

Pesterev AV, Bergman LA, Tan CA (2002) Pothole-induced contact forces in a simple vehicle model. J Sound Vib 256; 3:565–572

Pollock AJ, Craighead IA (1986) The selection of a criterion to evaluate ride-discomfort in off-road vehicles. Proc Polymodel 9, Newcastle upon Tyne, UK, pp 21–31

Potter TEC, Cebon D, Cole DJ, Collop AC (1995) An investigation of road damage due to measured dynamic tyre forces. Proc IMechE, J Automobile Eng, D1; 209:9–24

Rabinow J (1948) The magnetic fluid clutch. AIEE Trans 67, 1308

Rabinowicz E (1958) The intrinsic variables affecting the stick-slip process. Proc Physical Society of London 74; 4:668–675

Rabinowicz E (1965) Friction and wear of materials. Wiley, New York, NY

Rakheja S, Sankar S (1985) Vibration and shock isolation performance of a semiactive "on-off" damper. ASME J Vib, Acoust, Stress Reliab Des 107:398–403

Ramsbottom M, Crolla DA, Plummer AR (1999) Robust adaptive control of an active vehicle suspension system. Proc IMechE D 213:1–5

Rao SS (1995) Mechanical vibrations. Addison-Wesley, Reading, MA

Reynolds O (1886) On the theory of lubrication and its application to Mr Beauchamp tower's experiments, including experimental determination of the viscosity of olive oil. Phil Trans R Soc 177:157–234

Rice JR, Ruina AL (1983) Stability of steady frictional slipping. J Applied Mech 50:343–349

Rideout G, Anderson RJ (2003) Experimental testing and mathematical modeling of the interconnected hydragas suspension system. SAE paper 2003-01-0312

Sadeghi F, Sui PC (1989) Compressible elastohydrodynamic lubrication of rough surfaces. J Tribol 111; 1:56–62

SAE, Society of Automotive Engineers (1965) Ride and vibration data manual. J6a

Sain PM, Sain MK, Spencer BF (1997) Models for hysteresis and application to structural control. American Control Conference, Albuquerque, USA, pp 16–20

Sammier D, Sename O, Dugard L (2000) H_∞ control of active vehicle suspensions. Proc IEEE Int Conf on Control Applications, Anchorage, USA, pp 976–981

Sanliturk KY, Stanbridge AB, Ewins DJ (1995) Friction dampers: measurement, modelling and application to blade vibration control. Proc ASME Design Engineering Technical Conference, Boston, USA 84, 3:1377–1382

Satoh M, Fukushima N, Akatsu Y, Fujimura I, Fukuyama K (1990) An active suspension employing an electrohydraulic pressure control system. Proc 29th Conference on Decision and Control, Honolulu, USA 4:2226–2231

Sayers MW, Karamihas SM (1998) The little book of profiling - basic information about measuring and interpreting road profile. University of Michigan Transportation Research Institute, Ann Arbor, MI, USA

Sebesan I, Hanganu D (1993) Proiectarea suspensiilor pentru vehicule pe sine. Editura Tehnică, Bucharest, Romania

Sharp RS, Crolla DA (1987) Road vehicle suspension system design - a review. Vehicle Syst Dyn 16:167–192

Sharp RS, Hassan SA (1986) The relative performance capabilities of passive, active and semi-active car suspension systems. Proc IMechE, 200, D3

Shaw M (1999) Application of hydro-elastomer technology to vehicle suspensions, PhD Thesis, University of Bath, UK

Sireteanu T, Gündisch O, Părăian S (1981) Vibraţile aleatoare ale automobilelor. Confort şi aderenţă. Editura Tehnică, Bucharest, Romania

Sireteanu T, Balas C (1991) Free vibrations of sequentially damped oscillator. The Annual Symposium of the Institute of Solid Mechanics, Bucharest, Romania, pp 7–12

Sireteanu T, Giuclea M, Guglielmino E (2002) On the stability of semi-active control with sequential dry friction. Proc Rom Acad Series A, 3; 1/2:31–36

Sireteanu T, Stancioiu D, Giuclea M, Stammers CW (2003) Experimental identification of a magnetorheological damper model to compare semi-active control strategies of vehicle suspensions. Conf on Active Noise and Vibration Control Methods, Cracow, Poland

Sireteanu T, Stoia N (2003) Damping optimization of passive and semiactive vehicle suspension by numerical simulation. Proc Rom Acad: Series A, 4, 1

Sireteanu T, Stammers CW (2003) Semi-active vibration control with variable dry friction In: Topics in applied mechanics Vol I, Ch 12, pp 342-388, ed. Sireteanu T and Vladareanu L, Publishing House of Romanian Academy

Sireteanu T, Stammers CW, Ursu I (1997) Analysis of a sequential dry friction type semi-active suspension system, Active 97, Budapest, Hungary, pp 565–571

Slotine JJE (1984) Sliding controller design for nonlinear systems. Int J Control 40; 2:421–434

Slotine JJE, Li W (1992) Applied nonlinear control. Prentice-Hall, Englewood Cliffs, New Jersey

Smith OJM (1957) Closer control of loops with dead time. Chem Eng Progress 53:217–219

Smith DW (1977) Computer simulation of tractor ride for design evaluation. SAE paper 770704

Soong TT (1973) Random differential equations in science and engineering. Academic, New York, NY

Soong TT, Costantinou MC (1994) Passive and active structural vibration control in civil engineering. Springer-Verlag, Wien, New York

Spencer BF, Dyke SJ, Sain, MK, Carlson JD (1997) Phenomenological model of a magnetorheological damper. J Eng Mech 123:230–238

Spencer BF, Yang G, Carlson JD, Sain MK (1998) Smart dampers for seismic protection of structures: a full-scale study. Proc Second World Conference on Structural Control, Kyoto, Japan

Stammers CW, Sireteanu T (1997) Vibration control of machines by using semi-active dry friction damping. J Sound Vib 209; 4:671–684

Stammers CW, Guglielmino E, Sireteanu T (1999) A semi-active system to reduce machine vibration. 10th World Congress on the Theory of Machines and Mechanisms, Oulu, Finland, pp 2168–2173

Stammers CW, Sireteanu T (2004) Simulation of a state observer based control of a vehicle seat. 2nd Industrial Simulation Conference 2004, Malaga, European Simulation Society, pp 129–133

Stayner RM (1988) Suspensions for agricultural vehicles. Proc IMechE Conference on Advanced Suspensions, London, UK, pp 133–140

Steffens RJ (1966) Some aspects of structural vibration. Proc Symp Vibration in Civil Engineering, London, UK

Stribeck R (1902) Die Wesentlichen Eigenschaften der Gleit und Rollenlager. Zeitschrift des Vereines Deut Ing, 46, 38:1341–1348

Swevers J, AI-Bender E, Ganseman C, Prajogo T (2000) An integrated friction model structure with improved prediding behaviour for accurate friction compensation. IEEE Trans Autom Control 45; 4:675–786

Tan HS, Bradshaw T (1997) Model identification of an automotive hydraulic active suspension system. American Control Conference, Albuquerque, USA, 5:2920–2924

Thompson AG (1976) An active suspension with optimal linear state feedback. Vehicle Syst Dyn 5:187–203

Tiliback LR, Brood S (1989) Active suspension — the Volvo experience. SAE paper 890083

Titli A, Roukieh S, Dayre E (1993) Three control approaches for the design of car semi-active suspension (optimal control, variable structure control, fuzzy control). Proc IEEE Conf Decis Control 3:2962–2963

Tsampardoukas GI, Stammers CW (2007) Semi-active control of a heavy vehicle suspension to reduce road damage. Proc XII Int Symposium on Dynamic Problems of Mechanics, Ilhabela, SP, Brazil

Tsampardoukas GI, Stammers CW, Guglielmino E (2007) Hybrid balance control of magnetorheological truck suspension. To be published in J Sound Vib

Tsampardoukas GI, Stammers CW, Guglielmino E (2008) Ride & handling semi-active control of the passenger vehicle and human exposure to vibration. Proc IMechE, J Automobile Eng, 222, D3; 325–352

Tseng HE, Hedrick JK (1994) Semi-active control laws - optimal and sub-optimal. Vehicle Syst Dyn 23:545–569

Ursu I, Ursu F, Vladimirescu M (1984) The synthesis of two suboptimal electrohydraulic suspensions, active and semiactive, employing the receding horizon method. Nonlin Anal, Theory Appl 30, 4:1977–1984

Utkin VI (1992) Sliding modes in control optimization. Springer Verlag, Berlin

Valasek M, Kortum W, Sika Z, Magdolen L, Vaculin O (1998) Development of semi-active road friendly truck suspensions. Control Eng Practice 6:735–744

Vemuri V (1993) Artificial neural networks in control applications. Adv Comput 36:203–254

Visintin A (1994) Differential models of hysteresis. Springer, Berlin

Wang X, Noah S, Vahabzadeh H (1996) Analysis of a Coulomb friction damper for suppressing oscillations of drive shafts of vehicles. Proc ASME IMECE, Atlanta, USA 91:51–58

Watanabe Y, Sharp RS (1996) Controller design for an automotive active suspension using a learning method. Proc Symposium on Control, Optimization and Supervision in CESA IMACS Multiconference on Computational Engineering in Systems Applications, Lille France, 1:327–332

Watanabe Y, Sharp RS (1999) Neural network learning control of automotive active suspension systems. Int J Vehicle Des 21; 2/3:124–147

Wei L, Griffin M (1998a) Mathematical models for the apparent mass of the seated human body exposed to vertical vibration. J Sound Vib 212:855–874

Wei L, Griffin M (1998b) The prediction of seat transmissibility from measures of seat impedance. J Sound Vib 214:121–137

Weiss KD, Carlson JD, Nixon DA (1994) Viscoelastic properties of magneto- and electro-rheological fluids. J Intell Mater Syst Struct 5:772–775

Wen YK (1976) Method for random vibration of hysteretic systems. ASCE J Eng Mech 102, 2:249–263

Wereley NM, Pang L, Kamath GM (1998) Idealized hysteresis modeling of electrorheological and magnetorheological dampers. J Intell Mater Syst Struct 9; 8:642–649

Wettergren HL (1997) On the behaviour of material damping due to multi-frequency excitation. J Sound Vib 206; 5:725–735

Williams RA, Burnham KJ, Webb AC (1996) Developments for an oleo-pneumatic active suspension. Proc IEEE International Symposium on Computer-Aided Control System Design, Dearborn, USA, pp 44–49

Wills GJ (1980) Lubrication fundamentals. Marcel Dekker, New York, NY

Winslow WM (1947) Method and means for translating electrical impulses into mechanical forces. US patent 2417850

Winslow WM (1949) Induced fibration of suspensions. J Appl Phys 20:1137–1140

Wong JY (1993) Theory of ground vehicles. Wiley, New York, NY

Yagtz N, Özbulur V, Inanc N, Derdiyok A (1997) Sliding modes control of active suspensions. Proc 12th IEEE Int Symposium on Intelligent Control, Istanbul, Turkey, pp 349–353

Yang G, Spencer BF, Carlson JD, Sain MK (2002) Large-scale MR fluid dampers: modeling, and dynamic performance considerations. Eng Struct 24:309–323

Yi K, Song BS (1999) A new adaptive sky-hook control of vehicle semi-active suspensions. Proc IMechE, J Automobile Eng 213, D:293–303

Yokoya Y, Kizu R, Kawaguchi H, Ohashi K, Ohno H (1990) Integrated control system between active control suspension and four-wheel steering for the 1989 Celica. SAE paper 901748, 99, 6:1546–1561

Yoshimura T, Nakaminami K, Hino J (1997) A semi-active suspension with dynamic absorbers of ground vehicles using fuzzy reasoning. Int J Vehicle Des 18; 1:19–34

Yu LC, Khameneh KN (1999) Automotive seating foam: subjective dynamic comfort study. SAE paper 010588

Zeid A, Chang D (1991) Simulation of multibody systems for the computer-aided design of vehicle dynamic controls. SAE paper 910236

Zeller W (1949) Units of measurement for strength and sensitivity of vibrations. ATZ, 51:95

Ziegler JB, Nichols NB (1942) Optimum setting for automatic controllers. Trans ASME 65:433–444

Bibliography

Agapie A, Giuclea M, Fagarasan F, Dediu H (1998) Genetic algorithms: theoretical aspects and applications. Romanian J Inform Sci Tech 1; 1:3–21

Aracil J, Ollero A, Garcia-Cerezo A (1999) Stability indices for the global analysis of expert control systems. IEEE Trans Syst, Man Cyber 19; 5:998–1007

Bakker E, Pacejka HB, Lidner L (1989). A new tire model with an application in vehicle dynamics studies. SAE paper 890087, SAE World Congress, Detroit, USA

Bellizzi S, Bouc R, Campillo F, Pardoux E (1988) Contrôle optimal semi-actif de suspension de véhicule. In: Bensoussan A, Lions JL, Lectures Notes in Control and Information Science, pp 689–699, Springer, Berlin Heidelberg New York

Bose BK (1985) Sliding mode control of an induction motor. IEEE Trans Ind Appl, 21; 2:479–486

Bosso N, Gugliotta A, Somà A (2001) Multibody simulation of a freight bogie with friction dampers. ASME/IEEE Joint Rail Conference, Washington DC, USA

Cebon D (1993) Interaction between heavy vehicles and roads. Society of Automotive engineers, SAE Trans 930001, SP-951 (L. Ray Buckendale Lecture)

Cebon D, Besinger FH, Cole DJ (1996) Control strategies for semi-active lorry suspensions. Proc IMechE, D, J Automobile Eng 210; 2:161–178

Chapra SC, Canale RP (1988) Numerical methods for engineers. McGraw-Hill, New York

Chin Y, Lin WC, Sidlosky DM, Rule DS (1992) Sliding-mode ABS wheel-slip control. American Control Conference, Chicago, USA, pp 1–5

Cole DJ, Cebon D (1996) Truck suspension design to minimise road damage. Proc IMechE, D2, J Automobile Eng 210:95–107

Drazenovic B (1969) The invariance conditions in variable structure systems. Automatica 5:287–295

Driankov D, Hellendorn H, Reinfrank M (1993) An introduction to fuzzy control. Springer, Berlin Heidelberg New York

Dyke SJ, Spencer BF (1997) Semi-active control strategies for MR dampers: a comparative study. ASCE J Eng Mech 126; 8:795–803

Feng X, Lin CF, Yu TJ, Coleman N (1997) Intelligent control design and simulation using neural networks. AIAA Guidance, Navigation and Control Conference. A Collection of Technical Papers, Part 1, New Orleans, USA, pp 294–299

Ferri AA, Heck BS (1998) Vibration analysis of dry friction damped turbine blades using singular perturbation theory. J Vib Acoust 120; 2:588–595

Filippov AF (1960) Application of theory of differential equations with discontinuous righthand sides to nonlinear control problems. Proc 1st IFAC Congress, Russia, pp 923–927

Filippov AF, (1988) Differential equations with discontinuous righthand sides. Kluwer Academic, Dordrecht, The Netherlands

Fujita T (1994) Semi-active control of base isolated structures. The Active Control of Vibration, IUTAM Symposium, Bath, UK, pp 289–295

Gamble JB (1992) Sliding mode control of a proportional solenoid valve. PhD Thesis, University of Bath, UK

Gayed A, Benkhoris MF, Siala S, Le Doeuff R (1995) Time-domain simulation of discrete sliding control of permanent magnet synchronous motor. Proc 21st IEEE IECON Int Conf Industrial Electronics, Control, and Instrumentation, Orlando, USA 2:754–759

Genta G, Campanile P (1989) An approximate approach to the study of a motor vehicle suspensions with non-linear shock absorbers. Meccanica 24:47–57

Gillespie TD (1985) Heavy truck ride. SAE paper SP–607

Giuclea M, Sireteanu T (2000) Optimisation of seismic accelerograms by genetic algorithms. Proc 2nd Int ICSC Symposium on Engineering of Intelligent Systems, Paisley, Scotland, UK

Guglielmino E, Stammers CW, Stancioiu D, Sireteanu T (2002) Experimental assessment of suspension with controlled dry friction damper. The National Conference with International Participation, Automobile, Environment and Agricultural Machinery, Cluj-Napoca, Romania

Guglielmino E, Edge KA, Ghigliazza R (2002) Ammortizzatori ad attrito coulombiano controllato: un approccio alternativo nel progetto delle sospensioni di un veicolo. ATA review 55, 9/10

Guglielmino E, Edge KA (2002) Multivariable control of a friction damper based vehicle suspension. 5th JFPS International Symposium on Fluid, Nara, Japan

Guglielmino E, Edge KA (2003) The problem of controlling a friction damper: a robust control approach. American Control Conference, Denver, USA 4,4-6:2827–2832

Hagedorn P (1988) Non-linear oscillations. Clarendon, Oxford, UK

Handroos H, Liu Y (1998) Sliding mode control of a hydraulic position servo with different kinds of loads. Power Transmission and Motion Control, PTMC '98, Bath, UK, pp 379–392

Hashimoto H, Yamamoto H, Yanagisawa S, Harashima F (1988) Brushless servo motor control using a variable structure approach. IEEE Trans Industrial Applications 24 1:160–170

Hickson LR (1996) Design and development of an active roll control suspension. PhD Thesis, University of Bath, UK

Ibrahim IM, Crolla DA, Barton DC (2004) The impact of the dynamic tractor-semitrailer interaction on the ride behavior of fully-laden and unladen trucks. SAE paper 2004–01–2625

Jolly A (1983) Study of ride comfort using a non-linear mathematical model of a vehicle suspension. Int J Vehicle Des 43, 3:233–244

Karadayi R, Hasada GY (1986) A non-linear shock absorber model. Proc ASME-AMD Symposium on Simulation and Control of Ground Vehicles and Transportation Systems. Anaheim, USA 80:149–165

Kelso SP, Gordaninejad F (1998) Magneto-rheological fluid shock absorbers for off-highway, high-payload vehicles. SPIE Conference on Smart Material and Structures, San Diego, USA

Kwakernaak H, Sivan R (1972) Linear optimal control systems. Wiley-Interscience, New York, NY

Lane J, Ferri A, Heck B (1992) Vibration control using semi-active frictional damping. Proc ASME-DED Winter Annual Meeting, Friction-Induced Vibration, Chatter, Squeal and Chaos, New York, USA, 49:165–171

Lantto B, Krus P, Palmberg JO (1993) Robust control of an electrohydraulic pump using sliding mode planes. The Third Scandinavian Conference on Fluid Power, Linköping, Sweden 3:3–24

Lee R, Pradko F (1969) Analytical analysis of human vibration. SAE paper 680091

Lee HS, Choi SB (2000) Control and response characteristics of a magneto-rheological fluid damper for passenger vehicles. J Intell Mater Syst Struct 11; 1:80–90

Lizell M (1988) Semi-active damping. Proc IMechE Advanced Suspensions Conference, London, UK, pp 83–91

Lopez O, Garcia de Vicuna L, Castilla M, Matas J, Lopez M (1999) Sliding mode control design of a high power factor buck-boost rectifier. IEEE Trans Ind Electron, 46; 3:604–612

MacMahon TA (1984) Muscles, reflexes and locomotion. Princeton University Press.

Mastinu G (1988) Passive automobile suspension parameter adaptation. Proc IMechE Advanced Suspensions Conference, London, UK, pp 51–58

Otsuka K, Wayman L (1999) Shape memory materials. Cambridge University Press

Pandian SR, Leda K, Kamoyama Y, Kamawura S, Hayakawa Y (1998) Modelling and control of a pneumatic rotary actuator. Power Transmission and Motion Control '98, Bath, UK, pp 363–376

Peel DJ, Stanway R, Bullough WA (1996) Dynamic modelling of an ER vibration damper for vehicle suspension applications. Smart Mater Struct 5:591–606

Petzold L (1983) Automatic selection of methods for solving stiff and non-stiff systems of ODE's. SIAM J Sci Stat Comput 4:136–148

Pontryagin LS, Boltyanskii VG, Gamkrelidze D, Mishchenko A (1962) The mathematical theory of optimal processes. Interscience, New York

Potter TEC, Cebon D, Cole DJ, Collop AC (1996) Road damage due to dynamic tyre forces measured on a public road. Int J Vehicle Des, Heavy Vehicle Syst 3; 1/4: 346–362

Ramallo JC, Johnson EA, Spencer BF, Sain MK (1999) Semiactive building base isolation. Proc American Control Conference, San Diego, USA, pp 515–519

Richards CW, Tilley DG, Tomlinson SP, Burrows CR (1990) Type-insensitive integration codes for the simulation of fluid power systems. Proc ASME Winter Annual Meeting, Dallas, USA

Roberts JB, Spanos PD (1990) Random vibration and statistical linearization. Wiley, New York, NY

Ross-Martin T (1994) Low bandwidth active roll and warp control suspension. PhD Thesis, University of Bath, UK

Roxin E (1965) On generalized dynamical systems defined by contingent equations. J Differ Equ 1:188–205

Sarpturk SZ, Istefanopoulos Y, Kaynak O (1987) On the stability of discrete-time sliding mode control systems. IEEE Trans Autom Control 32; 10:930–32

Singh MP, Matheu EE, Suarez LE (1997) Active and semi-active control of structures under seismic excitation. Earthq Eng Struct Dyn 26; 2:193–213

Sireteanu T (1984) The effect of sequential damping on ride comfort improvement. Vehicle Noise Vib, C145/84, IMechE, pp 77–82

Sireteanu T, Giuclea M (2000) A genetic algorithm for syntesizing seismic accelerograms. Proc Romanian Academy, Series A, 1, 1:37–40

Sireteanu T, Balas C (1992) Forced vibrations of sequentially damped oscillator. The Annual Symposium of the Institute of Solid Mechanics, Bucharest, Romania, pp 1–6

Sireteanu T (1996) Effect of system nonlinearities obtained by statistical linearization methods. Euromech 2nd European Nonlinear Oscillation Conference, Prague 1:415–417

Sireteanu T, Stammers CW, Giuclea M, Ursu I, Guglielmino E (2000), Semi-active suspension with fuzzy controller optimized by genetic algorithms. ISMA 25 International Conference on Noise and Vibration Engineering, Leuven, Belgium

Sireteanu T, Stancioiu D, Stammers CW (2002) Use of magnetorheological fluid dampers in semi-active seat vibration control. Active 2002, Southampton, UK

Sireteanu T, Stancioiu D, Giuclea M, Stammers CW (2003) Experimental identification of a magnetorheological damper model to compare semi-active control strategies of vehicle suspensions. Conf on Active Noise and Vibration Control Methods, Cracow, Poland

Stammers CW (1995) Switchable dampers for vehicles: a switching demand problem. 9th World Congress on the Theory of Machines and Mechanisms, Milan, Italy, pp 2168–2173

Stancioiu D, Sireteanu T, Guglielmino E, Giuclea M (2003) Vehicle road holding and comfort improvement using controlled dampers. 7th International Conference on Economicity, Safety and Reliability of the Vehicle, Bucharest, Romania, pp 1195–2002

Tsypkin YZ (1974) Relay control systems. Cambridge University Press

Tunay I, Amin M, Beck A (1998) Robust control of a hydraulic valve for aircraft anti-skid operation, Proc IEEE International Conference on Control Applications, Trieste, Italy, pp 689–693

Uebing M (1999) Simulation and control of a multi-axes pneumatically actuated animated figure. PhD Thesis, University of Bath, UK

Utkin VI (1977) Variable structure systems with sliding modes. IEEE Trans Autom Control 22:212–222

Utkin VI, Young KKD (1978) Methods for constructing discontinuity planes in multidimensional variable structure systems. Autom Remote Control 39; 10:1466–1470

Ursu F, Sireteanu T, Ursu I (1989) On anti-chattering synthesis for active and semiactive suspension systems. 3rd IFAC International Workshop on Motion Control, Grenoble, France, pp 93–98

Ursu I, Sireteanu T (1998) A new learning-controlled sequential semiactive suspension system. FISITA World Automotive Congress, Paris, France

Ursu I, Ursu F, Sireteanu T, Stammers CW (2000) Artificial intelligence based synthesis of semi-active suspension system. Shock Vib Digest 3; 1:3–10

Vanduri S, Law H (1993) Development of a simulation for assessment of ride quality of tractor-semitrailers. SAE paper 962553

Vaughan ND, Gamble J (1991) Sliding mode control of a proportional solenoid valve. Fourth Bath International Fluid Power Workshop, Bath, UK, pp 95–108

Wang LX, Kong H (1994) Combining mathematical model and heuristics into controllers: an adaptive fuzzy control approach. Proc 33rd IEEE Conference on Decision and Control, Buena Vista, Florida, USA 4:4122–4127

Wang DH, Liao WH (2005) Modelling and control of magnetorheological fluid dampers using neural networks. Smart Mater Struct, 14:111–126

Wonham WM, Johnson S (1963) Optimal bang-bang control with quadratic performance index. 4th Joint Automatic Control Conference, Minneapolis, USA

Woodrooffe J (1995) Heavy truck suspension damper performance for improved road friendliness and ride quality. SAE paper 952636

Yager R, Filev DP (1994) Essentials of fuzzy modeling and control. John Wiley & Sons

Yang JN, Li Z, Wu JC, Young KKD (1993) A discontinuous control method for civil engineering structures. Proc 9th VPI & SU Symposium on Dynamics and Control of Large Structures, Blacksburg, USA, pp 167–180

Yiu Y, Waters TP, Brennan MJ (2005) A comparison of semi-active damping control strategies for vibration isolation of harmonic disturbances. J Sound Vib 208:21–39

Yoshimura T, Nakaminami K, Hino J (1997) A semi-active suspension with dynamic absorbers of ground vehicles using fuzzy reasoning. Int J Vehicle Des 18; 1:19–34

Young KD, Özgüner Ü (1999) Sliding mode: control engineering in practice. American Control Conference, San Diego, USA, pp 150–162.

Zadeh AL (1991) The calculus of if-then rules. Fuzzy engineering toward human friendly systems. Proc Int Fuzzy Eng Symp 1:11–12

Authors' Biographies

Dr Emanuele Guglielmino received his PhD in Mechanical Engineering from the University of Bath (UK) in 2001, and a master's degree in Electrical Engineering from the University of Genoa (Italy) in 1998. His doctoral research regarded the robust control of hydraulically actuated friction damper systems for vehicle applications. He subsequently joined Westinghouse Brakes (UK), where he worked as an R&D engineer on controlled braking systems. In 2004 he joined General Electric (Florence, Italy) where he held positions as a control engineer and an application engineer. In 2008 he joined the Italian Institute of Technology (IIT) in Genoa as a team leader.

He has authored over 25 publications in the fields of semi-active suspensions, fluid power systems, mechatronics and robust control, and co-authored a chapter in a book of applied mechanics and control. For his work he won the ASME Best Paper Award (Fluid Power Systems and Technology Division) in 2001. He was invited Guest Editor in a special issue on semi-active suspensions of the International Journal of Vehicle Design. He is also recipient of an award from IMechE, and an entrepreneurship award from the Italian Industrial Association for an outstanding business plan as a spin-off of a research project.

Dr Tudor Sireteanu graduated in Mathematics from the University of Bucharest (Romania) in 1966. He subsequently joined the Institute of Solid Mechanics of the Romanian Academy in Bucharest. In 1971 he was a Fulbright fellow at the California Institute of Technology, Pasadena.

In 1981 he was awarded a PhD in non-linear random vibration from the University of Bucharest. Since 1992 he has been a PhD advisor in applied mathematics. In 2000 he was awarded the Aurel Vlaicu Romanian Academy prize for a series of publications in the field of vibration control. At present he is Director of the Institute of Solid Mechanics and honorary member of the Academy of Technical Sciences of Romania.

His research interests include random vibration and semi-active damping systems, in particular friction dampers and magnetorheological dampers. He is co-author of two books on automotive random vibration and magnetorheological fluids and dampers, editor of three books of applied mechanics and author of over 100 publications in scientific journals and international conference proceedings.

Dr Charles W. Stammers worked at the Institute of Sound and Vibration, University of Southampton (UK) from 1963 to 1969, in 1968 being awarded a PhD for a thesis on the stability of rotor systems. He then joined Westland Helicopters Ltd. studying machine and rotor vibration problems. In 1973 he joined the Department of Mechanical Engineering at the University of Bath. Projects undertaken have included the manufacture of an ambulance stretcher suspension and a robot for disabled users. Recent work has centred on vibration control in machines and vehicles utilising smart semi-active control systems.

Since 1996 he has headed a collaborative programme with the Institute of Solid Mechanics in Bucharest, Romania, supported by the Royal Society of London. This collaboration has resulted in two books dealing with research topics in applied mechanics. Current work concerns experimental systems to protect historic buildings from seismic inputs. He has 100 publications (journals and international conferences).

Dr Gheorghe Ghita graduated in Aeronautical Engineering from Politehnica University of Bucharest (Romania) in 1975. After graduating he was employed by the aircraft company Aerostar, and since 1982 he has been a researcher in mechanical engineering. At present he is employed by the Institute of Solid Mechanics of the Romanian Academy in Bucharest. In 2003 he received his PhD in mechanical engineering from Politehnica University with a thesis in the field of semi-active vibration control.

His research interests focus on experimental methods, signal processing and applications of computational intelligence to semi-active vibration control. He is a co-author of a book on magnetorheological fluids and dampers and of over 50 journal and conference papers.

Dr Marius Giuclea graduated in Mathematics from the University of Bucharest, Romania in 1994 and obtained his MSc degree in Mathematics in 1995. Between 1994 and 2001 he worked as a researcher at the Institute of Microtechnology and from 2001 as a lecturer at the Academy of Economic Studies. In 2004 he was awarded a PhD in applications of intelligent techniques in dynamic systems control by the Institute of Mathematics, Bucharest. His research interests include intelligent techniques and their applications in modelling and control of dynamic systems. He is author of over 30 journal and conference publications.

Index